EFFIGIES TICHONIS BRAHE OTTONI[...]
ÆTATIS SVÆ ANNOS O. COMPLETO

QVO POST DIVTINVM IN PATRIA
EXILIVM LIBERTATI DESIDERAT
DIVINO PROVISV
[...]

TYCHO BRAHE

A PICTURE OF
SCIENTIFIC LIFE AND WORK IN THE
SIXTEENTH CENTURY

BY

J. L. E. DREYER, Ph.D., F.R.A.S.

GLOUCESTER, MASS.

PETER SMITH

1977

This Dover edition, first published in 1963, is an
unabridged and corrected republication of the work
first published by Adam and Charles Black in 1890.

Library of Congress Catalog Card Number: 63-19506

Reprinted, 1977, by
Peter Smith Publisher, Inc.

ISBN: 0-8446-1996-5

TO

RALPH COPELAND,

PH.D., F.R.S.E., &c.,

ASTRONOMER ROYAL FOR SCOTLAND,

𝔗𝔥𝔦𝔰 𝔅𝔬𝔬𝔨 𝔦𝔰 𝔇𝔢𝔡𝔦𝔠𝔞𝔱𝔢𝔡

BY HIS FRIEND

THE AUTHOR.

PREFACE.

ASTRONOMERS are so frequently obliged to recur to observations made during former ages for the purpose of supporting the results of the observations of the present day, that there is a special inducement for them to study the historical development of their science. Much labour has accordingly been spent on the study of the history of astronomy, and in particular the progress of the science in the sixteenth and seventeenth centuries has of late years formed the subject of many important monographs. The life of Copernicus has been written in considerable detail by Prowe, Hipler, and others. Of Kepler's numerous works we owe a complete edition to the patient industry and profound learning of the late Dr. Frisch of Stuttgart, while the life of Galileo, and particularly his persecution and trial, have called forth quite a library of books and essays. In the present volume I have attempted to add another link to the chain of works illustrating the birth of modern astronomy, by reviewing the life and work of Tycho Brahe, the reformer of observational astronomy.

Although not a few monographs have been published from time to time to elucidate various phases in the career of Tycho Brahe, while several popular accounts of his life (by Helfrecht, Brewster, &c.) have appeared, the only scientific

biography hitherto published is that of Gassendi. This
writer obtained valuable materials from some of Tycho
Brahe's pupils, and from the Danish savant Worm, but he
chiefly derived his information from a close scrutiny of
Tycho's own writings, never failing to make use of any
particulars of a biographical nature which might be recorded
in passing by Tycho. In studying Tycho's works, I have
repeatedly come across small historical notes in places where
nobody would look for such, only to find that Gassendi had
already noticed them. In 1745 a biography was published
in a Danish journal (Bang's *Samlinger*, vol. ii.), the contents
of which are chiefly taken from Gassendi, but which also
contains a few documents of interest. Of far greater importance is a collection of letters, royal decrees, and other
documents, published in 1746 by the Danish historian Langebek in the *Danske Magazin*, vol. ii., which still remains
the principal source for Tycho's life. A German translation
of this and the memoir in Bang's *Samlinger* was published
in 1756 by Mengel, a bookseller in Copenhagen, who wrote
under the high-flown pseudonym Philander von der Weistritz ; and as his book has naturally become more generally
known than the Danish originals, I have, when quoting
these, added references to Weistritz's book. During the
present century several Danish historians have brought to
light many details bearing on Tycho's life which will be
referred to in this volume; and in 1871 a Danish author,
F. R. Friis, published a popular biography in which were
given various hitherto unpublished particulars, especially
of Tycho's beneficiary grants and other endowments. The
same writer has also published a number of letters exchanged between Tycho and his relations, and various contemporary astronomers. Of great scientific interest is the
correspondence between Tycho and Magini, published and

commented by Professor Favaro of Bologna with the care
and learning by which the writings of this author are
always distinguished.   Some other letters from the last
years of Tycho's life have recently been published by Professor Burckhardt of Basle.   Lastly, we must mention the
meteorological diary kept at Uraniborg, which is of great
historical value as affording many interesting glimpses of
Tycho Brahe's home life.   It was published in 1876 by
the Royal Danish Society of Science.

Among other publications of importance for the study
of Tycho Brahe's life and activity must be mentioned the
biography of Kepler, by Frisch, in the last volume of
Kepler's *Opera Omnia*, and several papers by Professor
Rudolph Wolf of Zürich on Landgrave Wilhelm of Hesse-
Cassel, and his astronomers Rothmann and Bürgi.   Though
only indirectly bearing on Tycho (of whose merits Professor
Wolf on every occasion speaks somewhat slightingly), these
valuable papers throw much light on the state of science at
the end of the sixteenth century, and will often be found
quoted in the following pages.

Having for many years felt specially interested in Tycho
Brahe, it appeared to me that it would be a useful undertaking to apply the considerable biographical materials
scattered in many different places to the preparation of a
biography which should not only narrate the various incidents in the life of the great astronomer in some detail, but
also describe his relations with contemporary men of science,
and review his scientific labours in their connection with
those of previous astronomers.   The historical works of
Montucla, Bailly, Delambre, and Wolf have indeed treated
of the astronomical researches of Tycho Brahe, but as the
plans of these valuable works were different from that
adopted by me, I believe the scientific part of the present

volume will not be found superfluous, particularly as it is
founded on an independent study of Tycho's bulky works.
To these I have given full references for every subject, so
that any reader may find further particulars for himself
without a laborious search.   Many details, especially as to
the historical sequence of Tycho's researches, have been
taken from his original MS. observations in the Royal
Library at Copenhagen, which I was enabled to examine
during two visits to Copenhagen in 1888 and 1889.   On
the same occasions I also studied three astrological MSS.
of Tycho's, of which an account will be found in Chapter VI.   It may possibly be thought by some readers that
I have devoted too much space to the consideration of
the astrological fancies of the Middle Ages.   But my
object throughout has been to give a faithful picture of the
science of the sixteenth century, and for this purpose it is
impossible to gloss over or shut our eyes to the errors of
the time, just as it would be absurd, when writing the
scientific history of other periods, to keep silence as to the
phlogistic theory of combustion, the emission theory of
light, or the idea of the sun as having a solid nucleus.   If
the study of the history of science is to teach us anything,
we must make ourselves acquainted with the by-paths and
blind alleys into which our forefathers strayed in their
search for truth, as well as with the tracks by which they
advanced science to the position in which our own time
finds it.

With the exception of the astronomical manuscripts in
the Royal Library at Copenhagen (for facilities in using
which I was indebted to Dr. Bruun, chief librarian), I have
not made use of any unpublished materials; but the scanty
harvest reaped by modern searchers makes it extremely
unlikely that anything of importance remains to be found

among unpublished sources. I believe, however, that certain periods of Tycho Brahe's life in this volume will be found to appear in a light somewhat different from that in which previous writers have seen it. Especially it seems difficult to deny that Tycho's exile was almost entirely due to himself, and that there was no absolute necessity for his leaving Hveen, even though he had lost most of his endowments. As an amusing instance of the manner in which many incidents have been misunderstood by those who consider Tycho a martyr of science, we may mention that the trouble into which the minister of Hveen got with his superiors and with his parishioners (for his unwarranted interference with the Church ritual), has been described as a riot or fight, instigated by a wicked statesman, in which Tycho's shepherd or steward (pastor!) was injured.

I should scarcely have been able to write this book far from great libraries if I had not for many years taken every opportunity of acquiring books or pamphlets bearing in any way on the subject, or of making excerpts from such as could not be purchased. I have, however, been under great obligations to the Astronomer Royal for Scotland, who most kindly allowed me to consult the literary treasures on the star of 1572 in the Crawford Library of the Royal Observatory, Edinburgh. Hereby I have been enabled to examine even some writings on the new star which were unknown to Tycho Brahe.

That I have adopted the Latin form of the astronomer's name, by which he is universally known, instead of his real baptismal name of Tyge, scarcely requires an apology. It would indeed only be affectation to speak of Schwarzerd or Koppernigk instead of Melanchthon or Copernicus. The portrait of Tycho Brahe in this volume (about which see p. 264) has already appeared in Woodburytype in the

Memoirs of the Literary and Philosophical Society of Manchester, vol. vi., and in woodcut in *Nature*, vol. xv. Most of the other illustrations are taken from Tycho's own works. For photographs, from which the illustrations in Chapter XI. were made, I am indebted to Professor Safarik of Prague, who has also kindly communicated various particulars about Tycho's life in Bohemia.

J. L. E. DREYER.

THE OBSERVATORY, ARMAGH,
September 1890.

CONTENTS.

CONTENTS.

CHAPTER X.

TYCHO'S LIFE FROM HIS LEAVING HVEEN UNTIL HIS ARRIVAL AT PRAGUE.

CHAPTER XI.

TYCHO BRAHE IN BOHEMIA—HIS DEATH.

CHAPTER XII.

TYCHO BRAHE'S SCIENTIFIC ACHIEVEMENTS.

APPENDIX.

NOTES.

PLATES.

(The above by S. B. BOLAS & Co., London.)

ERRATA.

Page 66, Footnote 2, line 7 from end, *add:* That Hardeck speaks of the
comet of 1264, although he gives the year 1260, may be seen
from his references to Pope Clement IV. (1265-1268) and the
battle of Benevent (1266). According to Pingré, several writers
have been confused with regard to the year of this comet.

TYCHO BRAHE.

CHAPTER I.

THE REVIVAL OF ASTRONOMY IN EUROPE.

THE early part of the sixteenth century must always rank among the most remarkable periods in the history of civilisation. The invention of printing had made literature the property of many to whom it had hitherto been inaccessible, and the downfall of the Byzantine Empire had scattered over Europe a number of fugitive Greeks, who carried with them many treasures of classical literature hitherto unknown in the Western world, while Raphael, Michael Angelo, and other contemporaries of Leo X. revived the glory of the ancients in the realm of art. The narrow limits of the old world had vanished, and the Portuguese and Spanish navigators had led the way to boundless fields for human enterprise, while the Reformation revolutionised the spirit of mankind and put an end to the age of ignorance and superstition.

During this active period there were also signs of renewed vigour among the devotees of science, and the time was particularly favourable to a revival of astronomical studies. Students of astronomy were now enabled to study the Greek authors in the original language, instead of having to be content with Latin reproductions of Arabian translations from the Greek, which, through the Italian Univer-

sities, had been introduced into Europe during the Middle
Ages. Another impulse was given by the voyages of
discovery, as navigators were obliged to trust entirely to
the stars and the compass, and therefore required as perfect
a theory as possible of the motions of the heavenly bodies.
We see accordingly at the end of the fifteenth century
and the beginning of the sixteenth considerable stir in the
camp of science, but as yet only in Germany—a circumstance
not difficult to explain. Though divided into a great
number of semi-independent states, Germany bore still the
proud name of the Holy Roman Empire, and on account of
the claims represented by this name the Germans had for a
long time been in constant intercourse with Italy, the land
with the great past, and still, notwithstanding its political
misery, the leader of civilisation. It was an intercourse of a
peaceful and commercial as well as of a warlike character ;
but in both ways was this of benefit to the Germans, pro-
ducing among them much knowledge of foreign affairs, and
giving them greater facilities for taking up the scientific
work of the ancients than were found in other parts of
Europe.

The first astronomer of note was Georg Purbach (1423–
1461), who studied at the University of Vienna, and
afterwards for some time in Italy. His principal work
on astronomy (*Theoricæ Novæ Planetarum*) attempted to
develop the old hypothesis of material celestial spheres,
and was but a mixture of Aristotelean cosmology and
Ptolemean geometry ; but he was the first European to
make use of trigonometry, the principal legacy which
astronomers owe to the Arabs. Purbach endeavoured to
get beyond the rudiments of spherical astronomy, which
hitherto had formed the only subject for astronomical
lectures, and had been taught through the medium of a
treatise written in the thirteenth century by John Holy-

wood (Johannes de Sacrobosco) for use in the University of
Paris.   While lecturing at Vienna, Purbach's attention was
drawn to a young disciple of great promise, Johann Müller,
from Königsberg, a small village in Franconia, where he
had been born in 1436.   He is generally known by the
name of Regiomontanus, though he does not seem to have
used this name himself, but always that of Johannes de
Monteregio.   He entered heart and soul into his teacher's
studies of the great work of Ptolemy, which embodied all
the results of Greek astronomy, and the talented pupil soon
became an invaluable co-operator to Purbach.   They did
not confine themselves to theoretical studies, but, with such
crude instruments as they could construct, they convinced
themselves of the fact that the places of the planets
computed from the astronomical tables of King Alphonso X.
of Castile differed very considerably from the actual positions of the planets in the sky.[1]   In the midst of these
occupations the two astronomers had the good luck to
become acquainted with a man who was well qualified
to help them to carry out their greatest wishes.   This
man was Cardinal Bessarion, a Greek by birth, who, as
Bishop of Nicæa, had accompanied the Byzantine Emperor
on his journey to the Council of Ferrara in 1438, where
he tried to bring about a reconciliation between the Greek
and Roman Churches.   Bessarion remained in Italy and
joined the Roman Church, but he never forgot his old
country, and contributed very much to make the classical
Greek literature known in the West.   The translation of
the original Almagest (as Ptolemy's work was generally
called, from a corruption of the Arabic *Al megist*, in its

---

[1] The Tabulæ Alphonsinæ had been computed in the middle of the
thirteenth century by a number of Arabian and Jewish astronomers under
the personal direction of King Alphonso el Sabio.   They were founded on
the theory of Ptolemy and the observations of the Arabs, and were first
printed at Venice in 1483.

turn derived from μέγιστη σύνταξις) was a subject in which
he was particularly interested, and during his stay at
Vienna as Papal Nuncio he succeeded in communicating to
Purbach his own anxiety to make Ptolemy better known in
the scientific world. Purbach was on the point of starting
for Italy for the purpose of collecting Greek manuscripts,
when he died suddenly in 1461, but Regiomontanus succeeded to his place in the Cardinal's friendship, and set out
for Italy with Bessarion in the following year.

Regiomontanus stayed about seven years in Italy, visiting the principal cities, and losing no chance of studying
the Greek language and collecting Greek manuscripts. At
Venice he wrote a treatise on trigonometry, which branch
of mathematics he also, during the remainder of his life,
continued to develop, so that he constructed a table of
tangents (*tabula fecunda*), and probably only was prevented
by his early death from completing his treatise by introducing the use of tangents therein.[1] After his return to
Germany, he settled, in 1471, at Nürnberg. This city was
one of the chief centres of German industry and literary
life, and no other German city had such regular commercial
communication with Italy, from whence the produce of the
East was brought into the market, and nowhere did the
higher classes of citizens use their wealth so willingly in
support of art and science. The new art of printing had
recently been introduced at Nürnberg, where a regular
printing-press was now working—a circumstance of particular importance to the collector of Greek writings. A
wealthy citizen, Bernhard Walther (born 1430, died 1504),
became at once the friend and disciple of Regiomontanus,

---

[1] The treatise *De Triangulis Omnimodis*, libri v., was first published at
Nürnberg in 1533, while Regiomontanus himself printed the *Tabulæ Directionum* in 1475, containing both a table of sines for every minute, and the
above-mentioned table of tangents for every degree, extended to every
minute by Reinhold in a new edition in 1554.

and arranged an observatory for their joint use. Instruments, as fine as the skilful artisans of Nürnberg could make them, adorned the earliest of European observatories, and the two friends made good use of them (they observed already the comet of 1472), and originated several new methods of observing. But Regiomontanus did not forget the printing operations, and published not only Purbach's *Theoricæ Novæ* and trigonometrical tables, but also his own celebrated Ephemerides, the first of their kind, which, some years afterwards, were made known to the navigators through the German geographer Martin Behaim, and guided Diaz, Columbus, Vasco de Gama, and many others safely across the ocean. Nothing spread the fame of the astronomer like these Ephemerides, and the Pope was thus induced to invite Regiomontanus to Rome to reform the confused calendar. The invitation was obeyed in 1475, but Regiomontanus died in July 1476 very suddenly at Rome. He only reached the age of forty, and no doubt much might have been expected from him if death had not so early stopped his career; but he had rendered great service to science, not only by his endeavours to save the Greek authors from oblivion,[1] but by his Ephemerides, his development of trigonometry, and his observations. Walther survived him twenty-eight years, and continued his observations, which were published in 1544.

By Purbach and Regiomontanus the astronomy of the Alexandrian school had been introduced at the German Universities, and the increased demands which navigators made on astronomers continued to help forward the study of astronomy in Germany, which country, by having a sovereign in common with Spain, for a while had much intercourse with the latter country. Of the astronomers

---

[1] The Greek text of Ptolemy's work from the MS. brought home by Regiomontanus was published at Bâle in 1538.

who worked during the first half of the sixteenth century
we shall here mention Peter Apianus or Bienewitz, who
taught at the University of Ingolstadt. Besides other
works, he published in 1540 a large book, *Astronomicum
Cæsareum*, dedicated to Charles the Fifth. In this beautiful volume the author represented, by means of movable
circles of cardboard of various colours, the epicyclical motions
of the planets according to the Ptolemean system, and expected to be able in this way to find their positions without
computation. The book was received with much applause,
and is really in some ways to be admired, though one cannot
help agreeing with Kepler in regretting the "miserable
industry" of Apianus, which after all only produced a very
rough approximation to the real motions of the stars, but
which is eminently characteristic of the low state of science
at that time. Apianus deserves more thanks for having
paid much attention to comets, and for having discovered
the important fact that the tails of these bodies are turned
away from the sun. This was also pointed out about the
same time by Fracastoro of Verona in a work published in
1538, containing an elaborate attempt to revive the theory
of concentric spheres of Eudoxus, which had been pushed
into the background by the Ptolemean system of the world.

Only three years after Apian's volume appeared the great
work of Nicolaus Copernicus, *De Revolutionibus* (1543),
which was destined to become the corner-stone of modern
astronomy. We shall in the following so often have occasion
to refer to the labours of this great man, that a few words
will suffice in this place. Copernicus, who not only discovered the greatest truth in astronomy, but who even by his
opponents was admitted to be an astronomer worthy of being
classed with Hipparchus and Ptolemy, was born in 1473 at
Thorn, on the Vistula, a town which belonged to the Hansa
League, and a few years before had come under the suzerainty

of Poland. He studied first at the University of Krakau, where astronomy was specially cultivated, and at the age of twenty-four he proceeded to Bologna, where he enjoyed the teaching of Domenico Maria Novara. Thus Copernicus not only became acquainted with Ptolemy's work, but also acquired some familiarity with the astrolabe or astronomical circle, one of the few crude instruments then in use. From about the end of 1505 till his death in 1543, Copernicus lived in the diocese of Ermland, in Prussia, most of the time in the town of Frauenburg, where he held a canonry at the cathedral. It is much to be regretted that we are utterly unacquainted with the manner in which Copernicus came to design the new system of astronomy which has made his name immortal. But he had probably early perceived that, however valuable the labours of Regiomontanus had been, they had not improved the theory of celestial motion, so that the most important problem, that of computing beforehand the positions of the planets and accounting for their apparently intricate movements, was practically untouched since the days of Ptolemy. That great mathematician had completed and extended the planetary system of Hipparchus, and had in a wonderfully ingenious manner represented the complicated phenomena. But more than 1400 years had elapsed since his time, and the system, however perfect from a mathematical point of view, had long been felt to be too complicated, and not agreeing closely enough with the observed movements of the planets. This circumstance led Copernicus to attempt the construction of a new system, founded on the idea that the sun, and not the earth, is the ruler of the planets. But though Copernicus on the basis of this idea developed a theory of the planetary movements as complete as that of Ptolemy, he was unable to do more than to demonstrate the possibility of explaining the phenomena by starting from

the heliocentric idea. Having no materials from which to
deduce the true laws of the motion of the planets in elliptic
orbits, he was obliged to make use of the excentric circles
and epicycles of the ancients, by which he greatly marred
the beauty and simplicity of his system.[1] He did not
possess accurate instruments, and took but few observations
with those he had. The idea does not seem to have struck
him that it was indispensable to follow the planets through
a number of years with carefully constructed instruments,
and that only in that way could the true theory of planetary
motion be found.

There was much to be done yet ere the reform of astronomy could be accomplished. The pressing want of new
tables to take the place of the antiquated Alphonsine tables
was supplied a few years after the death of Copernicus by
Erasmus Reinhold, but though the positions of the planets
could be computed from them with greater accuracy than
from the old tables, the "Prutenic tables" (published in
1551) did not by this superiority offer any proof of the
actual truth of the Copernican principle.

A century had now elapsed since the study of astronomy
had commenced to revive in Italy and Germany, but as
yet the work accomplished had chiefly been of a tentative
and preparatory kind, Copernicus alone having attempted
to make science advance along a new path. Still, much
useful work had been done. The labours of the ancients
had now become accessible in the originals; the Arabs and
Regiomontanus had developed trigonometry, and thereby
greatly facilitated astronomical computations; Copernicus
had shaken the implicit conviction of the necessity of
clinging to the complicated Ptolemean system, and had
offered the world an alternative and simpler system, while
new tables had been computed to take the place of the

---

[1] We shall return to this subject in Chapter VII.

Alphonsine tables. But otherwise the astronomy of the ancients reigned undisturbed. No advance had been made in the knowledge of the positions of the fixed stars, those stations in the sky by means of which the motions of the planets had to be followed; the value of almost every astronomical quantity had to be borrowed from Ptolemy, if we except a few which had been redetermined by the Arabs. No advance had been made in the knowledge of the moon's motion, so important for navigation, nor in the knowledge of the nature of the planetary orbits, the uniform circular motion being still thought not only the most perfect, but also the only possible one for the planets to pursue. Whether people believed the planets to move round the earth or round the sun, the complicated machinery of the ancients had to be employed in computing their motions, and crude as the instruments in use were, they were more than sufficient to show that the best planetary tables could not foretell the positions of the planets with anything like the desirable accuracy.

No astronomer had yet made up his mind to take nothing for granted on the authority of the ancients, but to determine everything himself. Nobody had perceived that the answers to the many questions which were perplexing astronomers could only be given by the heavens, but that the answers would be forthcoming only if the heavens were properly interrogated by means of improved instruments, capable of determining every astronomical quantity anew by systematic observations. The necessity of doing this was at an early age perceived by Tycho Brahe, whose life and work we shall endeavour to sketch in the following pages. By his labours he supplied a sure foundation for modern astronomy, and gave his great successor, Kepler, the means of completing the work commenced by Copernicus.

CHAPTER II.

TYCHO BRAHE'S YOUTH.

TYCHO BRAHE belonged to an ancient noble family which had for centuries flourished not only in Denmark, but also in Sweden, to which country it had spread in the fourteenth century, when one of its members, Torkil Brahe, fled thither from Denmark to escape punishment for manslaughter. The family still exists in both countries. In the sixteenth century the Danish nobility was still of purely national origin, unmingled with the foreign blood which became merged in it in the course of the next two hundred years, when every new royal bride brought with her a train of needy adventurers, with empty purses and long titles, from the Holy Roman Empire. Like their foreign fellow-nobles, they were descended from men who had received grants of land on tenure of military service, and until about the end of the thirteenth century they can hardly be said to have formed a separate class, as their privileges and duties were not yet of necessity hereditary. They were untitled (till 1671), but all the same they were as proud and jealous of the privileges of their order as any Norman count or baron, and were called by the characteristic names of " free and well-born " or " good men." In the first half of the sixteenth century they had successfully resisted the attempts of King Christiern II. to curb their power, and had driven him from his throne; and when the lower orders afterwards had attempted to replace him on the throne rendered vacant

10

by the death of his brother and successor, the nobles had, after a hard struggle, crushed their opponents, though the latter were backed by the powerful Hansa city of Lübeck. The Reformation had broken the rival power of the Church, and the nobles had in consequence (though not to the same extent as in Germany and England) increased in wealth and possessions. And during the next fifty years they did not abuse their worldly advantages, but were, as a rule, faithful servants of their king and country, generous and kind to their tenants, fond of studies and learning. Most of them had in their youth travelled abroad, frequently for years at a time, and studied at foreign universities, where they acquired knowledge not only of books, but also of the world. At their country-seats many of them encouraged and protected men of learning, and kept up their acquaintance with classical literature, as well as with the more humble folk-lore which, in the shape of old epics and ballads (*Kjœmpeviser*), had been handed down from one generation to another among the humble as well as among the high-born. Almost every country-seat possessed what was at that time considered a fine library, so that it was quite natural that hardly a pamphlet or book was published without a dedication to some noble patron.

The father of the great astronomer was Otto Brahe, born in 1517, from 1562 or 1563 a Privy Councillor, and successively lieutenant of various counties, finally governor of Helsingborg Castle (opposite Elsinore), where he died in 1571. His wife was Beate Bille, whom he had married in 1544, and their second child and eldest son, Tyge, was born on the 14th December 1546 at the family seat of Knudstrup, in Scania or Skaane, the most southern province of the Scandinavian peninsula, which at that time still belonged to Denmark, as it had done from time

immemorial.[1] Tyge, or Tycho (as he afterwards Latinised
his name), had a still-born twin-brother, a fact alluded to
in a Latin poem which he wrote and had printed in 1572.[2]
Otto Brahe had in all five sons and five daughters (in
addition to the still-born son), the youngest being Sophia,
born in 1556, who will often be mentioned in the sequel.
Though he was the eldest son and heir to the family
estate of Knudstrup, Tycho did not remain under his
father's care for more than about a year, as his father's
brother, Jörgen (George) Brahe, who was childless, had been
promised by Otto, that if the latter got a son, he would let
Jörgen bring him up as his own.   The fulfilment of the
promise was claimed in vain; but Jörgen Brahe was not to
be put off so easily, and as soon as a second boy had been
born to Otto, the uncle coolly carried off his eldest nephew
by stealth as soon as he got an opportunity.   The parents
of Tycho gave way when the thing was done, knowing that
the child was in good hands, and doubtless expecting that
the foster-parents would eventually leave their adopted son
some of their wealth, which they also seem to have done.

We know nothing of Tycho's childhood except that he
was brought up at his uncle's seat of Tostrup, and was
from the age of seven taught Latin and other rudiments of
learning by a tutor.[3]   He acquired the necessary familiarity
with the only language which was then properly studied, so
that he was afterwards able not only to converse in and
write Latin, but also to write poetry in this language,
which was then and for a long time afterwards considered
a very desirable accomplishment for a learned man.   We
shall often have occasion to quote his poetry, some of which

---

[1] In several places in his writings Tycho alludes to the 13th December as
his birthday, but this is astronomically speaking, counting the day from noon,
as he was born between nine and ten o'clock in the morning.

[2] Reprinted in *Danske Magazin*, ii. p. 170 (Weistritz, ii. p. 23).

[3] Autobiographical note, *Astron. Inst. Mechanica*, fol. G.

is not without merit. Thus prepared, he was sent to Copenhagen in April 1559 to study at the University there.[1] This seat of learning had been founded in the year 1479 by permission of the Pope, but it had languished for a number of years for want of money and good teachers. The confiscation of the property of the monasteries enabled King Christian III. to commence improving it, and by the statutes of 1539 (which were still in force in Tycho Brahe's time) the number of professors was fixed at fourteen, three of Divinity, one of Law, two of Medicine, and eight in the Faculty of Arts, among whom were several whose names were honourably known outside their own country. Tycho now commenced his studies here, devoting himself specially to rhetorics and philosophy, as being the branches of learning most necessary to the career of a statesman, for which he was destined by his uncle, and probably also by his father, who had at first objected to his receiving a classical education.[2] But astronomy very soon claimed his attention. On the 21st of August 1560 an eclipse of the sun took place, which was total in Portugal, and of which Clavius has left us a graphic description. Though it was only a small eclipse at Copenhagen, it attracted the special attention of the youthful student, who had already begun to take some interest in the astrological predictions or horoscopes which in those days formed daily topics of conversation. When he saw the eclipse take place at the predicted time, it struck him " as something divine that men could know the motions of the stars so accurately that

---

[1] In those days students frequently entered a university at a very early age, and with an exceedingly slender stock of knowledge. At Wittenberg one of the professors in the Faculty of Arts was bound to teach the junior students Latin grammar, and one of the Wittenberg professors in his opening address pointed out how simple the rudiments of arithmetic were, and how even multiplication and division might be learned with some diligence. Prowe, *Nic. Coppernicus*, i. p. 116

[2] Gassendi, p. 4.

they could long before foretell their places and relative positions."[1] He therefore lost no time in procuring a copy of the Ephemerides of Stadius in order to satisfy his curiosity as to astronomical matters; and not content with the meagre information he could get from this book, he very soon made up his mind to go to the fountain-head, and at the end of November in the same year he invested two Joachims-thaler in a copy of the works of Ptolemy, published at Basle in 1551. This copy is still in existence, and may be seen in the University Library at Prague; there are many marginal notes in it, and at the bottom of the title-page is written in Tycho's own hand that he had bought the book at Copenhagen on the last day of November for two thaler. This book contained a Latin translation of all the writings of Ptolemy except the Geography, the Almegist being in the translation of Georgios from Trebizond. The study of this complete compendium of the astronomy of the day must have given the youthful student enough to do; indeed, it may well be doubtful whether he was at that time able to master it.

Tycho remained at Copenhagen for three years, chiefly occupying himself with mathematical and astronomical studies. We are not acquainted with any details as to this period of his life; all we know is that he formed an intimate friendship with one of the professors of medicine, Hans Frandsen, from Ribe, in Jutland (Johannes Francisci Ripensis), and especially with Johannes Pratensis, a young man of French extraction, who also afterwards became professor of medicine.[2] Jörgen Brahe now thought that the time had come to send his nephew to a foreign uni-

[1] Gassendi, p. 5.

[2] His father was Philip du Pré, from Normandy, who had come to Denmark with Queen Isabella, the wife of Christiern II. He afterwards became a Protestant and Canon of Aarhus Cathedral. N. M. Petersen, *Den Danske Literaturs His orie*, iii. p. 190.

versity, as was then customary. He probably hoped that, when removed from his friends at Copenhagen, the young worshipper of Urania might be induced to give up his scientific inclinations and devote himself more to studies which would in after years enable him to take the place in his native land to which his birth entitled him. The university he selected for his nephew was that of Leipzig. During the thirteenth and fourteenth centuries Danish students had followed the universal custom of the age and repaired to the University of Paris, where several of them had risen to great distinction, and even occupied the rectorial chair;[1] but gradually as the German universities improved they became more frequented by Danes than Paris. To accompany Tycho as tutor, Jörgen Brahe chose a young man of great promise, who, although only four years older than his pupil, was known to be steady enough to be intrusted with this responsible office. Anders Sörensen Vedel, son of a respected citizen of Veile, in Jutland, had been less than a year at the University, where he attended lectures on divinity, and at the same time devoted himself to the study of history. He became afterwards Royal Historiographer, and is particularly known by his translation of the Latin Chronicle of Saxo Grammaticus, an important source of Danish history from the end of the twelfth century, and also by his edition of the ancient national ballads or Kjæmpeviser.[2] Vedel was only too happy to accept the proposal of accompanying the young nobleman abroad, as there was at that time no Professor of History

[1] There were four times in the fourteenth century Danish Rectors of the University of Paris (N. M. Petersen, *Den Danske Literaturs Historie*, i. p. 74). Students from Denmark, Sweden, and Norway (provincia Daciæ) belonged to Anglicana Natio, one of the four *Nations* of the University.

[2] *Historiske Efterretninger om Anders Sörensen Vedel*, af C. F. Wegener. Appendix to a new edition of *Den Danske Krönike af Saxo Grammaticus*, translated by Vedel, Copenhagen, 1851, fol. This book is a valuable source for Tycho's early life.

in the Copenhagen University, while it was not uncommon
to find professors in German universities who combined the
chair of History with some other one.[1]

Vedel and his pupil left Copenhagen on the 14th February
1562, and arrived at Leipzig on the 24th March following.
They were at once installed in the house of one of the professors,[2] possibly on the recommendation of some of their
learned friends in Copenhagen, several of whom were in
constant communication with their colleagues at the Leipzig
University. They had at once their names entered in the
book of matriculation, where they may still be seen as
"Andreas Severinus Cimber" and "Tyho Brade ex Scan-
dria." There is, however, no sign whatever that Tycho
devoted himself to the study of law, while we know that he
at once sought the acquaintance of the professor of mathematics, Johannes Homilius; of his disciple, Bartholomæus
Scultetus (Schultz), and probably also of the "electoral
mathematician," Valentine Thau.[3] Homilius died on the
5th July, a little over three months after Tycho's arrival,
but we shall afterwards see that he had even in that
short time imparted valuable and practical knowledge to
the young student. Vedel did his best to carry out his
instructions by trying to keep Tycho to the study of jurisprudence, but Tycho would not allow himself to be hindered
in his favourite pursuits, and spent most of his money on

[1] Though the University of Leipzig did not get a chair of History till 1579,
Camerarius (about whom see next page) was to some extent considered as
being Professor of History, and is even once styled "Historiarum et utriusque
linguæ professor" (Wegener, *l.c.*, p. 31).

[2] His name is not known. Tycho only mentions him once in a note
among his observations: "1564, 14th Dec.—Sub cœnam Pfeffigerus, qui apud
doctorem nostrum hospitem convivabatur, dicebat . . ." (then follows an
account of the conversation in German).

[3] Thau is mentioned by Vedel as a friend of his own, and appears to have
been the inventor of an artificial car (rheda, viameter?). See Wegener, p.
32. Is he identical with Lucius Valentinus Otho, who edited the *Opus
Palatinum de Triangulis* of Rhäticus in 1596?

astronomical books and instruments, though he had to receive the money from his tutor and account to him for the way it was spent. He made use of the Ephemerides of Stadius [1] to find the places of the planets, having first learned the names of the constellations by means of a small celestial globe not larger than his fist, which he hid from Vedel, and could only use when the latter was asleep. Though this state of things at first produced some coolness between tutor and pupil, it appears that they soon renewed their friendly intercourse. Tycho could not but see that Vedel was only doing his duty, and Vedel gradually had to acknowledge that the love of astronomy had become so deeply rooted in his pupil that it was utterly impossible to force him against his will to devote himself to a study he disliked, or at least looked on with indifference. Another circumstance which was a bond of union between them was that the learned men whose society Vedel sought were to a great extent the same to whom Tycho looked for instruction. Thus the above-mentioned Valentine Thau had a great regard for Vedel, and even tried to get him to enter the service of the Elector, while Homilius was a son-in-law of Camerarius, the most renowned of the professors at Leipzig, and a man whom Vedel later in one of his writings mentions as his beloved teacher.[2] Drawn together through their intercourse with these and other men of learning, Vedel and Tycho laid the foundation of a warm friendship which lasted through life.

[1] Published at Cologne in 1556 for the years 1554–70, again in 1559 and 1560, being continued to the year 1576. Founded on the Prutenic tables.

[2] Joachim Liebhard (who changed his name to Camerarius because there had been several *Kämmerer* in his family), born at Bamberg in 1500, died at Leipzig in 1574; published the Commentary of Theon as an appendix to the edition of Ptolemy edited by Grynæus in 1538; wrote a book on Greek and Latin arithmetic (see Kästner, *Gesch. d. Mathem.*, i. p. 134), and published in 1559 a book, *De eorum qvi Cometæ appellantur, Nominibus, Natura, Caussis, Significatione,* in which he shows from history that comets sometimes announce evil, sometimes good events.

Though Tycho was during the greater part of the time he spent at Leipzig obliged to study astronomy in secret, he did not long content himself with the use of the Epheme- rides of Stadius, but procured the Alphonsine tables and the Prutenic tables.[1] We have already mentioned that the former were founded on the Ptolemean planetary system and the observations of the Arabs, as well as those made under the direction of Alphonso X. of Castile in the thirteenth century; while the latter, which got their name from being dedicated to Duke Albrecht of Prussia, were the work of Erasmus Reinhold, a disciple and follower of Copernicus. Tycho soon mastered the use of these tables, and perceived that the computed places of the planets differed from their actual places in the sky (even though he only inferred the latter from the relative positions of the planets and adjacent stars), the errors of the old Alphonsine tables being much more considerable than those of their new rivals. He even found out that Stadius had not computed his places correctly from Reinhold's tables. And already at that time, while Tycho was a youth only sixteen years of age, his eyes were opened to the great fact, which seems to us so simple to grasp, but which had escaped the attention of all European astronomers before him, that only through a steadily pursued course of observations would it be possible to obtain a better insight into the motions of the planets, and decide which system of the world was the true one. An astronomical phenomenon which took place in August 1563, a conjunc- tion of Saturn and Jupiter, which in those days was looked on as a very important one, owing to the astrological signifi- cance it was supposed to have, induced him to begin at once to record his observations, even though they were taken with the crudest implements only. A pair of ordinary

---

[1] In his observations from 1563–64 he also mentions the Ephemerides of Carellus (Venice, 1557).

compasses was all he had to begin with; by holding the
centre close to the eye, and pointing the legs to two stars
or a planet and a star, he was able to find their angular
distance by afterwards applying the compasses to a circle
drawn on paper and divided into degrees and half degrees.
His first recorded observation was made on the 17th August
1563,[1] and on the 24th of August in the morning he noted
that Saturn and Jupiter were so close together that the
interval between them was scarcely visible.[2] The Alphonsine tables turned out to be a whole month in error, while
the Prutenic ones were only a few days wrong as to the
moment of nearest approach. Tycho continued his observations, partly with the above-mentioned tool, partly using
eye-estimations as to which stars formed with a planet a
rectangular triangle, or which stars were in a right line
with it. But in the following year he provided himself
with a "radius," or "cross-staff," as it used to be called
in English, one of the few instruments employed by the
intrepid navigators who discovered the new worlds beyond
the ocean.[3] It consisted of a light graduated rod about three
feet long and another rod of about half that length, also
graduated, which at the centre could slide along the longer

[1] Die 17, H. 13, M. 15, Erat ♂ in 7 Gr. 8 lat. Mer. 3 Gr. ad fixas.

[2] "Intervallum ♄ et ♃ matutino tempore vix observatione oculari notari
potuit : in hac nocte enim uterque se invicem obumbrabat suis radiis sed
latitudo ipsorum diversa adhuc erat, ♄ enim meridionalior ipso ♃ erat. Die
27 (astron. 26) Mane vidi ♃ cum ♄ obtinere eandem alt. ab horizonte, hinc
licet conjicere eorum ♂ jam praeteriisse sed propius erant ab invicem dispositi
quam ante triduum : quare etiam tempus συζυγίας propinquius huic 27 Aug.
quam priori 24 fuisse manifestum est. In utroque autem die paulo ante
ortum ☉ distantiam ipsorum observavi."—MS. volume of observations, 1563–
1581 incl., in the Royal Library at Copenhagen (Gamle Kongelige Samlinger,
4to, No. 1824). The early observations (up to 1577) only exist in this copy,
the originals would seem to be lost, at least they are not at Copenhagen.

[3] In French called arbalète or arbalestrille, in Spanish ballestilla, in German
Jacobsstab. It seems to have been invented by Regiomontanus, and is described in his Problemata XVI. de Cometæ Longitudine, written in 1472, and
printed in 1531 in Schoner's Descriptio Cometæ.

one, so that they always formed a right angle. The
instrument could be used in two ways. Two sights might
be fixed at the ends of the shorter rod, and one at the
end of the longer rod, and the observer, having placed
the latter close before his eye, moved the cross-rod along
until he saw through its two sights the two objects of
which he wanted to measure the angular distance. Or
one of the sights of the shorter arm might be movable, and
the observer first arbitrarily placed the shorter arm at any
of the graduations on the longer one, and then shifted the
movable sight along until he saw the two objects through
it, and a sight fixed at the centre of the transversal arm.
In either case the graduations and a table of tangents
furnished the required angle. Tycho's instrument was of
the latter kind, and was made according to the directions of
Gemma Frisius. He got his friend Bartholomæus Scultetus
to subdivide it by means of transversals, which method of
subdividing small intervals was then beginning to be used,
and which Tycho ascribes to Homilius. The earliest observations stated to have been made with the radius are from
the 1st of May 1564, and Tycho says that he had to use it
while his tutor was asleep, from which we see that Vedel
had even at that time not given up his resistance to his
pupil's scientific labours. The observer soon found that the
divisions did not give the angles correctly, and as he could
not get money from Vedel for a new instrument, he constructed a table of corrections to be applied to the results of
his observations.[1] This is deserving of notice as the first
indication of that eminently practical talent which was in
the course of years to guide the art of observing into the
paths in which modern observers have followed. Kepler,
who more than any one else was able to appreciate his great

---

[1] *Astronomiæ Instauratæ Mechanica*, fol. G. 2. For a specimen of the
observations see Appendix A. at end of this volume.

predecessor, justly says that the "restoration of astronomy" was "by that Phœnix of astronomers, Tycho, first conceived and determined on in the year 1564."[1]

While occupied with the study of astronomy and occasional observations, Tycho, like everybody else at that time, believed in judicial astrology, and now and then worked out horoscopes for his friends. He even kept a book in which he entered these "*themata genethliaca.*" He mentions in a letter,[2] written in 1588 to the mathematician Caspar Peucer, the son-in-law of Melanchthon, that he had during his stay at Leipzig made out the nativity of Peucer, and found that he was to meet with great misfortunes, either exile or imprisonment, and that he should become free when about sixty years of age, through the agency of some "martial" person. This prediction chanced to turn out correct, as Peucer in 1574 was deprived of his professorship at Wittenberg, and kept in a rigorous imprisonment till 1586, being suspected of a leaning to Calvinism. From a lunar eclipse which took place while he was at Leipzig, Tycho foretold wet weather, which also turned out to be correct.[3]

Tycho left Leipzig on the 17th May 1565 with Vedel

[1] *Tabulæ Rudolphinæ,* title-page.

[2] Printed in *Resenii Inscriptiones Hafnienses* (Hafniæ, 1668), pp. 392 *et seq.;* and in Weistritz, *Lebensbeschreibung des T. v. Brahe,* i. pp. 239 *et seq.* (the matter referred to occurs on p. 259).

[3] In the volume of observations, 1563–81, there follow, after April 19, 1565, sixteen pages headed "Notationes interiectæ," of various contents. On a vacant quarter page is written in a different hand: "Duobus sequentibus annis nullæ extant observationes Brahei, sed earum loco sequebantur annotationes qualescunque in codice." Also in another hand is the following: "Tycho Brahe Tomo II. Epistolarum aliqvando excuso sed non edito fol. 54 scribit se hujus eclipsis tempore adhuc Lipsiæ studiorum causa commoratum, et pluvium tempus cum meteoris humidis ex hac eclipsi prædixisse." Among the notes is also "Observatio XII. dierum et noctium statim sequentium natalem Christi in Anno 1564 completo, pro constitutione et temperamento 12 mensium Anni 1565 proxime seqventis." The probable weather for January is concluded from the weather on December 26th; that for February from the 27th, and so on.

to return to Denmark.   During his absence from home war
had broken out between Denmark and Sweden, and his
uncle, Jörgen Brahe, who held the post of vice-admiral,
probably considered that his nephew's proper place at such
a time was in his native country.   Travelling by way of
Wittenberg, they reached Rostock on the 25th May, and
succeeded in crossing to Copenhagen without meeting any
hostile cruisers.   Whether the uncle had become reconciled
to the substitution of the study of astronomy for that of
law, is not known with certainty; but the two relatives did
not long enjoy each other's company, as Jörgen Brahe, who
had just returned from a naval engagement in the Baltic
(near the coast of Mecklenburg), died on the 21st June
1565.   It happened that King Frederick II., when riding
over the bridge which joined the castle of Copenhagen and
the town, fell into the water.   Jörgen Brahe was in his
suite, and hastened to help the king out; but a severe
cold he caught in consequence developed into an illness
which proved fatal in a few days.   After his uncle's death
there was nothing to keep Tycho at home.   Another uncle,
Steen Bille, maintained that he should be left to follow his
own inclinations; but, with this exception, all his relations
and other nobles looked with coldness at this young man
with his odd taste for star-gazing and his dislike to what
they considered sensible occupations.   He was, therefore,
glad to escape from these surroundings to others more
congenial, and early in 1566 he left Denmark for the
second time, and arrived at Wittenberg on the 15th April.
The University of Wittenberg had been founded in 1502,
and had then for nearly fifty years been one of the most
renowned in Europe, possessing great attractions for Protestant students.   Here Luther had lived, and from this
hitherto insignificant spot had shaken the spiritual despotism
under which the world had suffered so long; and a few years

had only elapsed since the death of Melanchthon had deprived the University of an accomplished scholar as well as a faithful and indefatigable worker for the Reformed faith. There were still many men of celebrity following in their footsteps, and keeping up the high reputation they had made for the University. Mathematics were specially cultivated at Wittenberg, because, as the Statutes stated, without them Aristotle, "that nucleus and foundation of all science," could not be properly understood. At the instance of Melanchthon, two chairs of Mathematics were founded, "Mathematum superiorum" and "inferiorum," the holder of the former having to lecture on astronomy—that of the latter on algebra and geometry. To Danish students Wittenberg had since the Reformation been a favourite resort, and, among a number of young countrymen, Tycho Brahe also found his former tutor, Vedel, who had arrived a few months before. We do not possess any information as to how Tycho spent his time at Wittenberg; all we know is that he had the advantage of studying under the above-mentioned Caspar Peucer, Professor of Medicine and Physician in ordinary to the Elector of Saxony. This man, who was distinguished both as a mathematician, a physician, and a historian, had been invested with unusual authority over the University.[1] In the history of astronomy Peucer is known as the author of a few treatises, among which is one on spherical astronomy. Tycho, however, did not profit very much from Peucer's instruction, as the plague broke out at Wittenberg, so that he was induced to leave it on the 16th September, after a stay of only five months,[2]

[1] As Præceptor primarius totius Academiæ. We have already mentioned Peucer's subsequent misfortune. He died in 1602.

[2] Tycho probably remembered that the well-known astronomer Erasmus Reinhold, author of the Prutenic tables and professor at Wittenberg, had in 1553 vainly tried to escape the plague by flying from Wittenberg to Saalfeld, where he died.

and to go to Rostock, where he arrived on the 24th September, and was matriculated at the University a few weeks later.[1]

Though not as celebrated as the University of Wittenberg, Rostock was also much frequented by Scandinavian students, a natural consequence of its being situated close to the shore of the Baltic, and within easy reach from the Northern countries. It can hardly have been the wish of studying astronomy under any of the professors at Rostock which induced Tycho to take up his abode there, for there was not at that time any savant attached to the University of Rostock who occupied himself specially with astronomy; and only one, David Chytræus, otherwise well known as a theological author, is very slightly known in the history of astronomy as one of the numerous writers on the new star of 1572. But if there were no astronomers at Rostock (and, indeed, they were not numerous anywhere), there were several men who devoted themselves to astrology and alchemy, in addition to mathematics or medicine. It must be remembered that it was at that time easy enough to be thoroughly acquainted with the little that was known in several sciences, and men frequently exchanged a professorship of medicine for one of astronomy or divinity. The connection of medicine in particular with astronomy was supposed to be a very intimate one, and as physicians, if they kept to what we should call their proper sphere, could do little but grope in the dark, they were only too glad to call in the aid of astrology to make up for the deficiency of their medical knowledge. The idea of a connection between the celestial bodies and the vital action of the human frame was a natural consequence of the

[1] As "Tycho Brahe, natus ex nobili familia in ea parte regni Danici quæ dicitur Scania." See G. C. F. Lisch, *Tycho Brahe und seine Verhältnisse zu Mecklenburg*, in *Jahrbücher des Vereins für Mecklenburgische Geschichte*, xxxiv., 1869 (Reprint, p. 2).

Aristotelean and scholastic views of the kosmos and of the dependence of the four elements of the sublunary region on the movements in the æthereal part of the universe. The dependence of vegetable life on the motion of the sun in the ecliptic, and the similarity of the period of the moon's orbital motion to that of certain phenomena of human life, were looked upon as proofs of the connection between the sublunary and the æthereal worlds; and as the human body was composed of the elements, it would, like these, be influenced by the forces, chiefly the planets, by which the celestial part of the kosmos exercised its power. Thus it was supposed that the state of the body was dependent on the positions of the planets among the signs of the zodiac, and that the power of the Deity over the fate of man was also exercised by the medium of the stars.[1] Galileo had not yet overthrown the Aristotelean system of Natural Philosophy, and Bacon had not yet taught us to look for the explanation of the phenomena of nature by seeking for the mechanically acting causes through observation and induction, instead of through metaphysical speculation. Until this was done, it is not to be wondered at that the greatest minds believed in astrology; and it only shows the narrowmindedness of some modern writers, and their ignorance of the historical development of man's conception of nature, when they, on every occasion, sneer at the greatest men of former ages for their belief in astrology.

Among the professors at Rostock was Levinus Battus, Professor of Medicine, born in the Netherlands, and originally a mathematician. He has left writings on alchemy, and was a follower of Paracelsus; so that it is likely enough that Tycho, who afterwards paid a good deal of attention to chemistry, attended his lectures. Tycho does not seem to have taken observations regularly at that time; at

[1] "Astra regunt hominem, sed regit astra Deus."

least we do not possess any made at Rostock earlier than January 1568. But shortly after his arrival, on the 28th October 1566, a lunar eclipse took place, and Tycho posted up in the college some Latin verses, in which he announced that the eclipse foretold the death of the Turkish Sultan. It was natural to think of him, as Soliman, who was about eighty years of age, had the year before startled Christendom by his formidable attack on Malta, which was heroically and successfully defended by the Knights of St. John. A few weeks later news was received of the Sultan's death ; but unluckily he had died before the eclipse, so that the praise Tycho received for the prophecy was not unmingled with sneers, while he defended himself by explaining the horoscope of Soliman, from which he had drawn his conclusions as to the Sultan's death.[1]

An event took place at Rostock soon after this, which was a good deal more unfortunate for Brahe, and which has become more widely known than many other and much more important incidents in his life. On the 10th December 1566 there was a dance at Professor Bachmeister's house to celebrate a betrothal, and among the guests were Brahe and another Danish nobleman, Manderup Parsbjerg. These two got into a quarrel, which was renewed at a Christmas party on the 27th, and finally they met (whether accidentally or not is not stated) on the 29th, at seven o'clock in the evening, "in perfect darkness," and settled the dispute with their swords. The result was that Tycho lost part of his nose, and in order to conceal the disfigurement, he

[1] In a marginal note in the volume of observations, 1563–81 (printed in *Danske Magazin*, ii. p. 177), Tycho states that Soliman died a few days before the eclipse. In reality he died on 6th September, while besieging the Hungarian fortress Szigeth, though his death was kept secret for more than a fortnight. There is a written pamphlet by Tycho, apparently intended to be printed, in the Hofbibliothek at Vienna, *De Eclipsi Lunari*, 1573, *Mense Decembri*, in which the eclipse of 1566 and the prediction of the Sultan's death are also treated of. Friis, in *Danske Samlinger*, 1869, iv. p. 255.

replaced the lost piece by another made of a composition of gold and silver. Gassendi, who recounts all these details, adds that Willem Jansson Blaev, who spent two years with Tycho at Hveen, had told him that Tycho always carried in his pocket a small box with some kind of ointment or glutinous composition, which he frequently rubbed on his nose.[1] The various portraits which we possess of Tycho show distinctly that there was something strange about the appearance of his nose, but one cannot see with certainty whether it was the tip or the bridge that was injured, though it seems to be the latter. A very venomous enemy of his, Reymers Bär, of whom we shall hear more farther on, says that it was the upper part of the nose which Tycho had lost.[2]

As already remarked above, Tycho does not seem to have taken many observations about this time, but on the 9th April 1567 an eclipse of the sun took place which he observed. At Rostock the eclipse was of about seven digits, but at Rome it was total, and the solar corona was seen by Clavius. In the summer of 1567 Tycho paid a visit to his native country, but he does not appear to have been altogether pleased with his reception there, and at the end of the year he returned to Rostock, where he arrived on the 1st January 1568. Already, at six o'clock on the following morning, he commenced to take observations, though he had not an instrument at hand, and therefore had to content himself with noting down the positions of Jupiter and Saturn among the stars. On the 14th he wrote a letter to

---

[1] Gassendi (p. 10) adds, that according to the Epistles of Joh. Bapt. Laurus (Protonotarius Apostolicus of Pope Urban VIII.), the dispute between Brahe and Parsbjerg was as to which of them was the best mathematician. But this is probably only gossip. They are said to have been very good friends afterwards. Towards the end of this book we shall see that Parsbjerg complained of the fight being referred to in Tycho's funeral oration.

[2] Delambre, *Astr. Moderne*, i. p. 297 ; Kästner, *Geschichte der Mathematik*, iii. p. 475.

a countryman Johannes Aalborg (whose acquaintance he had probably made at Rostock the previous winter), that he had since his arrival been staying at the house of Professor Levinus Battus, but that he hoped the same day to take up his residence in the College of the Jurists, where he would have a convenient place for observing. (We find that he commenced to use the radius or cross-staff there on the 19th.) In this letter, which is printed by Gassendi,[1] Tycho says that he intends remaining over the winter in his new abode, and adds : "But you, my dear Johannes, must keep perfect silence with regard to those reasons for my departure which I have confided to you, lest anybody should suspect that I complain of anything, or that there was some- thing in my native land which obliged me to leave it. For I am very anxious that nobody should think that I am com- plaining of anything, as in truth I have not much to com- plain of. I was indeed received better in my native land by my relations and friends than I deserved ; only one thing was wanting, that my studies should please everybody, and even that may be excused. There are many denunciators everywhere."

But though Tycho was dissatisfied with the want of sympathy which his countrymen showed for his love of the stars, it appears that there must have been those in Den- mark who appreciated the steady perseverance with which the young nobleman devoted himself to study, and the first sign of this appeared soon after. On the 14th May 1568, King Frederick II. granted him under his hand a formal promise of the first canonry which might become vacant in the Chapter of the Cathedral of Roskilde in Seeland.[2] To understand this, we must mention that the Danish cathedral

---

[1] Page 11, and reprinted in *Tychonis Brahei et ad eum doctorum Virorum Epistolæ*, Havniæ, 1876–86, p. 1.

[2] The letter may be seen in *Danske Magazin*, ii. p. 180 (Weistritz, ii. p. 45).

chapters were not abolished at the Reformation (1536), but that their incomes for more than a century were spent to support men of merit (or who were supposed to be such), and especially men of learning. The members were still called canons, and if they lived about the cathedral, they formed a corporate body and managed the temporalities of the cathedral and its associated foundations. Gradually the canonries became perfect sinecures, and the kings assumed the right to fill them, until their property, in the course of the second half of the seventeenth century, was taken possession of by the Crown. One of these sinecures was thus by royal letter promised to Tycho Brahe, who now might feel certain that means of following his favourite pursuit would not be wanting. He was possibly still at Rostock when this letter was issued, and it is not known when he left this town (his last recorded observation there is of the 9th February), but it must have been early in the year, as he was at Wittenberg some time in 1568,[1] and went to Basle in the course of the same year, where he was matriculated at the University.[2] He must have stayed at Basle till the beginning of the following year, when we find him at Augsburg, where he began to observe on the 14th April. On the way he had paid a visit to Cyprianus Leovitius (Livowski), a well-known astronomer, who lived at Lauingen, in Suabia,[3] who had published an edition of the trigonometrical tables of Regiomontanus (*Tabulæ Directionum*, 1552), various Ephemerides, and an astrological book on the signification of conjunctions of planets, eclipses, &c.[4] Leovitius thought the world was likely to

---

[1] *Mechanica*, fol. G. 2.

[2] R. Wolf, *Geschichte der Astronomie*, p. 271.

[3] Born in 1524 in Bohemia, died at Lauingen in 1574.

[4] The original edition (of 1564) is in German (see Pulkova Library Cat., p. 382), and not in Latin, as stated by Lalande. The London edition of 1573 (of which more below) is in Latin, and has the title given by Lalande.

come to an end in 1584, after the next great conjunction, and he was, on the whole, more of an astrologer than of an astronomer. Tycho asked him, among other things, whether he ever took observations, as he might thereby see that the Ephemerides, which he had with some trouble computed from the Alphonsine tables, did not agree with the heavens. To this Leovitius answered that he had no instruments, but that he sometimes "by means of clocks" observed solar and lunar eclipses, and found that the former agreed better with the Copernican (Prutenic) tables, the latter better with the Alphonsine, while the motion of the three outer planets agreed best with the Copernican, the inner ones with the Alphonsine tables.[1] It does not seem to have struck him, nor, indeed, any one before Tycho, that the only way to produce correct tables of the motions of the planets was by a prolonged series of observations, and not by taking an odd observation now and then.

In the ancient free city of Augsburg Tycho seems to have felt perfectly at home. Dear to all Protestants as being the place where the fearless Reformers had declared their faith, and where the Protestant princes and cities of Germany had signed the "Confession of Augsburg," the city possessed the further attraction of having many handsome public and private buildings and spacious thoroughfares, while the society of many men of cultured tastes and princely wealth (such as the celebrated Fugger family), made it an agreeable place of residence. Among the men with whom Tycho associated here were two brothers, Johann Baptist and Paul Hainzel, the former burgomaster, the latter an alderman (septemvir). Both were fond of astronomy, but Paul particularly so, and they were anxious to procure some good instrument with which to make observations at their country-seat at Göggingen, a village about an

---

[1] *Astr. Inst. Progymnasmata*, p. 708.

English mile south of Augsburg.[1] Tycho tells us in his
principal work, *Astronomiæ Instauratæ Progymnasmata*, at
some length, that he was in the act of making out how large
an instrument would have to be in order to have the single
minutes marked on the graduated arc, when Paul Hainzel
came in and a discussion arose between them on the subject.
Tycho was convinced that no good would result to science
from using " those puerile tools" with which astronomers
then observed, and he concluded that it was necessary to
construct a very large quadrant, so large that every minute
could readily be distinguished, and fractions of a minute
estimated ; " for he did not then know the method of subdividing by transversals." This last remark is curious, as
we have already seen that he attributed his acquaintance
with the method to Scultetus, but he evidently means that
it had not yet occurred to him to use this plan on an arc
as well as on a rectilinear scale. He spoke in favour of constructing a quadrant, as he had already made several of
three or four cubits radius (this is the only evidence we
have of this fact), and was sufficiently familiar with the
cross-staff to know that no accurate results were to be
expected from it. The outcome of this discussion was that
Paul Hainzel undertook to defray the expense of a quadrant
with a radius of 14 cubits (or about 19 feet). The most
skilful workmen were engaged, and within one month the
huge instrument was completed. Twenty men were scarcely
able to erect it on a hill in Hainzel's garden at Göggingen ;
it was made of well-seasoned oak ; the two radii and the arc
were joined together by a framework of wood, and a slip of
brass along the arc had the divisions marked on it. Unlike
all Tycho's later quadrants, it was suspended by the centre,

---

[1] The latitude of Göggingen is 48° 20′ 28″, and that of St. Ulrich's Church,
Augsburg = 48° 21′ 41″ (Bode's *Jahrbuch, Dritter Supplementband*, pp. 166–
167). Hainzel found, in 1572–73, the latitude of Göggingen = 48° 22′ with
the great quadrant (*Progymnasmata*, p. 361).

and was movable round it, the two sights being fixed on
one of the radii, and the measured altitude being marked
by a plumb-line. The weighty mass was attached to a
massive beam, vertically placed in a cubical framework of
oak, and capable of being turned round by four handles, so
as to place the quadrant in any vertical plane. The framework or base was strongly attached to beams sunk in the
ground. There was no permanent roof over it, but some
kind of removable cover. The instrument stood there for
five years, until it was destroyed in a great storm in
December 1574, and some observations made with it of
the new star of 1572 and other fixed stars are published in Tycho's *Progymnasmata*.[1] He does not himself
appear to have observed with it, although we possess his
observations made at Augsburg, with few interruptions,
from April 1569 to April 1570. Some of these are, as
formerly, mere descriptions of the positions of the planets,
stating with which stars they were in a straight line or in
the same vertical ; others are made with the cross-staff; others
again with a " sextant" or instrument for measuring angles
in any plane whatever, which he had designed about this
time. This instrument, which he presented to Paul Hainzel,
consisted of two arms joined by a hinge like a pair of compasses, with an arc of 30° attached to the end of one arm,
while the other arm could be slowly moved along the arc
by means of a screw.[2] We shall farther on describe this
instrument in detail.

In addition to these instruments, Tycho while at Augsburg
arranged for the construction of a large celestial globe five
feet in diameter, made of wooden plates with strong rings
inside to strengthen it. It was afterwards covered with

---

[1] Pages 360-367. The quadrant is figured ibid., p. 356; also in *Astron.
Inst. Mechanica*, fol. E. 5, and in Barretti *Historia Cœlestis*, p. cvii. About
its destruction, see *T. B. et ad eum Doct. Vir. Epistolæ*, p. 17.

[2] Figured in *Mechanica*, fol. E. 2.

thin gilt brass plates, on which the stars and the equator
and colures were marked. It was not finished when Tycho
left Augsburg; but Paul Hainzel, who was under great
obligations to him for having designed the quadrant and
given him the newly-constructed sextant, readily undertook
to superintend the completion of it.

At Augsburg Tycho made the acquaintance of Pierre de
la Ramée, or Petrus Ramus, Professor of Philosophy and
Rhetoric at the College Royal at Paris, who had been
obliged to leave France several times owing to his adherence
to the Huguenot party, and the odium he had drawn on
himself by his opposition to the then all-powerful Aristotelean
philosophy. He wanted to discourage the exclusive study
of this time-honoured system of philosophy, now worn to
a shadow, which had become a mere cloak for stagnation,
bigotry, and ignorance, and to introduce in its place the
study of mathematics in the University of Paris. But his
zeal only procured him much enmity and persecution; he
had to apologise for his abuse of the Peripatetic philosophy
before the Parliament of Paris, and by sentence of special
royal commissioners appointed to investigate the matter,
Aristotle was reinstated as the infallible guide to learning.
Ramus had therefore for a while withdrawn from France,
but, unluckily for himself, he returned in 1571, and perished
the following year in the massacre of St. Bartholomew.
This man, who was naturally inclined to hail with pleasure
a rising star in a science closely allied to his own, happened
to be at Augsburg in 1570, and became acquainted with
Tycho Brahe through Hieronymus Wolf, a man of great
learning, especially in the classical languages, and himself
drawn to Tycho by his love of astrology. Having been
invited by Hainzel to inspect the great quadrant, Ramus
expressed his admiration of this important undertaking, so
successfully carried out by a young man only twenty-three

years of age, and begged him to publish a description of it. In his work *Scholarum Mathematicarum Libri XXXI.*, published at Basle in 1569 (only a few months before his conversation with Tycho), Ramus had advocated the building up of a new astronomy solely by logic and mathematics, and entirely without any hypothesis, and had referred to the ancient Chaldeans and Egyptians as having had a science of this kind, which had gradually by Eudoxus and that terrible Aristotle been made absurd through the introduction of solid spheres and endless systems of epicycles. Ramus explained his views to Tycho (who has left us an account of this conversation[1]); but he answered that astronomy without an hypothesis was an impossibility, for though the science must depend on numerical data and measures, the apparent motions of the stars could only be represented by circles and other figures. But though Ramus could not bring over the young astronomer to his views, they could cordially agree in the desire of seeing the science of astronomy renovated by new and accurate observations, before a true explanation of the celestial motions was attempted; and it can hardly be doubted that the conversation of this rational, clear thinker (so different from a Leovitius, with his brain crammed full of astrology and other hazy and fanciful ideas) took root in the thoughtful mind of the young astronomer, and bore fruit in after years in that reformation of his science for which Ramus had hoped.

Tycho Brahe left Augsburg in 1570, but the exact month is not known, nor the route by which he travelled. We only know that he passed through Ingolstadt, and called on Philip Apianus,[2] a son of Peter Apianus (or Bienewitz), whose name is well known both by his having pointed

---

[1] In a letter to Rothmann, *Epist. Astron.*, p. 60 ; see also *Progymn.*, p. 359.
[2] *Progymn.*, p. 643.

out that the tails of comets are turned away from the sun, and
also by his work *Astronomicum Cæsareum*, to which we have
already alluded. He was probably called back to Denmark
by the illness of his father, Otto Brahe, for the first sign of
his having returned is an observation made on the 30th
December 1570 at Helsingborg Castle, where his father
was governor,[1] and it is known that his brother, Steen
Brahe, who was also abroad at that time, was called home
by the alarming state of his father's health.[2] Otto Brahe died
at Helsingborg on the 9th May 1571, only fifty-three years
of age, surrounded by his wife and family. Tycho has in
a letter to Vedel given a touching description of his last
moments.[3] His property of Knudstrup seems to have been
inherited jointly by his eldest two sons, Tycho and Steen,
the latter of whom was already the following year in a still
existing document styled " of Knudstrup."[4]

Tycho remained at home after his father's death, paying
occasional visits to Copenhagen, but spending most of his
time in Scania. He seems to have found it too lonely at
Knudstrup, and soon took up his abode with his mother's
brother, Steen Bille, at Heridsvad Abbey, about twenty
English miles east of Helsingborg, and not very far from
Knudstrup. Formerly there had been here a Benedictine
monastery, which, like several others in Denmark, was not
at once abolished at the Reformation; but in 1565 Steen
Bille had been ordered to take possession of the Abbey,
" because ungodly life was going on there," and to maintain

---

[1] It can hardly be called an observation : " ½ Hor. quasi post occasum ☉
vidi quod ☽ limbi illuminati extremitate distabat a ♃ per duplicem diametrum
sui corporis, habebatque eandem præcise cum ♃ latitudinem visam."

[2] *Danske Magazin,* ii. p. 182 (Weistritz, ii. p. 50).

[3] *T. B. et ad eum Doct. Virorum Epistolæ,* pp. 1–3. In this letter Tycho,
at Vedel's request, gives him a prescription against fever, and adds that he
could give him others, but will wait till he sees him, as he does not like to
put them in writing.

[4] *Danske Magazin,* 4th Series, vol. ii. pp. 324-325.

the Abbot, and to keep up divine service according to the Lutheran ritual, while he was to drive out "all superfluous learned and useless people." The Abbey does not appear to have been granted to him formally in fee till 1576.[1] We have already mentioned Steen Bille as the only one of Tycho's relations who appreciated his scientific tastes, and he seems indeed to have been a man of considerable culture, who took an interest in more than one branch of learning or industry. Tycho says that he was the first to start a paper-mill and glass-works in Denmark. Whether it was from living with this uncle, or from some other cause, that Tycho for a while devoted himself more to chemistry than to astronomy, is uncertain, but from the 30th December 1570 till November 1572 we do not possess a single astronomical observation made by him, while during this time he worked with great energy at chemical experiments, to which he had already paid some attention at Augsburg. His uncle gave him leave to arrange a laboratory in an outhouse of the Abbey, and was evidently himself much interested in the work carried on there.[2] Whether the object of this work was to make gold, as was most frequently the case with chemical experiments made in those days, there is no evidence to show; but even if this was not the case, there is nothing surprising in seeing an astronomer in the sixteenth century turn aside from the contemplation of the stars to investigate the properties of the metals and their combinations. We have already alluded to the idea of the universe as a whole, of which the single parts were in mystical mutual dependence on each other—an idea which had arisen among Oriental nations in the infancy of time, had thriven well owing to the mystical tendency of the Middle Ages, and had been gradually developed and formed into a complicated system by the speculations of philosophers

[1] Friis, *Tyge Brahe*, p. 31.   [2] *Progymn.*, p. 298.

of successive periods.   The planets were the rulers of the elementary world and of the *microcosmos*, the moon being represented among the metals by silver, Mercury by quick-silver, Venus by silver, the sun by gold, Mars by iron, Jupiter by gold or tin, and Saturn by lead.   It is therefore very probable that Tycho while working in the laboratory considered himself as merely for a while pursuing a special branch of the one great science, to the main branch of which he had hitherto felt specially attracted.   But if these mystical speculations had as yet some power over his mind, they would seem gradually to have been pushed into the background, while cool and clear reasoning took their place, and guided him safely to his great goal—the reformation of practical astronomy.

We have now followed Tycho through what may be con-sidered the first period of his life.   By study and intercourse with learned men he had mastered the results of the science of antiquity and the Middle Ages.   But though he had to some extent already, as a youth seventeen or eighteen years of age, perceived the necessity of a vast series of systematic observations on which to found a new science, he had hitherto shrunk from carrying out this serious undertaking himself, or had perhaps despaired of getting the means of doing so. But a most unusual and startling celestial phenomenon was now to occur, to rouse him to renewed exertion, and firmly fix in his mind the determination to carry out the plans he had so long entertained.

CHAPTER III.

THE NEW STAR OF 1572.

ON the evening of the 11th November 1572, Tycho Brahe
had spent some time in the laboratory, and was returning
to the house for supper, when he happened to throw his
eyes up to the sky, and was startled by perceiving an
exceedingly bright star in the constellation of Cassiopea,
near the zenith, and in a place which he was well aware
had not before been occupied by any star.   Doubtful
whether he was to believe his own eyes, he turned round
to some servants who accompanied him and asked whether
they saw the star; and though they answered in the
affirmative, he called out to some peasants who happened
to be driving by, and asked the same question from them.
When they also answered that they saw a very bright star
in the place he indicated, Tycho could no longer doubt his
senses, so he at once prepared to determine the position of
the new star.   He had just finished the making of a new
instrument, a sextant similar to the one he had made for
Paul Hainzel, and he was therefore able to measure the
distance of the new star from the principal stars in Cassiopea with greater accuracy than the cross-staff would
have enabled him to attain.[1]   In order to lessen the
weight, the instrument was not made of metal, but of
well-seasoned walnut-wood, the arms being joined by a
bronze hinge, and the metallic arc only 30° in extent, and

---

[1] The sextant is figured and described in *Astr. Inst. Progymnasmata*,
p. 337 *et seq.*

graduated to single minutes. The arms were four cubits, or about five and a half feet long,[1] three inches broad, and two inches thick; and to steady the instrument an undivided arc was attached to the arm which held the graduated arc, about eighteen inches from the centre, and passing loosely through a hole in the other arm, where it could be clamped by a small screw. This undivided arc and the long screw which served to separate the arms steadied the instrument, and kept its various parts in one plane. The graduated arc was not, as in his later instruments, subdivided by transversals, and the two sights were still of the usual kind, which he afterwards discarded, viz., two square metallic plates with a hole in the centre. The error of excentricity, caused by the unavoidable position of the observer's eye slightly behind the centre of the arc, was duly tabulated and taken into account.

With this instrument Tycho measured the distance of the star from the nine principal stars of Cassiopea. We can easily picture to ourselves the impatience with which he must have awaited the next clear night, in order to see whether this most unusual phenomenon would still appear, or whether the star should have vanished again as suddenly as it had revealed itself. But there the star was, and continued to be for about eighteen months, north of the three stars (now called β, a, γ Cassiopeæ) which form the preceding part of the well-known W of this constellation, and forming a parallelogram with them. It was only a degree and a half distant from a star (κ) of the $4\frac{1}{2}$ magnitude. Tycho continued, while the star was visible, to measure its distance from the other stars of Cassiopea; and in order to find whether it had any parallax, he repeated these measures from time to time during the night, and even left the

---

[1] One Tychonic cubit is = 16.1 English inches (D'Arrest, *Astr. Nachr.*, No. 1718, p. 219).

sextant clamped in the interval between two observations, to make sure that no change had taken place in the instrument in the meantime. The star being circumpolar for his latitude, he was able to follow it right round the pole, and he took advantage of this circumstance to observe its altitude at the lower culmination by the sextant, as he did not at that time possess a quadrant. He placed it in the plane of the meridian with the one arm, which we may call the fixed one, and to which he had now attached an arc of 60°, resting horizontally on a window-sill and a short column inside the room.[1] To ensure the horizontal position of the arm, it was moved until a plumb-line suspended from the end of the graduated arc touched a mark exactly at the middle of the arm,[2] and as the instrument might happen to be slightly moved while the observation was being made, a short graduated arc was traced at the middle of the arm, on which the plumb-line would immediately mark the small correction to be applied to the observed altitude. This simple but neat contrivance is highly characteristic of Tycho; we recognise here the modern principle of acknowledging an instrument to be faulty, and applying corrections for its imperfections to the results determined by it, a principle which we shall see he followed in the construction of all his instruments. From repeated observations he found the smallest altitude of the new star to be 27° 45′, and consequently, as he assumed the latitude of Heridsvad to be 55° 58′, the declination of the star was 61° 47′. He remarks that the declination was as constant as the distances from the neighbouring stars, and that the

---

[1] The sextant in this position is figured in *Progymnasmata*, p. 348; *Mechanica*, fol. E. 3 *verso*.

[2] Tycho considers it necessary to quote Euclid iv. 15 and i. 12 in explanation of this (*Progymn.*, p. 349). Euclid was apparently still considered an author with whose work but few were familiar (see H. Hankel, *Zur Geschichte der Mathematik im Alterthum und Mittelalter*, p. 355).

instrument was not perfect enough to show the change of
about a third of a minute which the precession of the
equinoxes made in the declination while the star was
visible, an amount which even his later and more perfect
instruments would hardly have been able to point out.[1]

With the sextant Tycho was not able to observe the
upper culmination of the star, at which it was only 6° from
the zenith. As a supplement to his own results, he therefore gives in his later work[2] the observations made with
the great quadrant at Augsburg by Paul Hainzel, which
give a value of the declination agreeing within a fraction
of a minute with his own.

The observations with the sextant must have occupied
Tycho during the winter of 1572–73, during which time
the brightness of the new star had already commenced to
decline considerably. When he first saw it, on the 11th
November, it was as bright as Venus at its maximum
brightness, and remained so during the month of November, so that sharp-sighted people could even see it in the
middle of the day, and it could be perceived at night
through fairly dense clouds. In December it was somewhat fainter, about equal to Jupiter; in January, a little
brighter than stars of the first magnitude; and in February and March, equal to them. In April and May it
was like a star of the second magnitude; in June, July,
and August, equal to one of the third, so that it was very

---

[1] *Progymn.*, p. 351. Individual results of observations are not given, and,
unluckily, the original observations of Nova are lost; at least they are not in
the volume of observations from 1563–81, repeatedly quoted. This only
contains the following observations of Nova :—" 1573 Die pentecostis 10
Maji inter flexuram Cassiopeæ et novam stellam 5ᵍ8', 5ᵍ0'. Inter supremam
cathedræ et novam stellam 5ᵍ28', 5ᵍ20'. Inter Schedir et novam stellam
8ᵍ5', 7ᵍ52'. Confide his observationibus subtracta tamen instrumenti parallaxi
(5 gradus habent parallaxin 8 minutorum, 8 gradus habent 13).—Augusti die
14 inter novam stellam et Polarem 25ᵍ9'."

[2] *Progymn.*, p. 360.

like the principal stars in Cassiopea. It continued to decrease during September, so that it reached the fourth magnitude in October, and was exactly equal to κ Cassiopeæ in November. At the end of the year and in January 1574 it hardly exceeded the fifth magnitude; in February it came down to the sixth magnitude, and about the end of March it ceased to be visible. At the same time the colour gradually changed; at first it was white, and by degrees became yellow, and, in the spring of 1573, reddish, like Betelgeuze or Aldebaran. About May 1573 it became like lead, or somewhat like Saturn, and seemed to remain so while the star was visible.[1]

About the time when the new star appeared Tycho Brahe had prepared an astrological and meteorological diary for the following year, giving the time of rising and setting of the principal stars, the aspects of the planets, and the phases of the moon, together with their probable influence on the weather. To this diary he added an account of his observations on the new star and its probable astrological signification. Early in 1573 he went to Copenhagen on a visit to his friend Professor Johannes Pratensis, and brought the manuscript with him. Pratensis had not yet heard of the new star, and would scarcely believe Tycho when he told him about it. Equally incredulous was another friend, Charles Dancey, French envoy at the Danish court, who invited Tycho and Pratensis to dinner as soon as he heard of the arrival of the former. During the dinner Tycho happened to mention the star, but Dancey thought he was joking and intending to sneer at the ignorance of Danish savants in astronomy, while Tycho only smiled, and hoped that the evening would be clear, so that they could see the star with their own eyes. The evening was favourable, and

---

[1] *Progymnasmata*, pp. 300-302, and p. 591, the latter place being a reprint of the preliminary account printed in 1573.

Pratensis and Dancey were as surprised as Tycho had been when they saw this new star, so utterly different from a comet (the only class of celestial bodies with which anybody thought of comparing it), and yet, according to Tycho's observations, more distant than the planets, and probably belonging to " the eighth sphere," which had always hitherto been considered the very picture of immutability.   Pratensis at once recollected the statement made by Pliny in the second book of his *Natural History* (on which he happened to be then lecturing at the University), that Hipparchus is said to have observed a new star; and perceiving the importance of the manuscript essay which Tycho had given him to read, he urged him to have it printed.   But Tycho declined, on the pretence that the essay had not received the final touches from his hand, but really because he was not quite free from the prejudice of some of his fellow-nobles, that it was not proper for a nobleman to write books.[1]

Tycho therefore returned to Scania with his manuscript. But when the spring came, and communication with Germany was reopened, he received from thence through Pratensis so many accounts of the star, both written and printed, containing a vast amount of nonsense, that he became inclined to let his own book be published, as it might serve to refute the erroneous statements circulated about the star.   During a second visit to Copenhagen he was entreated to publish the book, not only by Pratensis, but also by his kinsman, Peter Oxe, high treasurer of Denmark, whose sister had been the wife of Jörgen Brahe, and consequently had been a second mother to Tycho.   Shaken in his resolution by the persuasions of this intelligent man, who even suggested that he might hide his name under an anagram if he did not wish to put it on the title-page,

[1] *Progymnasmata*, p. 579.

Tycho finally yielded so far as to allow Pratensis to let the account of the star and the plan of the meteorological diary be printed, omitting the details of the latter. The book was therefore printed at Copenhagen in the year 1573, but very few copies appear to have been distributed or sent abroad, so that it afterwards became necessary to reprint the more important parts of it in the greater work, *Astronomiæ Instauratæ Progymnasmata*, on which Tycho was engaged during the last fourteen years of his life, and which was published after his death. The little book, *De Nova Stella*, is now extremely scarce, and does not appear to have been seen by any modern writer on the history of astronomy. It will therefore not appear inopportune if I give a somewhat detailed account of its contents in this place.[1]

On the back of the title-page is a versified address to the author from Professor Joh. Francisci Ripensis, one of his earliest friends. Then follows a letter from Pratensis, dated 3rd May 1573,[2] begging Tycho to print the book, at least the part relating to the star, the plan of the diary (if he should think the diary itself too long), and the forecast of the lunar eclipse. Tycho's answer comes next, dated "Knusdorp," 5th May. In this he remarks that

---

[1] The title given in Lalande's *Bibliographie* is erroneous. The complete title is: "Tychonis Brahe, Dani, De Nova et Nullius Aevi Memoria Prius Visa Stella iam pridem Anno a nato Christo 1572 mense Nouembrj primum Conspecta, Contemplatio Mathematica. Cui, præter exactam Eclipsis Lunaris, hujus Anni, pragmatian, Et elegantem in Vraniam Elegiam, Epistola quoque Dedicatoria accessit : in qua, nova et erudita conscribendi Diaria Metheorologica Methodus, utriusque Astrologiæ Studiosis, eodem Autore, proponitur. Cuius, ad hunc labentem annum, Exemplar, singulari industria elaboratum conscripsit, quod tamen, multiplicium Schematum exprimendorum, quo totum ferme constat, difficultate, edi, hac vice, temporis angustia non patiebatur." Hafniæ Impressit Lavrentius Benedictj, 1573. Printed in small 4to, 106 pp.

[2] Evidently written expressly for the book, as there is a previous letter from Pratensis in existence dated 16th April (*T. B. et ad eum Doct. Vir. Epist.*, p. 8), in which he says that the figures are being cut, but that there is some difficulty about the paper ; also that the word *lucubrationes* is not a good one for the title. Tycho must therefore have consented to the publication long before the 3rd of May.

the intricate diagrams and figures of the diary would be
very troublesome to reproduce, and the year is nearly half
over, so that it would not be worth while printing the
diary.   As to the star, he fears that the account of it is a
very immature one; still he will let it be published, partly
because his friend wishes it, partly because some of the
German accounts of the star place it at a distance of only
twelve or fifteen semi-diameters of the earth, while his own
observations of its distance from Schedir (α Cassiopeæ)
show that it is situated *in ipso cœlo*.   He has made his
observations with a new and exquisite instrument, much
better than a radius or any similar instrument, and the
horizontal parallax of three or four degrees, which the star
would have if it were as close to us as stated by the
German writers, would have been easily detected.   "O
cœcos cœli spectatores!"   Somebody had thought that it
was a comet with the tail turned away from the earth, but
that writer has forgotten what Apianus and Gemma Frisius
have taught us, that the tails of comets are turned away
from the sun, and not from the earth.   Others thought the
star was one of the tailless comets which the ancients
called *Crinitæ;* others again that it belonged to the class
called *Rosæ*, with a disc gradually fainter towards the edges.
But it looks exactly like other stars, and nothing like it
has been seen since the time of Hipparchus.   It does not
seem likely that it will last beyond September, or at most
October (1573), and it would be far more marvellous if
it remained, for things which appear in the world after
the creation of the universe ought certainly to cease again
before the end of the world.   As his own conclusions thus
differ so much from those of the German writers, he con-
sents to let his book be printed, and sends it herewith,[1]

[1] As mentioned above, the book was actually in the printer's hands when
this was written.

leaving him to settle the title, and only begging him to suppress the author's name or to hide it in an anagram, as many people are perverse enough to think it an *ingens indecus* for a nobleman to work in the free and sublime sciences. He has not had time to revise the manuscript owing to domestic affairs,[1] other studies, and social intercourse with friends. He next proceeds in some poetical lines [2] to declare his intention of seeing more of the world in order to increase his knowledge, as it will be time enough later on to return to the frigid North, and, like other nobles, waste his time on horses, dogs, and luxury, unless God should reserve him for something better. Having (in prose) assured Pratensis of his lasting friendship wherever they both may be, and reminded him that they shall at all events be contemplating the same sun, moon, and stars, he gives vent to his feelings in the following lines, which may serve as a specimen of Tycho's poetical effusions :—

> " Et quia disiunctis, Radios coniungere in unum
>   Non licet, et nosmet posse videre simul,
> Jungemus radios radiis radiantis Olympi,
>   Quando micant claro, sydera clara, Polo.
> Tunc ego, quam specto, figens mea lumina cœlo
>   Est quoque luminibus, Stella videnda, tuis.
> Sic oculos pariter Cœlum coniunget in unum,
>   Nostra licet iungi corpora Terra vetet."

Finally, he ends this lengthy introduction by asking Pratensis to urge the workmen who are making him a celestial globe and other instruments, so that they can all be ready when he comes over again.[3]

Next comes the account of the star, exactly as afterwards

---

[1] Perhaps this is an allusion to his having fallen in love about this time, as we shall see farther on.

[2] Reprinted in *Danske Magazin*, ii. p. 186 (Weistritz, ii. p. 59).

[3] This globe is mentioned at some length in the above-mentioned letter from Pratensis of April 16, 1573. It must have been a beautiful piece of work, the surface silvered with gilt stars, &c.

reprinted in his greater work, filling a little more than twenty-seven pages (A to second page of D2). As this is more generally accessible than the other parts of the book, a short abstract will suffice here. Having described how he first saw the star, he quotes the words of Pliny relating to the star of Hipparchus, which many had taken to be a comet; but as it would be absurd to fancy that a great astronomer like Hipparchus should not have known the difference between a star of the æthereal region and a fiery meteor of the air which is called a comet,[1] it must have been a star like the present one which he saw. Since that time no similar star has been seen till now, for the star of the Magi was not a celestial object, but something relating exclusively to them, and only seen and understood by them. How it was created he does not profess to offer an opinion about, but proceeds to treat of its position among the stars. This is illustrated by a diagram of the stars in Cassiopea, and the measured distances of the new star from a, β, and γ Cassiopeæ are given,[2] after which he shows how the rules of spherical trigonometry of Regiomontanus give the longitude and latitude of the star from these data. He adds that the accuracy of the co-ordinates deduced will, of course, depend on that of the positions of the fixed stars he has used; but as he has not any observations of his own to depend on, he is obliged to use the positions given by Copernicus, trusting that God will spare him and enable him to correct the accepted places of the fixed stars by new observations. In order to find the distance of the star from the earth, he has measured its angular distance from Schedir,

[1] The expression is remarkable, as it shows that Tycho had not yet shaken himself free from the old Aristotelean opinion of the comets as atmospheric phenomena.

[2] Respectively 7° 55', 5° 21, and 5° 1'; while *Progymnasmata*, p. 344, gives 7° 50'.5, 5° 19', and 5° 2'. Tycho remarks (ibid., p. 593) that the latter results are corrected for the error of excentricity, and were made with the improved pinnules which he afterwards adopted.

which passed the meridian nearly at the same time, both at
upper and lower culmination, and found no difference whatever; whereas he shows that there would have been a
parallax at lower culmination equal to $58\frac{1}{2}'$ if the star had
been as near to us as the moon is.[1] Therefore the star
could not be situated in the elementary region below the
moon, nor could it be attached to any of the planetary
spheres, as it would have been moved along with the
sphere in question in a direction contrary to that of the
daily revolution of the heavens, while his observations
show that it has since its first appearance remained immovable. Consequently, it must belong to the eighth
sphere, that of the fixed stars; and it cannot be a comet
or other fiery meteor, as these are not generated in the
heavens, but below the moon, in the upper regions of
the air, upon which all philosophers agree, unless we are to
believe Albumassar, who is credited with the statement
that he had observed a comet farther off than the moon,
in the sphere of Venus. Here again Tycho expresses the
hope that he will some time get a chance of deciding this
matter (as to the distance of comets); but anyhow, he
adds, this star cannot have been a comet, as it had neither
the appearance of one nor the proper motion which a comet
would have been endowed with.

The third paragraph deals with the magnitude and colour
of the star. The volume of a star is very considerable; the
smallest are eighteen times as great as the earth, those
of the first magnitude 105 times as great. Therefore the
new star must have been of immense size. He then
describes its gradual decline, until it "now, at the beginning
of May, does not exceed the second magnitude." It must,

---

[1] He assumed that the parallax would be $= 0$ at the upper culmination,
but in his later work he remarks that this is a mistake, and that it would be
nearly $7'$.

therefore, at first have been much more than a hundred times as large as the earth, but it has decreased in size. It twinkles like other stars, while the planets do not twinkle, which is another proof of its belonging to the eighth sphere. Having mentioned the change in colour, he finishes the astronomical part of the treatise on the star by remarking that the change in colour and magnitude does not prove it to be a comet or a similar phenomenon, for if it is possible that a new body can be generated in the æthereal region, as he has proved to be the case in opposition to the opinions of all philosophers, it must be considered far less impossible and absurd that this new star should change in brightness and colour. And if it could ever, beyond the ordinary laws of nature, have been seen in the heavens, it would not be more absurd if it should again cease altogether to be visible, though again in opposition to those laws.

Tycho now proceeds to give his opinion about the astrological effects of the new star.[1] These cannot be estimated by the usual methods, because the appearance of the star is a most unusual phenomenon. The only known precedent is the star said to have appeared at the time of Hipparchus, about B.C. 125. It was followed by great commotions both among the Jewish people and among the Gentiles, and there is no doubt that similar fatal times may be expected now, particularly as the star in Cassiopea appeared nearly at the conclusion of a complete period of all the Trigoni.[2] For in about ten years the watery Trigon will

---

[1] Fol. D. 2 *verso* to E. 3.

[2] As some readers may not be familiar with the phraseology of astrology, it may be well to mention here that each Trigonus consists of three signs of the Ecliptic, 120° from each other; the four Trigoni correspond to the four elements, and each of them is in turn the ruling one, until a conjunction of planets has taken place within one of its signs. In about 800 years the four Trigoni will all have had their turn, and a cycle is completed. See, *e.g.*,

end with a conjunction of the outer planets in the end of
the sign of Pisces, and a new period will commence with
a fiery Trigon. Referred to the pole (*i.e.*, according to
right ascension), the new star belongs to the sign of
Aries, where the new Trigon will also begin, and there will
therefore be great changes in the world, both religious and
political. The star was at first like Venus and Jupiter,
and its effects will therefore first be pleasant; but as it then
became like Mars, there will next come a period of wars,
seditions, captivity, and death of princes and destruction
of cities, together with dryness and fiery meteors in the
air, pestilence, and venomous snakes. Lastly, the star
became like Saturn, and there will therefore, finally, come
a time of want, death, imprisonment, and all kinds of sad
things. As it is not exactly known when the star first
appeared, he follows the example of Halus, a commentator
of Ptolemy (on the occasion of the appearance of a comet),
and assumes that it appeared at the time of new moon,
the 5th of November,[1] at 7h. 55m., for which moment he
finds that Mars was the ruling planet. The places most
affected by the star will be those in latitude 62° (in the
zenith of which the star could be); but as the star belongs
to Aries, its influence will be felt nearly over the whole of
Europe, and particularly after the great conjunction (of
April 1583) has added its great power to that of the
star.

It will be seen that this prediction is only expressed in
very vague terms, and we shall find, when we come to analyse

Cyprianus Leovitius, *De Conjunctionibus Magnis*, London, 1573; Kepler, *De
Stella Nova*, 1606, p. 13 (*Opera Omnia*, ii. p. 623); Ideler, *Handbuch der
Chronologie*, ii. p. 401. More about this in Chapter VIII.

[1] No doubt he was right, as this would be a capital day for a celestial
explosion to take place! The date and time of the first appearance was
required to prepare the horoscope of the star in the usual manner (see below,
Chapter VI.).

Tycho's later writings, that he afterwards modified and extended it.   When he wrote his preliminary treatise on the new star, he was evidently chiefly inclined to ascribe a direct physical and meteorological influence to the celestial bodies, though he was by no means blind to the difficulty of foretelling the results of this influence, but he became gradually more inclined to disregard the physical effect (dryness, pestilence, &c.), and solely to look to the effect of the stars on the human mind, and through that on the human actions. That an unusual celestial phenomenon occurring at that particular moment should have been considered as indicating troublous times, is extremely natural when we consider the state of Europe in 1573.   The tremendous rebellion against the Papal supremacy, which for a long time had seemed destined to end in the complete overthrow of the latter, appeared now to have reached its limit, and many people thought that the tide had already commenced to turn.   In the south of Germany and in Austria the altered tactics of the Church of Rome, due to the influence of the fast rising Society of Jesus, were stamping out the feeble attempts of Reformers; in France, the Huguenots were fighting their unequal battle with the fury of despair against an enemy who a year ago had attempted to end the strife by the infamous butchery of St. Bartholomew;[1] in the Netherlands, hundreds had suffered for their faith, while the country was being devastated with fire and sword in the vain efforts of the Spanish Government to make a free nation submit to their own sanguinary religion; in England, the hopes of Protestants might at any moment be seriously threatened if the dagger of an assassin should find way to the heart of their queen, or if her most formidable and venomous enemy should turn his dreaded power against her.   Who could

[1] The year 1572 was, according to the custom of the age, remembered by the line " LVtetIa Mater sVos natos DeVoraVIt."

doubt that fearful disturbances were in store for the generation that beheld the new star as well as for the following
one ?   The moderation of Tycho's astrological predictions is
therefore remarkable, and becomes more conspicuous if we
compare his opinions with the many silly ones expressed by
contemporary writers.

Before we say a few words about these, we shall, however, finish the review of the contents of Tycho's book.
We have already mentioned that he did not think it worth
while to print the astrological calendar for the year 1573,
of which the treatise on the new star originally formed a
part, but that he contented himself with publishing the
introduction, setting forth the principles on which the
calendar, or diary, as he calls it, had been constructed.
This fills sixteen and a half pages.   It begins with a good
deal of abuse of the ordinary prognostications, the absurdities of which he intends to expose in a book to be called
Contra Astrologos pro Astrologia.   This intention he does
not seem to have carried into effect, and two other treatises,
which he says were already written, seem not to have
been preserved.[1]   He remarks that both the Alphonsine
and the Prutenic Tables are several hours wrong with regard
to the time of the equinoxes and solstices, and it is useless to give the time of entry of the sun into any part of
the Zodiac to a minute, as the sun in an hour moves
less than 3', a quantity which cannot be observed with
any instrument.   Some writers are foolish enough to give
minutes and seconds when stating the time of any particular
position of a planet, although at the conjunction of Saturn
and Jupiter in 1563 the Alphonsine Tables were a whole

---

[1] One of these was *De variis Astrologorum in Cœlestium Domorum Divisione
Opinionibus, earumque Insufficientia*, in which he proposed a new plan of
dividing the heavens into "houses" by circles through the points of intersection of the meridian and horizon. The other treatise was *De Horis
Zodiaci inæqualibus, quas Planetarias vocant.*

month in error, while even the Prutenic Tables could hardly
fix the day correctly, not to speak of minutes or seconds.[1]
A calendar should contain the usual information as to the
aspects, time of sunrise and sunset, time of rising and
setting of the moon and planets, and the names of the
principal stars rising and setting at the same time. The
moon is of particular importance as it is nearest to the
elementary world, but even the planets must influence the
weather. Lastly, a calendar should give the probable
weather for every day, concluded from the configurations of
the celestial bodies. He would warn the reader not to expect
too much from the weather predictions, partly because much
remains yet to be done in exploring the motions of the
stars and their effect, partly on account of the fluidity of
the inferior matter, which sometimes delays, sometimes
hastens the effect produced by the stars. But any blame
should rest on him and not on the art. Besides terrestrial
influences must act differently in different parts of the earth,
so that one configuration of the stars cannot have the same
effect in several localities. Therefore he has undertaken
this work principally in order by observation to learn the
effect of the stars on this part of the earth, so that our
posterity may profit thereby, and in order to secure this
object he exhorts all meteorologists to take observations of
the weather.

The only part of the diary given in the book is that
relating to the total eclipse of the moon on December 8,
1573. It fills twenty-four pages, including two full-page
woodcuts—one of the progress of the eclipse, the other of
the earth, moon, and planets at that time. He gives first
the calculation of the eclipse by the Prutenic Tables, with all

[1] O audaces astronomos, O exquisitos & subtiles calculatores, qui Astronomiam in Tuguriis & popinis, vel post fornacem, in libris & chartis, non
in ipso cœlo (quod par erat) exercent. Plerique enim ipsa sidera (pudet dicere)
ignorant. Sic itur ad astra " (fol. G.).

the details step by step, and for the sake of comparison, the resulting time of the various phases (without details of calculation) by the Alphonsine Tables and Purbach's Tables; also the same data after correcting the places of the sun and moon by his own observations.[1] He adopts the meridian 35° *ab occasu*, by which he probably means 35° east of the peak of Teneriffe. He recommends observers to discard clocks of any kind, but to fix the time by observing altitudes of some stars not too far from the east or west horizon, but which, when on the meridian, would have a considerable altitude, and he gives the altitudes of a few stars for the beginning and end of the eclipse and of totality. The astrological significance he computes by the rules given by Ptolemy in the second book of his Tetrabiblion. Mercury, and in the second place Saturn, are the ruling planets. The former means robbery, stealing, and piracy; the latter want, exile, and grief. The regions chiefly affected by the eclipse are those which Ptolemy specially connected with the sign of Gemini, where the moon is. These are Hircania, Armenia, Cyrene, Marmaria, and Lower Egypt, to which later astrologers have added Sardinia, Lombardy, Flanders, Brabant, and Würtemberg. It has been observed that the sign of Gemini has a special significance for Nürnberg whenever an eclipse or a conjunction took place in it, and the Nürnbergers may therefore expect something, possibly pestilence, as Gemini is a "human sign,"

[1] It may not be without interest to insert these data here:—

	Tabulæ Prut.			Ex propria Motuum ratione.	
	H.	M.	S.	H.	M.
Initium primæ obscurationis . .	5	55	41	6	15
Initium totius obscurationis . .	6	59	50	7	20
Medium	7	51	29	8	10
Finis totius obscurationis. . .	8	43	8	9	0
Finis ultimæ obscurationis . .	9	47	17	10	5
Locus ☉ in Sagit . . .	26°	29′		26°	40′

and also on account of the positions of Mercury and Mars.
Countries whose rulers were born when Gemini was culminating may also be on the look-out, and generally speaking kings and princes are more affected by eclipses than
private people ("as I have observed myself"), because the
sun and the moon are the princes among the planets. The
effect will last as many months as the eclipse lasted hours,
and the beginning of it depends on the moon's distance
from the horizon at the commencement of the eclipse.
This eclipse will take effect from March till July 1574 (for
latitude 56°). As examples of this kind of prognostications he quotes several recent eclipses. First, the lunar
eclipse of April 3, 1558, after which Charles V. died; then
the solar eclipse of April 18, 1558, which did not begin
to take effect till the end of the year, and then Christian
III. of Denmark and Norway died; and shortly afterward
the deposed king, Christiern II., a captive for many years,
died also; and Tycho shows how beautifully this agreed
with their horoscopes. On November 7, 1565, a lunar
eclipse took place near the Plejades, a group of stars with
a moist and rainy influence, and consequently rainy weather
came on, as he had at the time predicted at Leipzig.
Similarly a lunar eclipse occurred on the 28th October
1566 close to Orion, and the effect should, according to
Ptolemy, begin at once; and so it did, and the whole winter
turned out wet, as he predicted himself at Rostock. He
does not say a word about the old Sultan!

The book is wound up with *In Vraniam Elegia Autoris*,
filling more than eight pages, and a page of verses by
Vedel. In the *Elegy*, Tycho promises soon to produce
something better, as neither the sneers of idle people nor
the hardships of study shall deter him; let others boast of
their achievements in war or of their ancient family, let
others seek the favour of princes or hunt for riches, or waste

their time gambling and hunting, he does not envy them,
for though sprung from ancient races both on the father's
and on the mother's side, he does not value it, and calls
nothing his but what has originated with himself.[1]   But
his mind is planning great things, and happy above all men
is he who thinks more of celestial than of earthly things.

I have given a very full account of the contents of Tycho's
little book, not only because it is now extremely scarce, but
also because it is very characteristic of him, and presents
us with a perfect picture of the young author, his plans and
his difficulties.   We see him thoroughly aware of the great
desideratum of astronomy, a stock of accurate observations,
without which it could not possibly advance a single step
further, and hoping that life and means might be granted
him to supply this deficiency; we see in him at the same
time a perfect son of the sixteenth century, believing the
universe to be woven together by mysterious connecting
threads which the contemplation of the stars or of the
elements of nature might unravel, and thereby lift the
veil of the future; we see that he is still, like most of
his contemporaries, a believer in the solid spheres and
the atmospherical origin of comets, to which errors of the
Aristotelean physics he was destined a few years later to
give the death-blow by his researches on comets; we see
him also thoroughly discontented with his surroundings, and
looking abroad in the hope of finding somewhere else the
place and the means for carrying out his plans.   At the
same time the book bears witness to the soberness of mind
which distinguishes him from most of the other writers on
the subject of the star.   His account of it is very short, but
it says all there could be said about it—that it had no
parallax, that it remained immovable in the same place,

---

[1] "Nil tamen his moveor.   Nam qvæ non fecimus ipsi
Et genus et proavos, non ego nostra voco."

that it looked like an ordinary star—and it describes the star's place in the heavens accurately, and its variations in light and colour.   Even though Tycho made some remarks about the astrological significance of the star, he did so in a way which shows that he did not himself consider this the most valuable portion of his work.   To appreciate his little book perfectly, it is desirable to glance at some of the other numerous books and pamphlets which were written about the star, and of most of which Tycho himself has in his later work (*Progymnasmata*) given a very detailed analysis, devoting nearly 300 pages to the task.   It would lead us too far if we were to follow him through them all, but it will not be without interest briefly to describe what some of the more rational of his contemporaries published about the star, and to what absurdities a fervid imagination led some of the common herd of scribblers.

At Cassel the star was observed by Landgrave Wilhelm IV., an ardent lover of astronomy, of whom we shall hear more in the sequel.   He did not hear of the star till the 3rd December, and took observations of its altitude in various azimuths from that date and up to the 14th March following.   From the greatest and smallest altitude Tycho found afterwards a value of the declination differing less than a minute from that found by himself.   From the azimuths and altitudes observed at Cassel Tycho deduced the right ascension and declination: the single results for the latter are in good accordance *inter se*, while those for right ascension differ considerably, the greatest difference being more than 2°.   Tycho justly concludes that this must be caused by the bad quality of the clock employed by the Landgrave, who merely gave the time of observation in true solar time, without furnishing the means of correcting for the error of his clock.   In a letter to Caspar Peucer, the Landgrave stated that the star might have a parallax not exceeding 3′,

as there was that difference between the polar distances
above and below the pole ; but his instruments had at that
time not reached the degree of accuracy which they did ten
years later, and the difference is not surprising. Peucer
and Wolfgang Schuler at Wittenberg found a parallax of
19′, which Tycho believed was a consequence of their having
used an old wooden quadrant ; and, in fact, when he learned
that the Landgrave had found little or no parallax, Schuler
had a large triquetrum constructed, and also found that the
star had no parallax, or at most a very small one.[1] Many
observers measured the distance of the new star from the
neighbouring ones, but the results found were generally
considerably in error. Thus the Bohemian, Thaddæus
Hagecius, physician to the Emperor, in an otherwise sensible
book,[2] gives a number of observed distances, some of which
are 7′ to 12′ (one is even 16′) wrong, and even the English
mathematician, Thomas Diggs (or Digges), who had made
a special study of the cross staff, and had his instrument
furnished with transversal divisions, differed 1½′ to 4′ from
Tycho,—possibly, as the latter thinks, because he did not
allow sufficiently for the error of excentricity.[3] Cornelius
Gemma, a son of the well-known astronomer, Gemma
Frisius, and professor of medicine at Louvain, had a great

---

[1] The triquetrum had been much in use from the time of Ptolemy. It consisted of two arms of equal length and movable round a hinge, while a third
and carefully graduated arm gave the means of measuring the angle between
the two former by the aid of a table of chords.

[2] "Dialexis de novæ et prius incognitæ Stellæ invisitatæ Magnitudinis et
splendissimi Luminis Apparitione et de eiusdem Stellæ vero loco constituendo.
Per Thaddæum Hagecium ab Hayck." Francofurti, a. M. 1574. 176 pp. 4to.
In an appendix are two papers on the star by Paul Fabricius and Corn.
Gemma. Some years after Hagecius sent Tycho a copy with many MS.
corrections and additions, which Tycho quotes extensively in his *Progymnasmata* (p. 505 *et seq.*). In this corrected copy the most erroneous measures had
been improved or struck out, whereby the greatest differences from Tycho's
results were reduced to 4′ or 6′.

[3] "Alæ seu Scalæ Mathematicæ, quibus visibilium remotissima Cœlorum
Theatra conscendi, et Planetarum omnia itinera novis et inauditis methodis
explorari . . . Thoma Diggeseo authore." Londini, 1573. 4to.

deal to say about the star, but most of his distance measures
are upwards of a degree wrong. On the other hand,
Michael Mœstlin, the teacher of Kepler, though he possessed
no instruments, determined the place of the star with fair
accuracy simply by picking out four stars so placed that the
new star was in the point of intersection of two lines drawn
through two and two of them. As the star did not move
relatively to these four stars during the daily revolution of
the heavens (of which he assured himself by holding a
thread before the eye, so that it passed through the three
stars), Mœstlin concluded that it had no parallax, and that
it was situated among the fixed stars, whose distance Copernicus, of whom he was a follower, had shown to be
extremely great. Digges tried the same method, using a
straight ruler six feet long, which he first suspended vertically until he found two stars which were in the same
vertical as the new star; six hours afterwards he tried
again, holding the ruler in his hand, whether the three
stars were still in a straight line. He found the star to be
exactly in the point of intersection of the line joining β
Cephei and γ Cassiopeæ, and the line joining ι Cephei and
δ Cassiopeæ, and concluded that it could not have a parallax
amounting to 2′. Tycho afterwards computed the place of
the star from these data, using his own accurate positions of
the four stars, and found the longitude only 2′ greater and
the latitude ½′ greater than what he had deduced from his
own observations.[1] Digges had hoped to test the Copernican theory of the motion of the earth by trying whether
the star had an annual parallax, but he could find none.

---

[1] *Digges*, l.c., chapter x., fol. K 3. By a mistake he says that the two lines
join δ Cassiopeæ, β Cephei, and ι Cephei, γ Cassiopeæ. Tycho remarks that
one can see at a glance that these two lines do not intersect each other between
the stars, but pretending not to see that it is merely a *lapsus calami*, he gravely
calculates places from these data, using his own distances, and of course gets
absurd results (*Progymn.*, p. 681), after which he interchanges the stars, and
gets the correct result given above.

The question of the star's distance from the earth being one
of special interest, all observers tried to determine the daily
parallax, but the results varied immensely according to the
skill of the observer.   While several writers, in addition to
those already mentioned, state that they could find no perceptible parallax,[1] others found a large one.   Thus Elias
Camerarius at Frankfurt on the Oder had at first thought
that he had found a parallax of 12', but in January 1573
he could only find one of $4\frac{1}{2}'$, from which he concluded that
the star had in the meantime receded from us in a straight
line (so that its apparent place was not altered), and that
this was the cause of its diminished brightness.   A German
writer of the name of Nolthius tried to find the parallax by
a method suggested by Regiomontanus from the hour angle,
the azimuth and the latitude of the observing station, comparing the altitude computed from these with the observed
one.   He chose, however, a bad time for the experiment,
when the altitude was very great (77°), and it is not to be
wondered at that he found an absurd result—39' for the
parallax—and it does not seem to have struck him that this
would correspond to a parallax equal to 2° 42' at the lowest
altitude of the star, which could not have escaped even
casual observers, as pointed out by Tycho.[2]

Of greater interest than these crude attempts are the
statements of the various writers as to the time when the
star first became visible.   Some writers say that the star
was already seen early in October, but none of them are
entitled to much credit.   The above-named Elias Camerarius
at Frankfurt on the Oder says that it appeared " in principio
Octobris Anni 1572 uesperi circa horam 10 prope Meridi-

---

[1] Thus Paul Fabricius at Vienna, Hainzel (using the great quadrant at
Augsburg), Reisacher at Vienna, Corn. Gemma (not stating how found),
Hieronimus Munosius at Valencia, Valesius from Covarruvias (physician to
Philip II. of Spain), and Johan Prætorius (Richter), professor at Wittenberg.

[2] *Progymn.*, i. p. 760.

anum;" but as he appears to be utterly unknown in the
history of science, too much weight ought not to be attached
to his unsupported statement.[1]   Annibal Raimundus of
Verona (of whom we shall hear more presently) tells us that
the star was seen "circa principium Octobris, a plurimis
Nobilibus et Ignobilibus, eruditis atque indoctis," but further
on he contradicts himself, saying that the star has now been
visible three months, and as he wrote at the end of January
1573, this would make the appearance of the star date
from the *end* of October or the beginning of November.[2]
A little French book, published in 1590, states that the
star was seen "au mois d'Octobre" in Spain by shepherds
keeping watch over their flocks, but this reminds one too
much of the words used by St. Luke, and is contradicted by
other testimony.[3]   According to Paul Fabricius at Vienna
it appeared "sub Octobris finem."[4]   All these statements
are contradicted by Munosius, professor in the University of
Valencia, who maintained that he was certain the star had
not yet appeared on the 2nd November, as he was showing
his pupils the constellations on that night, and could not
have failed to see it, and Spanish shepherds agreed with
him therein.   As Munosius took very fair distance measures
of the star, and wrote in a sensible strain, there is every
reason to believe him.[5]   The first trustworthy observation

[1] Ibid., p. 692. Tycho never saw the book, and only knew it from a MS.
abstract made for him by Hagecius. It is in the Poulkova Library, and
W. Struve mentions that the writer states in two places that he saw it
"principio Octobris." *Astron. Nachr.*, xix. p. 334. Elias Camerarius is not
mentioned in Jöcher's *Gelehrten Lexicon*, nor in any other historical work
that I have at hand.

[2] *Progym.*, i. pp. 721-723. Tycho remarks that Raimundus has forgotten
the proverb that liars should have a good memory.

[3] *La novvelle Estoille apparve svr tous les Climats dv Monde: Et de ses effects.*
Paris, 1590. 28 pp. small 8vo. This book was not known to Tycho Brahe.

[4] *Hagecii Dialexis*, p. 129. Tycho remarks (*Progym.*, p. 548) that if it had
been visible in October, Mœstlin (who saw it "the first week in November")
would probably have noticed it.

[5] *Progymn.*, i. pp. 565, 566.

seems to have been made by Wolfgang Schuler at Wittenberg, who says that he saw it at six o'clock in the morning
on the 6th November.[1]   On the 7th at 6 P.M. it was seen
by Paul Hainzel,[2] and the same evening by Bernhard
Lindauer, minister at Winterthur in Switzerland.[3]   Maurolycus, the well-known astronomer at Messina, and David
Chytræus at Rostock, saw it on the 8th.[4]   Many writers
have quoted the words of Cornelius Gemma, stating that the
star appeared first on the 9th November, and that it had
not been visible on the previous evening in clear weather,[5]
but they have overlooked the fact that Gemma, in his
book *De Naturæ Divinis Characterismis, seu raris & admi-
randis Spectaculis*, Libri ii. (Antwerp, 1575, 2 vols. 8vo),
tells quite a different story, viz., that some people had
already seen it before the end of October.   He does not
say when he first saw it himself, but he did not begin to
observe its position till the 26th November, as he thought
it idle talk when he first heard of a new star.[6]   Gemma's

---

[1] *Progymn.*, i. p. 621.   [2] Ibid., p. 536.

[3] Rudolf Wolf in *Astr. Nachr.*, lxv. p. 63.

[4] About the observation of Maurolycus, see *Nature*, xxxii. p. 162 (June 18,
1885).   About Chytræus, see R. Wolf, *Geschichte der Astronomie*, p. 415.

[5] "Nona Nouembris, die Dominico vesperi, cum tamen obseruantibus proxi-
mum cœli locum die octauo, etiam sereno æthere non apparuerit" (*Hagecii
Dialexis*, p. 137), also in his separate pamphlet, "Stellæ Peregrinæ iam
primum exortæ et Coelo constanter hærentis φαινόμενον .... per D. Cornelium Gemmam." Lovanij, 1573.   13 pp. 4to (fol. A2).   There is one reprint (s.a.e.l.) of this, with some omissions, and coupled with a paper by
Postellus, and another coupled with a reprint of a paper by Cyprianus Leovitius.   Among writers who have quoted Gemma may be mentioned Newton
(*Principia*, iii., ed. Le Seur and Jacquier, p. 670), who thought that Gemma
himself had looked at the sky on the 8th without seeing it; but this was a
mistake, as we have just shown above.

[6] Gemma's book is a very curious one.   The first volume is about terrestrial
curiosities, Siamese twins, and much queerer beings (well illustrated); vol. ii.
is about celestial wonders, comets, &c., chapter iii. being "De prodigioso
Phænomeno syderis noui" (pp. 111-155).   Page 113 : "Sed qui se primos ob-
seruasse voluerunt, nonum diem pro initio tradiderunt : cum tamen interea
conuenerim plures, quorum alij diem secundum aut tertium annotarint,
plerique vel ante Octobris finem ferant etiam a vulgaribus obseruatum. . . .
Primum observationis tempus fuit nobis die Nou. 26."

testimony is therefore worth nothing, and it may safely be
assumed that the star became visible between the 2nd and
the 6th of November, and was seen by an apparently trustworthy observer on the morning of the 6th.

That many different attempts should be made to explain
the nature of the new star and the cause of its sudden
appearance is very natural.   Most writers contented themselves by saying that it was some sort of a comet, though
not of the usual kind, as these, according to Aristotle,
were sublunary, while the star was far beyond the moon.
That it did not in the least look like a comet was generally
not considered an objection to this theory, as instances
could be quoted of comets having appeared without tails;[1]
a greater difficulty was the absence of motion relatively to
the other stars in Cassiopea, as only very few writers had the
hardihood to maintain that it had actually moved before it
disappeared.[2]   Gemma sought to explain this by supposing,
with Elias Camerarius, that the star was moving in a
straight line away from us,[3] but this could not account
for the sudden appearance of the star with its maximum
brightness.   Others thought it more probable that the star
was not a new one, but merely an old and faint star, which
had become brighter through some sudden transformation

[1] In a pamphlet, "La Declaration d'vn comete ou estoille prodigieuse
laqvelle a commencé a nous apparoistre à Paris, en la partie Septentrionale
du ciel, au mois de Nouembre dernier en l'an present 1572, & se monstre
encores auiourd'huy.   Par I. G. D. V.," Paris, 1572, 4to, 8pp., it is said that
people who had good sight could see several rays, of which the longest, which
might be called the tail of the comet, was always turned to the east ! Its
distance from the pole-star, when above the pole, was "le plus souvent"
25° 30', but afterwards it became 24° 40', and below the pole 24° 30', which
the author takes to be the effect of parallax ! The author was probably Jean
Gosselin de Vize, librarian to the King. The pamphlet was not known to
Tycho ; it is not in Lalande's *Bibliographie*.

[2] Leovitius, writing in February 1573, says it seems to him that the star
had during the last month moved three degrees towards the north !

[3] The English astronomer, John Dee, was of the same opinion (*Progymn.*,
p. 691).

of the air between it and the earth, or a condensation of part of one of the spheres through which its light had to pass. The principal reason why some writers (*e.g.*, Reisacher and Vallesius) adopted this explanation was, that God had ceased creating on the sixth day, and nothing new had been made since then. Reisacher had at first thought that the star was identical with κ Cassiopeæ, which had merely become brighter, but when the light of the star had become less dazzling he perceived that κ was still in the heavens, and that he had merely failed to see it hitherto owing to the overpowering light of the new star. More obstinate was Raimundus of Verona, who in two publications maintained that it was nothing but κ. He seems to have done so with unnecessary heat, and using contemptuous expressions about people who thought differently, as Tycho in reviewing his writings uses stronger language than usual, and Hagecius thought it necessary to publish a refutation full of the most violent invectives and written in a very slashing style.[1] Another Italian, Frangipani, also took the star to be κ Cassiopeæ, and as its place did not agree with that assigned to the latter by Ptolemy, he calmly assumed that the old star must have moved. He quotes the old story about the seventh star of the Plejades (Electra) having disappeared after the destruction of Troy, and asserts that the pole-star did the same for a while after the taking of Constantinople by the Turks.[2] All this is, however, very tame compared with the fancies of a German painter, Georg Busch, of Erfurt, who wrote two pamphlets "Von dem Cometen." According to him it was a comet, and these bodies were formed by the ascending from the earth of human sins and wickedness, formed into a kind of gas,

[1] "Thaddæi Hagecij ab Hayck, Aulæ Cesareæ Medici, Responsio ad virulentum et maledicum Hannibalis Raymundi Scriptum," &c.   Pragæ, 1576. 4to.

[2] *Progymnasmata*, p. 743.

and ignited by the anger of God. This poisonous stuff falls down again on people's heads, and causes all kinds of mischief, such as pestilence, Frenchmen (!), sudden death, bad weather, &c. Perhaps it was the night of St. Bartholomew which made Busch think of Frenchmen in this connection.

The question as to whether new stars had ever appeared before was touched on by several writers, who referred to the star of Hipparchus and the star of Bethlehem. Landgrave Wilhelm IV., in his letter to Peucer, also alludes to the star stated by Marcellinus to have appeared A.D. 389.[1] Cyprianus Leovitius states that similar stars appeared in the same part of the heavens in the years 945 and 1264, the " comet " of the latter year being without a tail and having no motion, and says that this information was taken from an old manuscript.[2] It is certainly a very suspicious circumstance that real comets appeared both in 945 and in 1264, and the absence of tail and motion might merely be subsequent embellishments by the writer of the manuscript referred to by Leovitius; but, on the other hand, it is quite possible that new stars may have appeared in those

---

[1] About this star see *Calvisii Opus Chronologicum*, p. 413 (second edit., 1620), and *The Observatory*, vii. p. 75.

[2] "De Nova Stella. Judicium Cypriani Leovitii a Leonicia, Mathematici, de nova stella siue cometa, viso mense Nouembri ac Decembri A.D. 1572. Item mense Januario & Februario A.D. 1573. Lavingæ ad Danubium, 1573." 4to, 8 fol. :—"Historiæ perhibent tempore Ottonis primi Imperatoris, similem stellam in eodem fere loco Coeli arsisse, A.D. 945. Vbi magnæ mutationes plurimaque mala, uarias Prouincias Europæ peruaserunt, potissimum propter peregrinas gentes infusas in Germaniam. Verum multo locupletius testimonium in historijs extat de A.D 1264, quo Stella magna & lucida in parte Coeli Septentrionali circa Sydus Cassiopeæ apparuit, carens similiter crinibus, ac destituta motu suo proprio." In the margin, opposite the date 1264, is : "Descriptio huius Cometæ desumpta est ex antiquo codice, manu scripto. Euentus hi congruent cum significationibus stellæ propositæ: quod bene notandum est : videoque hic aliquid insigne." Tycho has reprinted the whole pamphlet (pp. 705-706), leaving out the "Judicium breve" at the end, and also the marginal notes. The latter are also omitted from a reprint published (s. l.) in 1573, together with a reprint of Gemma's pamphlet.

years without being noticed by other chroniclers, as science was then at it its lowest ebb in Europe, and a new star of perhaps less than the first magnitude and of short duration (like the stars of 1866 and 1876) could easily escape detection.[1] The only other contemporary author who alludes to the years 945 and 1264 is Count Hardeck, who in 1573 was Rector of the University of Wittenberg; but as his little book is dated the 1st May 1573, and that of Leovitius the 20th February, he would have had time to copy from Leovitius, and in any case it is certain that he speaks of a real comet of the year 1264, as he mentions its tail, while it is doubtful whether he means a comet or a star when speaking of 945.[2] It has been repeatedly suggested that the star of Cassiopea might be a variable star, with a period of about three hundred years, in which case it should again become visible about the present time, but it is needless to say that the vague assertions of Leovitius form a very slender foundation on which to build such a

---

[1] According to Klein, *Der Fixsternhimmel*, p. 102, the Chronicle of Albertus Stadensis (Oldenburg) mentions a bright star in Capricornus in 1245 (not alluded to elsewhere), as bright as Venus, but more red, and which lasted for two months.

[2] "Orationes duae. Vna de legibus et disciplina. Altera de Cometa inter Sidera lucente in mensem septimum, continens commonefactionem de impendentibus periculis. A Joh. Comite Hardeci. Wittenberg, 1573." 8vo. Fol. C., p. 2:—"Reperimus Cometas qui ante hæc tempora in eodem octaui orbis loco fulserunt, fere gentes concitasse Boreas, suis excitas sedibus, ad quærendas nouas. Qui Honorij principatu conspectus est, cuius meminit Claudianus, haud dubie finem Imperio occidentis cum tristi ac horribili ruina attulit . . . Qui Ottone primo imperante ad eandem Cassiopæam flagrauit Cometa, Vngaros in Germaniam, Ottonem in Italiam impulit . . . Qui anno a nato Christo sexagesimo supra millesimum ducentesimum ibidem luxit interregni tempore, coma ad coeli medium usque dispersa, Carolum Andegauensem e Gallia, per furiosa & scelerata consilia Clementis Pontificis attraxit in Italiam." This book is not mentioned by Tycho Brahe. In his *Cometographia*, p. 817, Hevelius quotes Christianus as mentioning the star of 945. This may seem to some readers to refer to the Chronicle of Christianus of 1472 (*Pulkova Cat.*, p. 76), but, as Professor Copeland has pointed out to me, it is merely a quotation of *D. Christiani Tractatus de Cometarum Essentia*, 1653, and therefore it does not prove anything as to the correctness of the statement of Cyprianus.

theory. All the same, it is desirable that the place where the star of 1572 appeared should be examined from time to time. Argelander has, from a discussion of all Tycho's distance-measures, found the most probable position of the star for the equinox of 1865 to be: RA = 0h. 17m. 20s., Decl. = +63° 23′.9. This position agrees remarkably well with that of a small star of the 10.11 magnitude, No. 129 of D'Arrest's list of stars in the neighbourhood of Tycho's Nova, which is for 1865: 0h. 17m. 19s. +63° 23′.1.

Whether this small star is variable or not must be left for the future to decide. Argelander stated in 1864 (speaking from memory) that he had about forty years previously failed to see any star in the place with the transit instrument at Åbo (of $5\frac{1}{2}$ inches aperture), and that he had also later—probably in 1849—been unable to see anything with the transit circle at Bonn.[1] There is thus a possibility that D'Arrest's star may have increased in light of late years, and observations made at Twickenham by Hind and W. E. Plummer in 1872–73, and at Prague by Safarik in 1888–89, seem to indicate that it is subject to very slight fluctuations of light.[2] The map of all the stars in the neighbourhood, prepared by D'Arrest (which is complete down to the fifteenth or sixteenth magnitude, within a radius of 10′ from the place of the Nova) may in future be compared with photographs of this interesting spot, which deserves to be watched from time to time.

---

[1] D'Arrest, *Oversigt over det kgl. Danske Vidensk. Selskabs Forhandlinger*, 1864, p. 1, where a list of stars near the place and a map are given. Micrometric observation of the star No. 129 in *Astr. Nachr.*, vol. lxiv. p. 75. Argelander, *Ueber den neuen Stern vom Jahre* 1572, *Astr. Nachr.*, vol. lxii. p. 273.

[2] *Monthly Notices, R. Astr. Soc.*, xxxiv., p. 168; *Astr. Nachr.*, vol. cxxiii. p. 365. D'Arrest in 1863–64 found no variability. The place was already examined by Edward Pigott between 1782 and 1786, but without finding any variable star (*Phil. Trans.*, 1786); it was first photographed by Mr. Roberts in 1890 (*Monthly Notices*, L. p. 359).

I shall not here enter into a lengthy examination of the
various prognostications and more or less wild speculations
to which the new star gave rise in 1572. As remarked
by Tycho, the usual methods of astrology were of no avail
in this exceptional case, and there is therefore little to be
gained even to the student of the history of astrology (a
subject of considerable interest) by an examination of the
literature on the star. I shall only point out a few curious
particulars. That the star portended great events, possibly
of an evil character, seemed evident to most writers, and
the star of Bethlehem was frequently referred to as a
phenomenon of a similar nature. As the star seen by the
wise men foretold the birth of Christ, the new one was
generally supposed to announce His last coming and the end
of the world. This was already suggested by Wilhelm IV.
in his above-mentioned letter to Peucer, and among others
who declared their belief in this idea was the successor of
Calvin at Geneva, Theodore Beza, who announced it in a
short Latin poem.[1] He even says that it is the very same
star which was seen by the Magi; but, as Tycho remarks,
perhaps that was only said " poetica quadam festivitate."
Gemma, in his book on the comet of 1577, points out the
great disturbances which followed the star seen by Hipparchus, and expects similar ones to occur now; Tycho
justly remarks that it looks as if Gemma had copied all
this from his own little book.[2] Catholic authors naturally
thought that the star foretold the victory of their Church;
among these is Theodore Graminæus, Professor of Mathe-

---

[1] Published in the above-mentioned reprint " De Nova Stella Judicia Dvorum
Præstantium Mathematicorum, D. Cypriani Leovitii et D. Cornelii Gemmæ,"
1573, s.l. ; perhaps also elsewhere. Reprinted by Tycho, *Progymn.*, p. 327.

[2] " De Prodigiosa Specie Naturaque Cometæ . . . 1577. Per D. Cornelium
Gemmam, Antwerp, 1578," 8vo, p. 42 (compare *Progymnasmata*, p. 565). There
is a curious picture in this book of Belgica weeping amidst the burning ruins
of a city, while the paternal government of Philip II. is represented by gibbets
and wheels in the background, and the comet is blazing overhead.

matics at Cologne, author of a book in which there is
nothing astronomical, but a great deal about old prophecies.[1]
According to one of these, dating from 1488 and founded
on the conjunction of Jupiter and Saturn in 1484, a false
prophet was soon to arise, who, of course, turned out to be
Luther, and a picture is given of the prophet dressed like
a monk, with a shrivelled little devil sitting astride on his
neck, and followed by a small monk or choir-boy.   Unluckily Luther was not born in 1484, but in 1483, and
not on the 22nd October, as assumed by the mathematician Cardan, who worked out his horoscope (in what spirit
may easily be conceived), but nineteen days later.   The
above-mentioned French pamphlet of 1590, printed at a
time when Henry IV. had not yet come to the conclusion
that " Paris vaut bien une messe," also declares that the
star meant the victory of the Church and the King, but
the latter must not be a heretic, but *fide plenus*.   The
author also states that the star disappeared the 18th
February 1574, " qui fut le propre iour que le feu Roy
Henry de Valois feist son entree en Cracouie." [2]   Doubtless
the star expired from grief at seeing this charming creature
bury himself so far from his admiring country.   Strange that
it did not light up again with joy when he bolted from his
Polish kingdom a few months later !

After this digression we shall now return to Tycho, deferring to a later chapter an account of the researches and
speculations on the subject of the new star which he made
in after-years, and which it would not be possible to describe
in this place without a serious break in the continuity of
our narrative.

[1] " Erklerung oder Auslegung eines Cometen. . . . Durch Theodorum
Graminæum Ruremundanum. Cöllen am Rhein, 1573, 4to." Tycho mentions
him as "Autor Stramineus, Graminæus volebam dicere " (*Prog.*, p. 778).
[2] This beautiful remark is also made in Gosselin's *Historia Imaginum
Coelestium*, Paris, 1577, 4to, p. 11.

CHAPTER IV.

TYCHO'S ORATION ON ASTROLOGY AND HIS TRAVELS IN 1575.

AFTER the publication of the book on the new star Tycho Brahe had intended to go abroad for some time, and it appears, even, that he was inclined to leave his native land for ever, but the journey had to be put off owing to an attack of ague, which continued during the greater part of the summer of 1573. Another circumstance which doubtless contributed to keep him at home, was that he had formed an attachment to a young girl some months before. Her name was Christine, but otherwise nothing is known about her; some authors say she was the daughter of a farmer on the Knudstrup property, others that she was a servant-girl; others, again, believed her to have been the daughter of a clergyman. At any rate, it is certain that she was not of gentle blood, and this contributed greatly to estrange his proud relations from him, as they, of course, considered the connection a disgrace. Tycho had no scruples in this respect, and probably considered that a quiet and domestic woman was more likely to be a suitable companion for him through life than a high-born lady to whom his scientific occupations, perhaps, might be distasteful. It is nowhere expressly stated that he and she were united by a Church marriage, and it is almost certain that this was not the case, as it is stated in several contemporary genealogies that Tycho was not married, but that he had children by

an "unfree woman." [1]   Twenty-nine years after his death
his sister Sophia and several others of his relations signed
a declaration stating that Tycho's children were legitimate,
and that their mother (though his inferior in rank) had been
his wife, adding that he would not have been allowed to
live with an unmarried woman in Denmark for twenty-six
years. But this does not in the least prove that Christine
had been formally married to Tycho.   According to the
ancient Danish law, a woman who publicly lived with a man
and kept his keys and ate at his table was after three
winters to be considered as his wife.   In this rule the
Reformation made no change, as Luther and his followers
did not consider a Church ceremony necessary to legalise a
marriage, but adopted the old rule of canonical law, that
the consent of the parties made the marriage, which, therefore, really dated from the betrothal (*matrimonium in-
choatum*), though the full consequences only began when
the parties went to live together or were married (*matrimonium consummatum*).   A natural result of these views
was, that the parties frequently began to live together
immediately after the betrothal, as they did not see the
necessity of the Church ceremony, which could make no
difference as to the legal effects of the connexion.   Gradually a change took place in these views, as the Church
could not look with indifference at this setting aside of its
authority; but though in Denmark betrothed people about

---

[1] *Danske Magazin*, ii. p. 192 (Weistritz, ii. p. 70).   The English traveller,
Fynes Moryson, tells us that Tycho was said "to liue vnmarried, but keeping
a Concubine, of whom he had many children, & the reason of his so liuing
was thought to be this ; because his nose hauing been cut off in a quarrell,
when he studied in a Vniversity of Germany, he knew himselfe thereby dis-
abled to marry any Gentlewoman of his own quality.   It was also said that
the Gentlemen lesse respected him for liuing in that sort, and did not acknowledge his sonnes for Gentlemen."   Moryson heard this at Elsinore in 1593 ;
see his "Itinerary of his ten Yeeres Travell through the twelve Domjnions of
Germany, Bohmerland, &c."   London, 1617, p. 59.

the year 1566 began to be punished if they commenced living together before the wedding, and an ordinance of 1582 declares that a formal betrothal before a minister and witnesses shall precede a wedding, it was not yet expressly ordered that a Church ceremony was the only way of legalising a marriage, and, in fact, this was not done till a hundred years later.[1] Tycho Brahe lived just at a time when the law of the land was still formally unaltered, and it is therefore intelligible how his children might be considered legitimate, and the companion of his life have been looked upon as his lawful wife. Doubtless the only fault anybody had to find with her was her low origin, and if she had been his equal in rank nobody would have thought that she was anything but his wife.[2]

Tycho's eldest child, a daughter of the name of Christine, was born in October 1573, but died in September 1576. His other children were Magdalene (born in 1574), Claudius (born in January 1577, died six days after), Tyge (born in August 1581), Jörgen (born 1583), and three other daughters, Elizabeth, Sophia, and Cecily, as to the dates of whose birth nothing is known.[3]

The ague seems to have left Tycho in August 1573, as

---

[1] By the *Danske Lov* of 1683 and the Church ritual of 1685. See an article in the *Historisk Tidsskrift*, fifth series, vol. i. 1879, by the Danish Minister of Justice, J. Nellemann.

[2] Early in the sixteenth century a Danish nobleman, Mogens Lövenbalk, brought a young Scotch lady, Janet Craigengelt (on the female side said to have been related to the Grahams of Montrose), home to his castle, Tjele, in Jutland, where she lived for many years and bore him two children. Her son tried in vain to obtain recognition as his father's legitimate heir, and his claims were set aside chiefly because his mother had clearly not been treated as the mistress of the house, but rather as a dependent. On the other hand, the University of Wittenberg declared in favour of the legitimate birth of the children, evidently guided by the then ruling principle of canonical law, that a long intercourse with all the outer resemblance of wedlock had the same legal weight as a formal marriage.

[3] The eldest daughter was buried in Helsingborg church. In the epitaph she is called *filiola naturalis*, which has made Langebek doubt whether she had the same mother as the other children (*D. Magazin*, ii. p. 194); but this

we still possess a couple of observations from the 14th
of that month.   The lunar eclipse of the 8th December,
which he had computed in the book on the new star, was
duly observed, and he was on that occasion assisted by his
youngest sister Sophia, at that time a girl seventeen years
of age, highly educated, and not only conversant with classical literature, but also well acquainted with astrology and
alchemy, and therefore in every way fit to assist her great
brother.   She was the only one of his relations who showed
any sympathy with his pursuits, and was a frequent visitor
in his home.   In March, April, and May 1574 Tycho
observed at Heridsvad, but the remaining part of the year
he chiefly spent at Copenhagen, where his daughter Magdalene was born.[1]   In the capital his rising fame had now
attracted considerable attention, and some young nobles
who were studying at the University requested him to
deliver a course of lectures on some mathematical subject
on which there were no lectures being given at that time.
His friends Dancey and Pratensis urged him to consent to
this proposal, but Tycho was not inclined to do so, until
the King had also requested him to gratify the wishes of
the students, and at the same time to give the University
a helping hand.   He then yielded, and the lectures were
commenced on the 23rd September 1574, with an oration
on the antiquity and importance of the mathematical
sciences.   This was printed after his death, but has long
ago become very scarce, for which reason we shall give an
abstract of the contents.[2]

very expression, which originated in the Roman jurisprudence, shows that
the humble companion of Tycho's life was her mother (see Gibbon's *Decline
and Fall of the Roman Empire*, chapter xliv.).

 [1] He had also observed at Copenhagen on the 24th April.   Nearly all these
observations are distance-measures of planets from fixed stars, doubtless with
the sextant, "satis exquisite, subtracta instrumenti parallaxi;" but a small
quadrant is also mentioned.

 [2] "Tychonis Brahei de Disciplinis mathematicis oratio publice recitata in

The oration begins with an allusion to his having been requested to lecture, not only by his friends, but also by the King, and then goes on to describe the various branches of mathematics cultivated by the ancients. Geometry has a higher purpose than merely measuring land, and the divine Plato turned all those away from his teaching who were ignorant of geometry, as being unfit to devote themselves to other branches of philosophy. To this he attributes the high degree of learning reached by the ancient philosophers, as they were imbued with geometry from their childhood, " while we, unfortunately, have to spend the best years of our youth on the study of languages and grammar, which those acquired in infancy without trouble." Astronomy is a very ancient science, and, according to Josephus, it can be traced back to the time of Seth, while Abraham from the motions of the sun, moon, and stars perceived that there was but one God, by whose will all was governed. It was next studied by the Egyptians; while we owe our knowledge, above all, to Hipparchus, Ptolemy, and more recently to Nicolaus Copernicus, who not without reason has been called a second Ptolemy, and who, having by his own observations found both the Ptolemean and the Alphonsine theories insufficient to explain the celestial motions, by new hypotheses deduced by the admirable skill of his genius, restored the science to such an extent, that nobody before him had a more accurate knowledge of the motions of the stars. And though his theory was somewhat contrary to physical principles, it admitted nothing contrary to mathematical axioms, such as the ancients did in assuming the motions

Academia Haffniensi anno 1574, et nunc primum edita . . . studio et opera Cunradi Aslaci Bergensis. Hafniæ, 1610, 4to." Dedicated to Tycho's brother, Sten Brahe of Knudstrup, and the editor has added some of his own speeches. Second edition, Hamburg, 1621, to the title is added "in qua simul Astrologia defenditur et ab objectionibus dissentientium vindicatur. Cum Præloquio Joach. Curtii." Both editions are very scarce.

of the stars in the epicycles and eccentrics to be irregular
with regard to the centres of these circles, "which was
absurd." The lecturer next alludes to the beauty of the
celestial phenomena, and shows that we must distinguish
between the casual contemplation of the heavens and their
scientific examination, as only the latter will detect the
variation in the moon's distance from us, the revolutions of
the planets, &c. The utility of astronomy is easy to perceive, as no nation could exist without means of properly
dividing and fixing time, while the science exalts the
human mind from earthly and trivial things to heavenly
ones. A special use of astronomy is, that it enables us to
draw conclusions from the movements in the celestial regions
as to human fate. The remainder of the lecture is devoted
to considerations on the importance and value of astrology,
and tries to answer the objections which philosophers and
theologians had made against it. It is evident, from the
detailed manner in which this is done, how important
Tycho considered this subject to be. We cannot, he says,
deny the influence of the stars without disbelieving in the
wisdom of God. The importance of the sun and moon is
easy to perceive, but the five planets and the eighth sphere
have also their destination, as they cannot have been
created without a purpose, but were placed in the sky and
given regular motions to show the wisdom and goodness
of the Creator. The sun causes the four seasons, while
during the increase and decrease of the moon all things
which are analogous to it, such as the brain and marrow of
animals, increase and decrease similarly. The moon also
causes the tides, and its influence on these becomes greatest
when that of the sun is joined to it at new-moon and
full-moon. Sailors and cultivators of the soil have noticed
that the rising and setting of certain stars cause stormy
weather, and more experienced observers know that the

configurations of the planets have also great influence on the weather. Conjunctions of Mars and Venus in certain parts of the sky cause rain and thunder, those of Jupiter and Mercury storms, those of the sun and Saturn turbid and disagreeable air. The most ancient writers on agriculture, as well as poets and astrologers, have observed that the rising and setting of the more conspicuous stars simultaneously with the sun produced rain, wind, and other atmospheric changes, particularly when the planets joined their effect to that of the stars.[1] The sun and stars move in the same manner from year to year, but this is not the case with the planets, and the weather of one year cannot, therefore, be like that of another. Among planetary conjunctions, he mentions that of Jupiter and Saturn in 1563, in the beginning of the sign of Leo near the hazy stars of Cancer (Præsepe), which Ptolemy already considered pestilential. This conjunction was in a few years followed by an outbreak of the plague. While many people admitted the influence of the stars on nature, they denied it where mankind were concerned. But man is made from the elements, and absorbs them just as much as food and drink, from which it follows that man must also, like the elements, be subject to the influence of the planets; and there is, besides, a great analogy between the parts of the human body and the seven planets. The heart, being the seat of the breath of life, corresponds to the sun, and the brain to the moon. As the heart and brain are the most important parts of the body, so the sun and moon are the most powerful celestial bodies; and as there is much reciprocal action between the former, so is there much mutual dependence between the latter. In the same way the liver corresponds

---

[1] " Habent se enim stellæ fixæ in coelo veluti matres, quæ nisi a septem errantibus stellis stimulentur et impregnentur, steriles sunt et nihil in hac inferiori natura progignunt" (*Oratio*, p. 20).

to Jupiter, the kidneys to Venus, the milt to Saturn, the gall to Mars, and the lungs to Mercury, and the resemblance of the functions of these various organs to the assumed astrological character of the planets is pointed out in a manner similar to that followed by other astrological writers. He believes experience to have shown that those who are born when the moon is affected by the evil planets (Saturn and Mars) and is unluckily placed, have a weak brain and are under the influence of passions, while those in whose case the sun was influenced by those planets suffered from palpitation of the heart. But if both luminaries are in unlucky aspects, those born at that time are of weak health and intellect. Those people at whose birth Saturn, the highest planet, was favourable, are inclined to sublime studies, while those whom Jupiter has influenced are attracted to politics. The solar influence makes people desire honour, dignities, and power; that of Venus makes them devote themselves to love, pleasures, and music; while Mercury encourages people to mercantile pursuits, and the moon to travelling.

Many philosophers and theologians, continued Tycho in his lecture, have contended that astrology was not to be counted among the sciences, because the moment of birth was difficult to fix, because many are born at the same moment whose fates differ vastly, because twins often meet with very different fortunes, while many die simultaneously in war or pestilence whose horoscopes by no means foretold such a fate. It had also been maintained that a knowledge of the future was useless or undesirable, and theologians added that the art was forbidden in God's Word and drew men away from God. To these objections Tycho answered, that even if there was an error of an hour in the assumed time of birth, it would be possible from subsequent events to calculate it accurately. With regard to war or pestilence, prudent astrologers always made a reservation as to public calamities which proceed from universal causes. Difference of education, mode of life, and similar circumstances explained the different fates which people born at the same time met with; and as to twins, they were not born exactly at the same moment, and one was always naturally weaker than the other, and this the stars could not correct. Astrology was not forbidden in the Bible, but sorcery only.

So far Tycho's astrological ideas are in accordance with those of contemporary and previous writers on such subjects, but towards the end of his discourse he shows more distinctly than most of these, that he did not consider the fate of man to be absolutely settled by the aspect of the stars, but that God could alter it as He willed. Nor was man altogether bound by the influence of the stars, but God had so made him that he might conquer that influence, as there was something in man superior to it. The objection to astrology, that it was a useless art, as knowledge of the future was undesirable, would only hold good if it were impossible to resist the influence of the stars; but being forewarned, we might try to avert the threatening evils, and in this way astrology was of great use.

In conclusion, Tycho stated that as the doctrine of the *primum mobile* (spherical astronomy) was very easy, and was frequently lectured on in the University, he had thought it more advisable to take for his subject the motions of *secundum mobile*, explain the method of calculating the motions of the seven planets by the Prutenic tables, which were the most accurate ones, and describe the circles by means of which the tables had been computed.

Early in 1575 these lectures were finished, and Tycho Brahe shortly afterwards started on the long-deferred

journey.[1] Leaving his family at home until he had decided
where he would finally settle down, he went first to Cassel
to make the acquaintance of the distinguished astronomer,
Landgrave Wilhelm IV. of Hesse. Wilhelm was born in
1532, and was the son of Landgrave Philip the Magnani-
mous, one of the most determined champions of the Reformation, who, after the disastrous battle of Mühlberg (1547),
had surrendered to the Emperor, and had been kept a close
prisoner for five years, during which anxious time his
dominions had been governed by Wilhelm. When Philip
became free in 1552, Wilhelm gladly turned back to the
learned occupations, to which he had already for some years
been devoted. By accident he came across the curious work
of Peter Apianus, *Astronomicum Cæsareum*, in which the
orbits of the planets are represented by movable circles of
cardboard, and he became so much interested in the subject,
that he had circles of copper made for the same purpose.
Having afterwards studied Purbach's planetary theory and
the other principal works of the time, he became, like Tycho,
convinced of the necessity of making systematic observations,
as he found considerable errors in the existing star catalogues. In 1561 he built a tower on the Zwehrer Thor at
Cassel, of which the top could be turned round to any part
of the sky, and here he observed regularly up to 1567, when
the death of his father and his own consequent accession to
the government of his dominions gave him less leisure for
scientific occupations. As yet he had not any astronomer

[1] Shortly before starting he had occasion to show his friendship for his
former tutor Vedel and his patriotism. Vedel was just in the act of finishing
his translation of the Danish Chronicle of Saxo Grammaticus (from the end
of the twelfth century), but the cost of the paper necessary for so large a work
was so great, that Vedel's friends feared that the work might remain un-
printed. Tycho wrote a Latin poem to encourage his friend, calling on the
Danish women to sacrifice some of their linen, and to send it to the paper-
mill in Scania, lest the deeds of their ancestors should be buried in oblivion
(Wegener's *Life of Vedel*, p. 83).

to assist him, and the work at his observatory had for a
long time made little or no progress, when Tycho Brahe
arrived at Cassel in the beginning of April 1575. The
Landgrave was well pleased to receive the young astronomer
as his guest, and they conversed by day about their favourite
science, and observed the heavens by night together, the
Landgrave with his own quadrants and unwieldy torqueta,
Tycho with some portable instruments, among which was
probably his sextant. Among other observations they determined the position of Spica Virginis.[1] Naturally they discussed the nature and position of the new star, and the
Landgrave told Tycho how he had once been so intent on
determining the greatest altitude of the star, that he had
not even desisted when he was told that part of the house
was on fire, but had calmly finished the observation before
leaving the observatory. Tycho was also interested to learn
that the Landgrave had remarked how the motion of the sun
became retarded when it approached the horizon at sunset,
which might be seen by watching a sun-dial. Tycho recollected having read the same in the observations of Bernhard
Walther (before whom, however, Alhazen had recognised in
this phenomenon an effect of refraction), and he determined
to follow up the matter by-and-by, so as to be able to correct observations made at low altitudes for refraction.[2]

More than a week had elapsed in thus exchanging ideas
and opinions, when a little daughter of the Landgrave died,
and Tycho, who did not wish to intrude his company on the

---

[1] *Tychonis Epist. Astron.*, Dedication. In his *Progymn.*, p. 616, Tycho
states that the Landgrave on this occasion gave him a copy of his own catalogue of improved star-places. Tycho prints as specimens the places of Aldebaran, Betelgeux, and Sirius; but though superior to the positions given by
Alphonsus and Copernicus, those of the Landgrave were *as yet* very inferior
to Tycho's. We shall, farther on, see how the observations made at Cassel
afterwards became much more accurate than they were at the time of Tycho's
visit.

[2] *Gassendi*, p. 29.

afflicted father, took his departure. He never saw the Landgrave again, but the visit of the young enthusiast had renewed the wavering scientific ardour of his host, and the
friendship thus commenced was revived in after-years
by frequent correspondence and the interchange of observations.

From Cassel, Tycho went to Frankfurt-on-the-Main,
where he purchased some books at the half-yearly mart,
particularly some of the numerous pamphlets on the new
star. He went thence to Basle, where he had already spent
some time in the winter of 1568–69, and where he now
found his stay so agreeable that he thought seriously of
settling down there. The University of Basle was one of
the most important centres of learning in Europe, and Tycho
might hope to find the same refined tastes and culture
among the scientific men living there which, some sixty
years before, had decided Erasmus to take up his residence
at Basle. The central situation of the city, between Germany and France and not far from Italy, seemed also very
convenient.[1] Deferring, however, for the present the final
step of returning home for his family, Tycho went through
Switzerland to Venice, and spent some days there, after
which he retraced his steps back to Germany, and went in
the first instance to Augsburg. The friendships with the
brothers Hainzel and Hieronimus Wolf formed during his
former visit had in the meantime not been forgotten, and
several letters had been exchanged between them. Thus,
Paul Hainzel had in March 1574 written to express his
warmest thanks for a copy of Tycho's book on the new star,
and in March 1575 both he and Wolf had written to tell
Tycho that they had succeeded in procuring for him from
Schreckenfuchs of Freiburg a zodiacal sphere constructed
according to the description of Ptolemy as formulated by

---

[1] *Astr. Inst. Mechanica*, fol. G. 2.

Copernicus.[1] The money sent by Tycho had been stolen by
the carrier, who had never since been heard of, but the
instrument had now arrived, and would be forwarded.[2] Tycho
can hardly have received these letters before starting from
home, and was therefore possibly still ignorant of another
piece of news contained in them, namely, that the great
quadrant at Göggingen, which he had designed six years
before, had in the previous December been blown down and
destroyed in a great storm.[3] The great globe which he had
ordered to be made during his former visit was now nearly
completed, and was the following year brought to Denmark
with great trouble. At Augsburg, Tycho on this occasion
made the acquaintance of a painter, Tobias Gemperlin, and
induced him to go to Denmark, where he afterwards painted
a number of pictures for Uraniborg and the royal castles.

At Ratisbon great numbers of princes and nobles from
all parts of the empire were just then gathering to witness
the coronation as King of the Romans of Rudolph the Second,
King of Hungary and Bohemia, on the 1st November.
Tycho also betook himself thither in the hope of meeting
the Landgrave, and perhaps some other scientific men. He
was, however, disappointed as to the Landgrave, who did
not appear; but he had the consolation of meeting, among
others, the physician-in-ordinary to the Emperor, Thaddæus
Hagecius or Hayek, a Bohemian, whose name we have
already met with among the writers on the new star. He gave

---

[1] Erasmus Oswald Schreckenfuchs (1511–1579), Professor of Mathematics,
Rhetorics, and Hebrew, first at Tübingen, afterwards at Freiburg in Breisgau;
editor of the works of Ptolemy (Basle, 1551), and author of commentaries on
the writings of Sacrobosco, Purbach, and Regiomontanus.

[2] By Petrus Aurifaber, "cum supellectile sua" (Was he the maker of the
globe?) These letters are published in *T. Brahe et ad eum doct. vir. Epist.*,
pp. 11 *seq.* Whether Tycho ever got the sphere is not known.

[3] It may have been re-erected later, as Joh. Major wrote to Tycho Brahe in
1577 that it was still in use; but, on the other hand, P. Hainzel wrote in 1579
that he had not observed the comet of 1577 for want of convenient instruments (*T. B. et doct. vir. Epist.*, pp. 42 and 46).

Tycho a copy of a letter he had received from Hieronimus Munosius of Valencia on this subject, and Tycho tried in vain to dissuade him from publishing an answer to the scurrilous and absurd assertions of Raimundus of Verona.[1] Another and most precious gift which Hagecius bestowed on Tycho on this occasion was a copy of a MS. by Copernicus, *De Hypothesibus Motuum Cœlestium Commentariolus*, an account of the new system of the world, which its author had written for circulation among friends some ten years before the publication of his book, *De Revolutionibus*, but which had never been printed.[2] In after years Tycho communicated copies of this literary relic to various German astronomers. Probably he presented to Hagecius in return a copy of his own paper on the star, as the latter is quoted in Hagecius' reply to Raimundus.[3]

From Ratisbon Tycho returned home viâ Saalfeld and Wittenberg. At the former place he visited Erasmus Reinhold, the younger, a son of the author of the " Prutenic Tables," who showed him his father's manuscripts, among which were extended tables of the equations of centre of the planets for every 10' of the anomaly.[4] At Wittenberg he inspected the wooden triquetrum with which Wolfgang

---

[1] *Progymn.*, pp. 567 and 734 ; see also above, p. 64.

[2] *Progymn.*, p. 479. Though the description there given of the MS. ought to have attracted attention, and have led to a search for copies of it, the *Commentariolus* remained perfectly unknown till the year 1878, when it was noticed that there was a copy of it in the Hof-Bibliothek at Vienna, and immediately afterwards another copy was found at the Stockholm Observatory. The Vienna MS. had been presented by Longomontanus on his departure from Prague in 1600 to another of Tycho Brahe's disciples, Joh. Eriksen, and it is therefore doubtless a copy of the MS. belonging to Tycho. See Prowe, *Nicolaus Coppernicus*, i. part ii. p. 286.

[3] *Danske Magazin*, ii. p. 196, quotes Thomasini *Elog. Viror. Illustr.*, according to which, Tycho, in his younger days, received an offer of an appointment at the Emperor's court. There is no confirmation anywhere of this statement ; but if the offer was ever made, it was probably done at Ratisbon in 1575.

[4] *Progymn.*, p. 699.

Schuler and Johannes Prætorius had observed the new star.[1]

About the end of the year Tycho returned home, apparently intending very soon to leave his native land for ever in order to reside at Basle.   He had, however, not yet confided his intentions to anybody, but luckily King Frederick II. had his attention specially drawn to Tycho through an embassy to Landgrave Wilhelm, which happened to return to Denmark from Cassel about that time.   The Landgrave had requested the members of the embassy to urge the king to do something for Tycho, so as to enable him to devote himself to his astronomical studies at home; as these would do much credit to his king and country, and be of great value for the advancement of science.[2]   When Tycho paid his respects to the king, the latter offered him various castles for a residence, but Tycho declined these offers. King Frederick was, however, fond of learning, and anxious to retain in the kingdom so promising a man; and he shortly afterwards sent off a messenger with orders to travel day and night, until he could deliver into Tycho's own hands the letter of which he was the bearer.   On the 11th of February, early in the morning, as Tycho was lying in bed at Knudstrup, turning over in his mind his plan of emigrating, the royal messenger, a youth of noble family and a connection of Tycho's, was announced, and was at once brought to his bedside to deliver the king's letter.   In this Tycho was commanded immediately to come over to Seeland to wait on the king.   He started the same day, and arrived in the evening at the king's hunting-lodge at Ibstrup, near Copenhagen.[3]   The king now told him that

---

[1] *Progymn.*, p. 636; see also above, p. 58.

[2] *Epist. Astron.*, Dedication, fol. 2.

[3] Afterwards called Jaegersborg, about five English miles north of Copenhagen; it was demolished long ago.   The present king's summer residence, Bernstorff, is close to the place.

one of his courtiers had understood from Tycho's uncle, Sten
Bille, that Tycho was thinking of returning to Germany,
and asked him whether he had perhaps refused to accept a
royal castle because he feared to be disturbed in his studies
by affairs of court and state.   The king next told him how
he had lately been at Elsinore, where he was building the
castle of Kronborg, and that his eye had fallen on the little
island of Hveen, situated in the Sound, between Elsinore
and Landskrona in Scania, and that it had occurred to him
that this lonely little spot, which had not been granted in
fee to any nobleman, might be a suitable residence for the
astronomer, where he might live perfectly undisturbed;
adding, that he believed he had heard from Sten Bille
before Tycho went to Germany that he liked the situation.
The king offered him the island and promised to supply him
with means to build a house there.   He finally told Tycho
to think the matter over for a few days, and give his final
answer at the castle of Frederiksborg; if he accepted the
offer, the king would immediately give the necessary orders
for payment of a sum of money for the building.

Having returned home, Tycho at once wrote a long
letter to his friend Pratensis, telling him in detail all that
had happened, and confessing his former intention of
leaving Denmark.   He asked Pratensis to show the letter
only to Dancey, and requested them both to advise him in
the matter.[1]   They both strongly urged him to accept the
king's offer, which he accordingly did, and already on the
18th of February the king by letter granted Tycho "five
hundred good old daler" annually until further orders.[2]

---

[1] *T. B. et Doct. Vir. Epist.*, p. 21 *et seq.*   In the letter of February 14,
Tycho asks Pratensis to tear up or burn the letter as soon as Dancey had seen
it, and in his reply next day, Pratensis writes that he had destroyed it.   Tycho
must therefore have kept a copy.

[2] About £114, but of course this represented at that time a much greater
sum.   In Denmark the first Joachimsthaler had been coined in 1523, exactly
of the same value as those first issued in North Germany in 1519, which value

Four days after, on the 22nd February 1576, Tycho paid
his first visit (at least as far as we know) to the little
island which was destined to become famous through him,
and the same evening took his first observation there of a
conjunction of Mars and the moon.[1] If he could have
foreseen that he was destined to furnish the means of cir-
cumventing the tricks of the inobservable Sidus (as Pliny
called Mars), and himself to add more to our knowledge of
the moon's motion than any one had done since Ptolemy,
he would certainly by this coincidence have been confirmed
in his belief in astrology. On the 23rd May a document
was signed by the king of which the following is an exact
translation: [2]—

"We, Frederick the Second, &c., make known to all men,
that we of our special favour and grace have conferred and
granted in fee, and now by this our open letter confer and
grant in fee, to our beloved Tyge Brahe, Otte's son, of
Knudstrup, our man and servant, our land of Hveen, with
all our and the crown's tenants and servants who thereon
live, with all rent and duty which comes from that, and is

the Danish *daler* retained nearly unaltered, though the name changed, first
to species (from *in specie*, or in one piece), then to rigsdaler species. The
coinage had greatly deteriorated during the war with Sweden, hence doubtless
the expression "good old daler."

[1] Februarii die 22. Existente in M. C. ultima in capite Hydræ quæ est
versus ortum, et sola juxta collum, apparebat visibilis conjunctio ☽ et ♂ admodum partilis, adeo ut ☽ inferiore et meridionaliore cornu fere attingeret
corpus ♂ distans saltem ab eo parte sexta sui diametri accipiendo distantiam
hanc ab inferiori cornu limbi. Erat autem circa idem tempus per observationem alt. lucidiss. in pede Orionis 11 g. 20 m.—H. 9 M. 30. Infimus vero ☽
limbus circa quem ♂ conspiciebatur elevari visus est 10 g. 50 m. Observatio
hæc facta i. Huennæ. Langebek in *Danske Mag.*, ii. p. 194 (Weistritz, ii. p.
73), refers to this visit to Hveen as made in the year 1574. In the original
the year is not given, and the observation follows after one of May 19, 1574.
But on February 22, 1574, the moon was only a few days old, and Mars was
at the other side of the heavens, while they were very close together on the
same date in 1576.

[2] *Danske Magazin*, ii. p. 198.

given to us and to the crown, to have, enjoy, use and hold, quit and free, without any rent, all the days of his life, and as long as he lives and likes to continue and follow his *studia mathematices*, but so that he shall keep the tenants who live there under law and right, and injure none of them against the law or by any new impost or other unusual tax, and in all ways be faithful to us and the kingdom, and attend to our welfare in every way and guard against and prevent danger and injury to the kingdom. Actum Frederiksborg the 23rd day of May, anno 1576.

" FREDERICK."

The same day the chief of the exchequer, Christopher Valkendorf, was instructed to pay to Tycho Brahe 400 daler towards building a house on the island of Hveen, for which Tycho was himself to provide building materials. This money was paid on the 27th May.[1] Just at the moment when everything was settled and Tycho's prospects in life were most brilliant, he had the grief to lose his friend Johannes Pratensis, who died suddenly on the 1st June from a bleeding of the lungs while lecturing in the University. He was only thirty-three years of age, and had been professor of medicine since 1571. Tycho had promised to write a Latin epitaph over his friend in case he should survive him, and he had it printed in 1584 at his own printing office at Uraniborg. He also caused a monument to be erected to the memory of Pratensis in the Cathedral of Copenhagen.[2]

---

[1] Friis, Tyge Brahe, p. 58.

[2] The epitaph is reprinted in *Danske Magazin*, p. 199 (Weistritz, ii. 84), and in *T. B. et ad eum Doct. Vir. Epist.*, p. 28.

CHAPTER V.

THE ISLAND OF HVEEN AND TYCHO BRAHE'S OBSERVATORIES AND OTHER BUILDINGS—HIS ENDOWMENTS.

IN the beautiful scenery along the coast of the Sound between Copenhagen and Elsinore, the isle of Hveen with its white cliffs, rising steeply out of the sea, forms a very conspicuous feature. It is about fourteen English miles north of Copenhagen, and about nine miles south of Elsinore, rather nearer to the coast of Scania than to that of Seeland. The surface is a nearly flat tableland of about two thousand acres, sloping slightly towards the east, and of an irregularly oblong outline, the longest diameter extending from north-west to south-east and being about three miles long. From time immemorial it was considered an appendage to Seeland, but in 1634 it was placed under the jurisdiction of the court of justice at Lund in Scania, because the inhabitants had complained of the long distance to their former court of appeal in Seeland.[1] In consequence of this change the island was ceded to Sweden in 1658, when the Danish provinces east of the Sound were conquered by the king of Sweden. Though there are no considerable woods on the island and the surface is but slightly undulated, the almost constant view of the sea in all directions, studded with ships and bounded by the well-wooded coasts

---

[1] There is still extant a Latin poem written by T. Brahe in 1592, "In itinere a Ringstadio domum," in which he charges the judge who had tried some lawsuit of his with injustice. *Danske Magazin*, ii. p. 279. About the change of jurisdiction see Bang's *Samlinger*, ii. p. 265 (Weistritz, ii. p. 226, and i. p. 56).

of Seeland and Scania in the distance, helps to form very attractive scenery, which adds to the peculiar charm the island has for any one who is interested in the great memories connected with it. One can understand why Tycho calls it "Insula Venusia, vulgo Hvenna," as if it were worthy of being called after the goddess of beauty. Another name, which Tycho mentions as sometimes applied to the island by foreigners, is "Insula Scarlatina," and with this name a curious and probably apocryphal story is connected, which is told by the English traveller, Fynes Moryson (who was in Denmark in 1593), in the following words : "The Danes think this Iland of Wheen to be of such importance, as they have an idle fable, that a King of England should offer for the possession of it, as much scarlet cloth as would cover the same, with a Rose-noble at the corner of each cloth. Others tell a fable of like credit, that it was once sold to a Merchant, whom they scoffed when he came to take possession, bidding him take away the earth he had bought." [1]

The island forms one parish, and the church, which is the only building to be seen with the naked eye from the Danish coast, is situated at the north-west corner of the island, close to the edge of the cliff. As already mentioned, the island is a table-land, with steep cliffs round nearly the whole circumference, through which narrow glens in several places form the beds of small rivulets, the prettiest one being Bäkvik, on the south-east coast. At the time of Tycho Brahe the inhabitants lived in a village called Tuna (i.e., town, Scottice, "the toun"), towards the north coast; there were about forty farms, and the land was tilled in common. From the map in Blaev's *Grand*

[1] An Itinerary written by Fynes Moryson, &c., London, 1617, fol., p. 60. The story also occurs in *P.D. Huetii Commentarius de Rebus ad eum pertinentibus*, Amsterdam, 1718, 8vo, p. 85. Tycho merely mentions the name Scarlatina, *Astr. Inst. Mech.*, fol. G. 2, and *De Mundi Aeth. Rec. Phaen.* ii. Preface.

Atlas it appears that most of the land in the south-eastern half of the island was only used for grazing. We give here a reduced copy of Blaev's map, which agrees well with

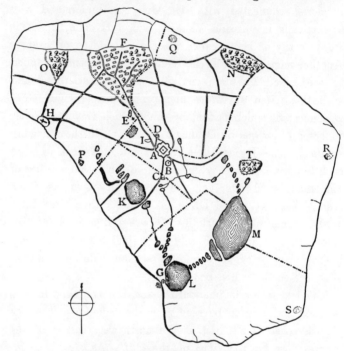

HVEEN AT THE TIME OF TYCHO BRAHE.

EXPLANATION OF THE MAP.

A. Uraniborg.
B. Stjerneborg Observatory.
C. Farm.
D. Workshop.
E. Windmill.
F. Village.
G. Paper-mill.

H. Church.
I. Hill where Petty Sessions were held.
K, L, M. Fish-ponds.
N. Grove of nut-trees.
O. Morass with alder trees.
P, Q, R, S. Ruins of old forts.
T. A small wood.

Tycho's own map,[1] except that we have slightly altered the contour of the island, in accordance with modern maps.

[1] *Astr. Inst. Mechanica*, fol. I. 2, and *Epist. Astron.*, p. 264. There is a small copy of it on the frontispiece of Kepler's *Tabulæ Rudolphinæ*.

[2] I possess another large map (18 in. by 13 in.), with one page letterpress on the back, "Topographia Insulæ Huenæ in Celebri Porthmo Rcgni Daniæ

Neither before Tycho's time nor afterwards has this little island played any part in the history of Denmark, and yet tradition points to a time long ago when even this little spot is supposed to have been the scene of heroic deeds. On the map appear the ruins of four castles or forts, which are supposed to have been destroyed in 1288, when the Norwegian king, Erik the Priesthater, ravaged the coasts of the Sound. Nowadays a few stones and a slight rise of the ground scarcely marks the site of each fort, but in Tycho's time there were more distinct traces of them left.[1] Their names were Nordborg, on the north coast; Sönderborg, on the south-west coast; Hammer, at the north-east, and Carlshöga, at the south-east corner. Tycho's friend and former tutor, Vedel, published a collec-lection of ancient Danish popular ballads and romances,

quem vulgo Oersunt uocant. Effigiata Coloniæ, 1586." I believe it belongs to Braunii *Theatrum Urbium*. There are very few details on it, and the coast-line is very incorrect, but the plans and views of Uraniborg in the corners of the map, and the descriptive letterpress on the back, are of value, as they contain some particulars not to be found elsewhere, and the author has evidently got reliable information, probably from A. S. Vedel, who is known to have contributed to the work. Willem Janszoon Blaev (1571-1638) had himself lived at Hveen with Tycho. The following particulars from the description of the island in his son's *Grand Atlas, ou Cosmographie Blaviane* (Amsterdam, 1663, vol. i. p. 61), are of interest :—" Elle est fertile en bons fruits et n'a aucune partie qui soit sterile, elle abonde en toutes sortes de gros bestail, nourrit des daims, lievres, lapins et perdrix en quantité. La pesche y est de tous costez : elle a un petit bois de couldriers, noisettiers, dont jamais les noix ne sont mangées des vers ny vermoluës. Il ne s'y trouve aucun loir ny taulpe. . . . Cette isle n'a point de riviere, mais quantité des ruisseaux et fontaines d'eau douce. Vne entre autres qui ne gele jamais, ce qui est tres-rare en ces quartiers." A similar account is given in Wolf's *Encomion Regni Daniæ*, Copenhagen, 1654, p. 525.

[1] The Swedish antiquarian, Sjöborg, who visited the island in 1814, mentions a place close north of the south-east ruin, called Lady Grimhild's grave, of which he could find no trace. On the north-east coast there was another ruin, apparently a quadrangular building, 80 feet by 24, with a walled-in enclosure in front. It was called the Monks' Kirk, but nothing is known about it, and it is not mentioned by Tycho. See Sjöborg, *Samlingar för Nordens Fornälskare*, T. iii., Stockholm, 1830, pp. 71–82. About the four castles see also Braun's map, where it is stated that there were (in 1586) no ruins left, but only traces of the foundations.

among which are three which give the following account
of the traditions about these ruins.[1]

Lady Grimhild, who owned the whole island, made a
festival at Nordborg, to which she, among others, invited
her brothers, Helled Haagen and Folker, the minstrel, both
well-known figures in Danish mediæval ballads. She intended, however, to slay the two brothers, with whom she
was at enmity, but they accepted her invitation, though
they were warned while crossing the Sound, first by a
mermaid and next by the ferryman, both of whom were
beheaded by Helled Haagen as a punishment for the evil
omen. On arriving at Nordborg they were well received
by Grimhild, who, however, soon persuaded her men to
challenge the brothers to mortal combat. She was specially
infuriated against Helled Haagen, and enticed him into
promising that he would confess himself defeated if he
should merely stumble. To bring about this result, she
had the lists covered with hides, on which peas were strewn,
and of course Helled Haagen slipped on these, and, true
to his vow, remained lying and was slain. His brother,
Folker, was likewise killed. But one of Grimhild's maids,
Hvenild, after whom the island got its name, bore a son
who was called Ranke, and who afterwards avenged the
death of his father Helled Haagen. The poems merely
mention the revenge, without going into details, but in his
introduction Vedel tells how Ranke enticed Grimhild into a
place in Hammer Castle, where he said his grandfather,
Niflung or Niding, had hidden his treasure, but when she
had gone inside, he ran out and bolted the door, leaving
her to die of hunger. The resemblance of this story to
the principal events of the Niebelungenlied is striking,
and doubtless the story is, both in the German epic and

---

[1] I take the following account from the Danish poet Heiberg's delightful
article on Hveen and its state in 1845, in his year-book, *Urania*, for 1846.

in the Scandinavian tradition, derived from a common source.[1]

Nearly in the centre of the island, 160 feet above the level of the sea,[2] Tycho selected a site for his new residence and observatory, which he very appropriately called Uraniburgum or Uraniborg, as it was to be devoted to the study of the heavens. The work was at once commenced, and on the 8th August 1576 the foundation-stone was laid. The French minister Dancey had asked to be allowed to perform this ceremony, and had provided a handsome stone of porphyry with a Latin inscription, stating that the house was to be devoted to philosophy, and especially to the contemplation of the stars. Some friends and other men of rank or learning assembled early in the morning, "when the sun was rising together with Jupiter near Regulus, while the moon in Aquarius was setting; libations were solemnly made with various wines, success was wished to the undertaking, and the stone was put in its place at the south-east corner of the house at the level of the ground."[3] The building operations were now steadily proceeded with under the direction of the architect, Hans van Stenwinchel from Emden, but Tycho doubtless superintended the work himself, as he seems to have almost constantly resided in the island. We find, at least, that

---

[1] According to another tradition mentioned by Sjöborg (*l. c.*, p. 74), Ranke threw the keys of Hammer Castle into the sea, and bewitched the castle so that it sank into the earth or into the sea; but if there shall ever be three posthumous men in the island at the same time, each called after his father, then Hammer Castle shall again stand in its old place, and the keys be found. Other traditions say that Hvenild was a giantess (Jettekvinde), who carried pieces of Seeland in her apron over to Scania, where they formed the hills of Runeberga, but as her apron-strings burst on the way, she dropped a piece in the sea, which formed the island of Hveen. The hill close to Uraniborg, Hellehög, where in Tycho's time the local court was held, is evidently called after Helled Haagen.

[2] According to Picard 27 toises (*Ouvrages de Mathematique*, p. 71).

[3] *Astron. Inst. Mechanica*, fol. H. 6.

he took observations pretty regularly from December 1576. On his birthday, the 14th December, he commenced a series of observations of the sun, which were steadily continued for more than twenty years.[1] Having now plenty of occupation, Tycho thought it best to decline an offer made to him the following year by the professors of the University, who on the 18th May 1577 unanimously paid him the compliment of electing him Rector of the University for the ensuing year, although it had not, since the Reformation, been customary to elect anybody to this post who was not a professor. Tycho replied on the 21st May, expressing his appreciation of the proffered honour and his regrets that the building operations and other business obliged him to decline the post offered him.[2]

Although the house was probably soon sufficiently advanced to enable Tycho to take up his residence in it, it does not appear to have been completed till the year 1580. Uraniborg was situated in the centre of a square enclosure, of which the corners pointed to the four points of the compass. The enclosure was formed by earthen walls, of which the sides were covered with stones, about 18 feet high, 16 feet thick at the base, and 248 feet from corner to corner.[3] At the middle of each wall was a semicircular bend, 73 feet in diameter, and each enclosing an arbour.[4] At the east

---

[1] "Die 14 qui mihi est natalis feci primam observationem Hvenæ ad Solem circa ipsum Solstitium hybernum et inveni alt. ⊙ meridianam minimam quæ illic potest 10° 43'." Previous to this date there is only an observation of Mars on the 22nd October.

[2] Tycho's answer is printed in *Danske Magazin*, ii. p. 202, see also Rördam, *Kjöbenhavns Universitets Historie*, Copenhagen, 1872, vol. ii. p. 174.

[3] Here and in the following, English measures are always used. Tycho expresses all his measures in *feet*, of which one is = 0.765 French foot = 0.815 English foot, or in cubits of 16.1 English inches. See D'Arrest's paper on the ruins of Uraniborg in *Astron. Nachrichten*, No. 1718.

[4] On the figure on Braun's map (see above, p. 90 note) the four walls are perfectlystraight, and the four arbours are in the middle of the flower-gardens. The semicircular bends were therefore later improvements.

and west angles gates gave access to the interior of the
enclosure, and in small rooms over the gateways English
mastiffs were kept, in order that they might announce the
arrival of strangers by their barking. At the south and
north angles were small buildings in the same style as the
main edifice, and affording room respectively for the printing

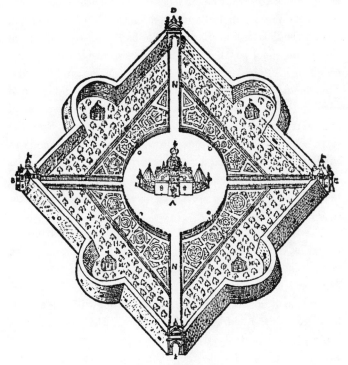

URANIBORG AND GROUNDS.

office and for the domestics. Under the latter building was
the castle-prison, probably used for refractory tenants.[1] Inside the walls were first orchards with about three hundred
trees, and inside these, separated from them by a wooden

[1] See letterpress on Braun's map. This cellar is one of the very few rem-
nants now left of Tycho's buildings.

paling, flower-gardens. Four roads ran through the orchards
and gardens from the four angles of the enclosure to the
open circular space in the middle, where the principal building was situated on a slightly higher level than the surrounding grounds. Uraniborg was built (apparently of red
bricks with sandstone ornaments) in the Gothic Renaissance
style, which towards the end of the sixteenth century was
becoming more generally adopted in the North of Europe,
where the heavier mediæval style had hitherto still been
the ruling one, so that Tycho Brahe's residence became
epoch-making in the history of Scandinavian architecture.
The slender spires and tastefully decorated gables and
cornices were indeed in better harmony with the peaceful
and harmonious life of a student of the heavens than the
more severe and dry Gothic style which the Renaissance was
superseding; and the pictures, inscriptions, and ornaments
of various kinds profusely scattered through the interior
reminded the visitor at every step of the pursuits and tastes
of the owner.

The woodcut below (which, like the previous and following ones, is a reduced copy of a figure in Tycho's own
description) gives a general idea of the aspect of the edifice
from the east, and by comparison with the plan of the
ground-floor on the next page, the reader will get a clear
idea of this remarkable structure.[1] The base of the
principal and central part was a square, of which each side
was 49 feet long, and to the north and south sides of this
there were round towers 18 feet in diameter, surrounded
by lower outhouses for fuel, &c., while narrow towers on the
east and west sides contained the entrances. Including the
towers, the entire length of the building from north to south

---

[1] The buildings and instruments are described in *Epist. Astron.*, p. 218 *et
seq.*, and *Astron. Inst. Mech.*, fol. H. 4 *et seq.* Some short Latin inscriptions,
with which various places in the house were ornamented, are given in Resenii
Inscriptiones Hafnienses (1668), p. 334, reprinted in *Weistritz*, i. p. 225.

was about 100 feet.   The central part was surmounted by
an octagonal pavilion, with a dome with clock-dials east and
west, and a spire with a gilt vane in the shape of a Pegasus.
In the pavilion there was an octagonal room with a dial in
the ceiling, showing both the time and the direction of the
wind, and round the pavilion ran an octagonal gallery, north
and south of which were two smaller domes with allegorical

ORTHOGRAPHIA

PRAECIPVAE DOMVS   ARCIS VRANIBVRGI IN
INSVLA PORTHON DANICI   HVÆNNA Astronomiæ inſtaüran-
da gratia arcis annum 1580   à TYCHONE BRAHE
tradi-   *fuata*

URANIBORG FROM THE EAST.

figures on the top.   The height of the walls of the central
building was 37 feet, and the Pegasus was 62 feet above
the ground.   The two towers north and south were about 18
feet high, and had each a platform on the top surmounted by
a pyramidal roof made of triangular boards, which could be
removed to give a view of any part of the sky.   North and
south of these observatories were two smaller ones, each
standing on a single pillar, and communicating with the

larger ones; they were also covered with pyramidal roofs, and were not built till after the completion of the house. Galleries around the towers gave the means of observing with small instruments in the open air, and on the east and west side of each gallery there was a large globe to serve as a support for a sextant. When not in use these globes were protected by pointed covers, of which the two eastern ones are visible on the figure. This also shows the founda-

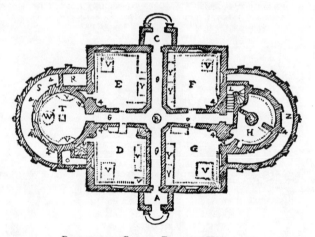

PLAN OF THE GROUND FLOOR OF URANIBORG.

A. East entrance.	P. Stairs to laboratory.
B. Fountain.	R, O. Aviaries.
C. West entrance.	T. Library.
A. Passages.	W. Great globe.
D. Sitting-room in winter.	S. Cellar for charcoal for the laboratory.
E, F, G. Guest-rooms.	Z. Wood cellar for kitchen.
H. Kitchen.	V. Tables.
K. Well.	Y. Beds.
L. Stairs to upper storey.	4. Chimneys.

tion-stone in the south-east corner, and next to it a door leading down to the basement.

The south-east room on the ground-floor was the sitting-room of the family in winter; later on it was enlarged by pulling down the wall between it and the passage west of it. The three other rooms were guest-rooms, but the south-west room, in which a large quadrant was attached to the west wall, was probably also used as a study. In the storey above there were the red room to the north east, the blue room to the south-east, the yellow room (a small octagonal one) over the porch on the east side, and on the west side one long room, the green one, with the ceiling covered with pictures of flowers and plants. Tycho specially mentions the beautiful view from this room of the Sound, with its numerous sails, particularly in summer. Above the second storey there were eight little rooms or garrets for students and observers. The south tower contained in the basement a chemical laboratory with furnaces, &c., above that on the ground floor was the library, and above that the larger southern observatory. In the north tower the centre of the basement was occupied by a deep well built round with masonry, which reached to the kitchen above.[1] Over the kitchen was the larger northern observatory.

In the library the great globe from Augsburg was mounted. It was five feet in diameter, the inside made of wooden rings and staves firmly held together. When returning to Augsburg in 1575, Tycho found that it was not perfectly spherical and showed some cracks, but after it had in the following year been brought to Denmark, the cracks were stopped and the sphericity made perfect by covering it with numerous layers of parchment. It was then left to dry for two years, and as the figure remained perfect, it was covered with brass plates, on which two great circles were engraved to represent the equator and the zodiac, divided into single degrees, and by transversals into minutes. Gradually the stars and constellations were laid down on it as their positions resulted from the observations, and not

---

[1] This well, from which the water could be pumped up and sent to the various rooms by concealed pipes, is still in existence.

till about twenty-five years after the construction of the globe had been commenced was it completely finished.[1] It was mounted on a solid stand, with graduated circles for meridian and horizon, and a movable graduated quadrant for measuring altitudes. On the horizon was the unavoidable inscription stating how the great work of art was made. A hemispherical cover of silk could be lowered over it from the ceiling to protect it from dust. In addition to this great globe, the library or museum contained four tables for Tycho's assistants to work at, also his collection of books and various smaller knicknacks, portraits of astronomers and philosophers, among whom Hipparchus, Ptolemy, Albattani, Copernicus, and the Landgrave figured conspicuously. There was also a portrait of George Buchanan, who played so important a part in the religious and political revolutions in Scotland, and whose acquaintance Tycho had probably made in 1571 when Buchanan was in Denmark. This portrait ha been presented to Tycho by Peter Young.[2] Under the pictures were versified inscriptions composed by Tycho.[3]

We may form some idea of the elegance and taste which pervaded Tycho's residence by examining the large picture which adorned his great mural quadrant. This instrument was, as already mentioned, mounted on the wall in the south-west room on the ground-floor, and consisted of a

[1] *Astr. Inst. Mechanica*, fol. G. The globe must have been quite finished about 1595; it is said to have cost Tycho about 5000 daler (Gassendi, p. 135).

[2] Young had been the first tutor to James VI., and became afterwards his almoner. He was several times in Denmark. Tycho had sent his little book about the new star to Buchanan, who thanked him for it in a letter dated Stirling, the 4th April 1575. In this letter Buchanan (who was then Lord Privy Seal) expresses his regret that he has not had leisure to finish his poem on the sphere (it was published after his death), and praises Tycho's book for having refuted popular errors. *T. Brahei et ad eum Doct. Vir. Epist.*, p. 18.

[3] As specimens Tycho prints the poems on Ptolemy and Copernicus. *Epist. Astron.*, pp. 239–240.

MURAL QUADRANT.

brass arc of $6\frac{3}{4}$ feet radius, 5 inches broad and 2 inches thick, fastened to the wall with strong screws, and divided in his usual manner by transversals; it was furnished with two sights, which could slide up and down the arc. At the centre of the arc there was a hole in the south wall, in which a cylinder of gilt brass projected at right angles to the wall, and along the sides of which the observer sighted with one of the sliding sights. This was one of the most important instruments at Uraniborg, and was much used. It is, therefore, no wonder that Tycho (who claimed it as his own invention) wished to fill the empty space on the wall inside the arc with a picture of himself and the interior of his dwelling. Tycho is represented as pointing up to the opening in the wall, and he says the portrait was con- sidered a very good likeness; at his feet lies a dog, "an emblem of sagacity and fidelity." In the middle of the picture is a view of his laboratory, library, and observatory, and on the wall behind him are shown two small portraits of his benefactor, King Frederick II., and Queen Sophia, and between them in a niche a small globe. This was an automaton designed by Tycho, and showing the daily motions of the sun and moon and the phases of the latter. The portrait was painted by Tobias Gemperlin of Augsburg, whom Tycho had encouraged to come to Denmark; the views of the interior of Uraniborg by its architect, Sten- winchel; and the landscape and the setting sun by Hans Knieper of Antwerp, the King's painter at Kronborg. The picture bears the date 1587, but the quadrant itself had been in constant use since June 1582.[1]

Another instrument on which Tycho found room for a picture was his smallest quadrant, one of the earliest in- struments constructed at Uraniborg. The radius of the

[1] There are a few meridian altitudes of Spica observed in April 1581, "per magnum instrumentum," which probably were also made with this quadrant.

quadrant was only 16 inches (one cubitus); the divided arc
was turned upwards, and within it were forty-four concentric
arcs of 90° to subdivide the single degrees according to the
plan proposed by Pedro Nunez.   On the empty space between the centre and the smallest of these arcs was a small
circular painting, representing a tree, which on the left side
is full of green leaves and has fresh grass under it, while on
the right side it has dead roots and withered branches.
Under the green part of the tree a youth is seated, wearing
a laurel wreath on his head and holding a star-globe and a
book in his hands.   Under the withered part of the tree
is a table covered with money-boxes, sceptres, crowns, coats
of arms, finery, goblets, dice, and cards, all of which a
skeleton tries to grasp in its outstretched arms.   Above is
the pentameter, " Vivimus ingenio, cætera mortis erunt,"
pointing out the vanity of worldly things, while only earnest
study confers immortality.   The first part of the sentence
is over the green part, the second over the withered part of
the tree.   In another place [1] Tycho had a similar picture,
in which there appeared among the green leaves symbols
of the life and doctrine of Christ, while the symbols of
philosophy are moved over to the withered side under the
dominion of Death, and the inscription is changed to
" Vivimus in Christo, cætera mortis erunt," so that the two
pictures showed the superiority of the noble efforts of the
human mind over trivial occupations, and yet the insuffi-
ciency of either except man turns to the Redeemer.

This small instrument does not seem to have had any
fixed place, and was afterwards removed to the subterranean
observatory, but the larger ones were all erected in the
observatories at the north and south ends of Uraniborg.

[1] It is not stated where. I conclude from the description in *Epistolæ
Astron.*, p. 254, that these were two different pictures, and not one picture
seen from two points of view, as one might almost conclude from *As'r. Inst.
Mech.*, fol. A.

In each of the two small observatories there was an equatorial armillary sphere, of which the northern one was ornamented with pictures of Copernicus and Tycho himself.[1] In the large southern observatory were the following instruments. A vertical semicircle (eight feet in diameter) turning round a vertical axis, and furnished with a horizontal circle for measuring azimuths (fol. B. 5); a triquetrum, or, as Tycho calls it, "instrumentum parallacticum sive regularum" (fol. C.); a sextant for measuring altitudes with a radius of $5\frac{1}{2}$ feet (fol. A. 5); and a quadrant of two feet radius with an azimuth circle (fol. A. 4). In the large northern observatory were another triquetrum of peculiar construction, with an azimuth circle 16 feet in diameter, resting on the top of the wall of the tower (fol. C. 2); a sextant of 4 feet radius for measuring distances (fol. E.); and a double arc for measuring smaller distances. Probably the last two instruments were removed or used on the open gallery when the triquetrum was erected, as the latter must have been large enough to fill the whole room, and, indeed, even in the southern observatory there cannot have been much elbow-room for the observers. In the northern observatory was also preserved an interesting astronomical relic, the triquetrum used by Copernicus, and made with his own hands.

By degrees, as Tycho's plans for collecting observations became extended and a greater number of young men desired to assist him, he felt the want of more instruments and of more observing rooms, in which several observers could be engaged at the same time without comparing notes. In 1584 he therefore built an observatory on a small hill about a hundred feet south of the south angle of the enclosure of Uraniborg, and slightly to the east. In

[1] *Astr. Inst. Mechanica*, fol. C. 5 (north one) and D. (south one). The first observations, "per armillas astrolabicas," are from 1581. References to the descriptions and figures of the other instruments are given above in the text.

this observatory, which he called Stellæburgum (Danish,
Stjerneborg), the instruments were placed in subterranean
rooms, of which only the roofs rose above the ground, so
that they were well protected from the wind.   As shown
by the view and plan on p. 106, there were five instrument
rooms, with a study in the centre, and the entrance to the
north.   The north-east and north-west rooms were built
somewhat later than the others, and were nearly at the

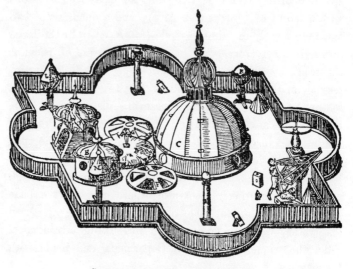

STJERNEBORG, SEEN FROM THE WEST.

level on the ground.[1]   The whole was surrounded by a low
wooden paling, forming a square with semicircular bends at
the middle of each side, and the sides facing north, south,

[1] This appears from the stone steps leading *up* to the crypt E., found in
1823, as we shall see in the Appendix.   The above figure also shows that not
only the roofs, but most of the walls of crypts D. and E. were above ground.
The quadrant in the crypt D. was erected in December 1585, twelve months
after Tycho had placed in position the stone on which the lower end of the
axis of the instrument in crypt C. was supported.   When he had built the
three crypts, he perhaps regretted having sunk them in the earth, and therefore built the two new ones higher.

east, and west.   The enclosure was 57 feet square, and the
diameter of the semicircles was 20 feet.   The entrance
was on the north side, and a door and some stone steps led
down to the study.   Over the portal were three crowned
lions hewn in stone, with the appropriate inscription—

"NEC FASCES NEC OPES
SOLA ARTIS
SCEPTRA PERENNANT."

Below this and over the door was the coat of arms of the
Brahe family, and some other allegorical figures.

On the back of the portal, towards the south was a large
tablet of porphyry, with a long inscription in prose, stating
that these crypts had, like the adjoining Uraniborg, been
constructed for the advancement of astronomy, at incredible
labour, diligence, and expense, and charging posterity to
preserve the building for the glory of God, the propagation
of the divine art, and the honour of the country.   Going
down the steps to the "Hypocaustum," another slab over
the door exhibited a versified inscription, expressing the
surprise of Urania at finding this cave, and promising even
here, in the bowels of the earth, to show the way to the stars.
The study was about 10 feet square, and only the vaulted
roof and the top of the walls were above the ground.   The
vault was sodded over to look like a little hill, "representing Parnassus, the mount of the Muses," and on the middle
of it stood a small statue of Mercury in brass, cast from a
Roman model, and turning round by a mechanism in the
pedestal.[1]   The study was lighted by four small windows
just above the ground, and contained a long table, some
clocks, &c., and on the wall hung a semicircle in brass, 8
feet in diameter, for measuring distances of stars, and which,

[1] In addition to this, Tycho possessed several other automata, which startled
the peasants of the island, and made them believe him to be a sorcerer.
Gassendi, p. 196.

when required, could be placed on a stand outside, similar to those which Tycho used for his sextants. On the ceiling was represented the Tychonian system of the world, and on the walls were portraits of eight astronomers, all in a reclining posture, namely, Timocharis, Hipparchus, Ptolemy,

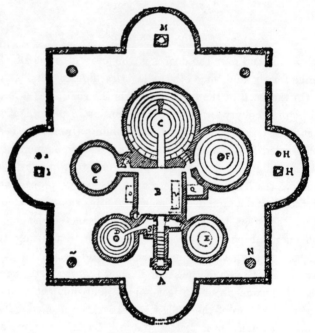

PLAN OF STJERNEBORG.

A. Entrance.
B. Study.
C. Crypt with large armillæ.
D. ,, ,, azimuthal quadrant.
E. ,, ,, zodiacal armillæ.
F. ,, ,, azimuthal quadrant.
G. ,, ,, sextant.
H, I. Stone piers for portable armillæ.

K, L, N, T. Globular stands for sextants.
M. Stone table.
O. Tycho Brahe's bed.
P. Fireplace.
V. Table.
Q. Bedroom for assistant.
S. Unfinished subterranean passage towards Uraniborg.

Albattani, King Alphonso, Copernicus, Tycho, and lastly Tychonides, an astronomer who is still unborn. Under each portrait was the name, approximate date, and a distich

setting forth the merits of each.   While that under Tycho's
picture leaves posterity to judge his work, the lines under
the picture of his hoped-for descendant are less modest,
expressing the hope that the latter might be worthy of his
great ancestor.   Tycho was represented as pointing up to
his system of the world, while his other hand held a slip of
paper with the query, "Quid si sic?"

In the centre of each crypt was a large instrument, the
floor rising gradually by circular stone steps up to the
walls.[1]   The instruments were an azimuthal quadrant
(*quadrans volubilis*) of $5\frac{1}{2}$ feet radius, with an azimuth
circle at the top of the wall (*Mechanica*, fol. B. 2), a zodiacal
armillary sphere (C. 4), a large quadrant of brass (radius 7
feet) enclosed in a square of steel, and likewise furnished
with an azimuth circle on the wall (B. 4); a sextant of $5\frac{1}{2}$
feet radius for measuring distances (D. 5), and in the largest
southern crypt a large equatorial instrument, consisting of
a declination circle of $9\frac{1}{2}$ feet diameter, revolving round a
polar axis, and a semicircle of 12 feet diameter, supported
on stone piers, and representing the northern half of the
Equator (D. 2).   In addition to these fixed instruments,
there were various smaller portable ones kept at Stjerneborg, which could be mounted on the pillars and stands outside or held in the hand; namely, a portable armilla 4 feet
in diameter,[2] a triquetrum, a small astrolabium or planisphere, the small quadrant described above, and two small
instruments made by Gemma Frisius, namely, a cross-staff
and a circle (*annulus astronomicus*), both of brass.   With
the exception of these two and the triquetrum of Copernicus,
all the instruments in Tycho's possession were made in his

---

[1] The number of steps in each crypt may be seen on the plan above.   The
floor of the crypt *G* (where the sextant was placed) was flat.

[2] This was placed either at *H* or at *I*, and served to measure declinations of
stars near the horizon which could not be got at with the subterranean instruments.   See *Epist. Astron.*, p. 229.

own workshop, which was situated close to the servants'
house, about 100 feet to the west.[1]

Tycho would scarcely have been able to construct these
magnificent instruments if he had not continually been provided with new sources of income through the liberality of
his royal patron. We have seen that Tycho, from February
1576, enjoyed an annual pension of 500 daler (£114). In
addition to this, the king granted him on the 28th August
1577 the manor of Kullagaard, in Scania, to be held by him
during the king's pleasure. Kullagaard is situated near
the north-western extremity of Scania, on the mountain of
Kullen, which forms a steep promontory in the Kattegat.
The king's letter stated expressly that Tycho Brahe should
not, like his predecessor, be bound to keep the lighthouse
of Kullen in order; but apparently the king soon found
that this exemption was a mistake, and already on the 18th
October 1577 a second royal letter was issued, in which it
was stated that as the late holder of the benefice had received
it for the purpose of keeping the light going, in order that
seafaring men should have no cause for complaint, the same
should be done by Tycho, if he wished to continue to hold
the manor.[2] This obligation was apparently not to the taste
of Tycho; at least he must have been negligent in seeing
that the light was regularly attended to, for already in the
autumn of 1579 the governor of Helsingborg Castle was
ordered to take possession of Kullagaard manor, in order to
keep the lighthouse properly attended to, as complaints had
frequently been made about it. As Tycho, however, begged
to be allowed to keep the manor, on the plea that he had no
other place from which to get fuel for Uraniborg, the king
again granted him the manor on the 13th November 1579,
on condition that the light was regularly lighted.[3] In May

---

[1] In 1577 Tycho had employed a smith at Heridsvad, but he was not able
for the work. *T. B. et Doct. Vir. Epist.*, p. 42.

[2] Friis, *Tyge Brahe*, pp. 80–81.   [3] Ibid., p. 96.

1578 he had also been granted the use of eleven farms in the county of Helsingborg, free of rent, to be held during the king's pleasure. These and the Kullen manor he lost again for a while in August 1580, probably because he had in the meantime been granted other sources of income; but he received them again in June of the following year, "to enjoy and keep, free of rent, as long as he shall continue to live at Hveen," with the repeated injunction to keep the light at Kullen in order. On the 27th October 1581 the customs officers at Elsinore were instructed, that whereas the light was in future to be kept burning in winter as well as in summer, they were out of the increased lighthouse fees received from navigators to pay Tycho Brahe 300 daler a year for the increase of trouble. This seems, however, to have been more than the additional fees amounted to, and on the 9th July 1582 the order about the 300 daler was cancelled by a royal decree, in which it was stated that Tycho was already in receipt of sufficient payment for keeping the lighthouse.[1] In 1584 the governor of Helsingborg Castle and the chief magistrate of Scania were ordered to proceed to Kullen, together with Tycho, to examine the lighthouse, which was said to be very dilapidated. The tower was ordered to be rebuilt in August 1585 at the public expense, and at the same time the indefatigable generosity of the king dictated a letter to the customs officers at Elsinore, commanding them until further orders to pay Tycho 200 dalers annually, in order that the light might be kept burning summer and winter as long as navigation lasted.[2]

We have seen that Tycho Brahe already in 1568 received the king's promise of the first vacant canonry in the cathedral of Roskilde. In 1578 this promise was more distinctly renewed, as by royal letter, dated Frederiksborg the 18th May, Tycho was appointed to succeed to the pre-

---

[1] Friis, *Tyge Brahe*, pp. 116–117.   [2] Ibid., p. 148.

bend attached to the chapel of the Holy Three Kings [1] in the
said cathedral whenever the holder of it should die.   In the
meantime he was to enjoy the income of the Crown estate
of Nordfjord in Norway, with all rent and duty derived from
it.[2]   He had not to wait very long for the prebend, as
Henrik Holk, who had held it since the Reformation, died
in 1579, on the 5th June of which year the canonry was
conferred on Tycho.   In the patent all the temporalities of
the · above-mentioned chapel were granted to Tycho during
pleasure, including the canon's residence, farms, and other
property belonging thereto, on the condition that hymns
were daily to be sung in the chapel to the praise of God,
and that for this purpose two poor schoolboys were to be
kept in food and clothes in order to assist the vicars-choral
in the daily service.   Furthermore, he was to maintain two
poor students at the University of Copenhagen, and to see
that these, as well as the two choir-boys, were diligent, and
fit to devote themselves to learned pursuits.   The chapel
and residence were to be kept in proper repair, and the
tenants to be dealt with according to law and justice, and
not to be troubled by any new tax or other impost.[3]

About a month after the prebend had been granted to
Tycho, he was ordered, in accordance with the rules of the
chapter, to allow the widow of his predecessor and the
University of Copenhagen to enjoy *annum gratiæ* of the
rents and other income of the prebend.   With characteristic
coolness the astronomer seems to have turned a deaf ear
to this injunction, and he even went so far as to forbid
the tenants to pay anything to the widow.   On the 3rd
December 1579 the king therefore found it necessary to send

---

[1] Anglice, the Three Wise Men of the East.   The chapel is an excrescence
on the south side of the cathedral, built in 1464.   Among other royal tombs,
that of Tycho's patron, Frederick II., is in this chapel.

[2] Royal letter, printed in *Danske Magazin*, ii. p. 203 (Weistritz, ii. p. 92.)

[3] Ibid., p. 204 (Weistritz, ii. p. 94).

a second and peremptory order to pay to the widow and the University what was due to them.[1]　Three years later Tycho thought that he saw a chance of making the heirs of Henrik Holk disgorge some of the money he had been obliged to let them have, for it appears that some repairs had to be made to the chapel, and that Tycho demanded payment for these from the heirs. But here again the king showed that, however favourably disposed he was to the renowned man of learning, he would have no injustice done to anybody; and in July 1582 he directed that the repairs were to be paid for out of public funds, but that in future Tycho, or whoever else might hold the prebend, was to pay for them.[2]　We shall afterwards see that the possession of this prebend gave rise to more serious troubles to Tycho Brahe.

It was mentioned above that Tycho obtained a grant of the Crown estate of Nordfjord on the west coast of Norway, to be held by him during the time that he was waiting for the vacancy in the prebend. But when he got possession of the latter, the king did not deprive him of the Nordfjord estate, but granted it to him again on the 13th June 1579, during pleasure, free of rent, and merely with the usual stipulation that he was to keep the tenants under the laws of Norway, and not injure any of them, nor was he to cut down any of the woods on the estate.[3]　This benefice may only have been intended to indemnify Tycho for the year of grace which he was to pay out of the Roskilde prebend, for on the 10th August 1580 the king's lieutenant at Bergen was ordered to receive the Nordfjord estate from Tycho Brahe, and in future to account to the king's exchequer for the income of the same.　Tycho must, however, have persuaded the king that he could ill afford to lose this income, for already,

---

[1] *Danske Magazin*, ii. p. 208 (Weistritz, ii. p. 100).

[2] E. C. Werlauff, *De hellige tre Kongers Kapel i Roeskilde Domkirke* (Copenhagen, 1849), p. 17.

[3] *Danske Magazin*, ii. p. 206 (Weistritz, ii. p. 97).

on the 11th November 1580, a new grant of the estate was
made to Tycho in exactly the same terms as the previous
one, and two months afterwards the lieutenant at Bergen was
directed to hand over the estate to Tycho Brahe, and to re-
fund all money received from it during the time he had been
deprived of it.[1]   The king evidently now thought that he had
done enough for Tycho, for on the 29th March 1581 he wrote
to him that although Tycho had applied to have the pension of 500 daler continued, still, as he had been provided
for in other ways, the pension was to be paid for the past
year, but was then to cease.   The same day the chief of the
exchequer, Valkendorf, received instructions to this effect ;
but already six months after he was directed again to pay
the pension to Tycho, who seems to have received it without
interruption till 1597.[2]

Tycho continued in undisturbed possession of the Norwegian estate till March 1586, when he and several other
tenants of Crown estates in Norway received notice to
surrender them, as " fish and other victuals " which they
produced were wanted for the navy.   It was, however,
stated that they were not to consider this as a sign of
disgrace, but that they would be indemnified in other ways.
Thus Tycho got in the first instance 300 daler from the
treasury,[3] and on the 11th September following he was
informed that he would, until further notice, receive an
annual sum of 400 daler from the customs paid at Elsinore.
This grant was renewed on the 4th June 1587, the
money to be paid annually on the 1st May.[4]   The estate

[1] Both letters to the lieutenant are printed in *Danske Magazin*, ii. pp.
211–212 (Weistritz, ii. p. 106).

[2] Letter to Tycho Brahe of March 29th, printed in Friis, *Tyge Brahe*, p.
114 ; letter to Valkendorf of same date, in *Danske Magazin*, ii. p. 217 (Weistritz, ii. p. 117).

[3] Friis, *Tyge Brahe*, p. 161.

[4] Letters of 11th September 1586 and 4th June 1587, printed in *Danske
Magazin*, ii. pp. 244–245 (Weistritz, ii. p. 165–166), where is also a letter from

of Nordfjord was restored to Tycho in June 1589, and the grant was renewed in June 1592, when the allowance from the Sound duties was discontinued.[1]

It would not at the present time be easy to form an accurate opinion as to the actual amount of income enjoyed by Tycho Brahe during the years he lived at Hveen (though we may mention here that, according to his own statement, it was about 2400 daler a year), and on the other hand we have no way of knowing exactly how much he spent on his instruments and buildings.[2] But at any rate, it will be evident from the above account of the various grants of land and money that King Frederick II. had very amply provided for his wants, and never forgot the promises made to Tycho when the latter was prevailed on to settle in his native country. The circumstances which gradually led to his being deprived of most of these grants will be detailed in a future chapter; but we may here mention that Tycho, shortly after the death of King Frederick II., in 1588, represented to the new Government that his great expenses in connexion with the scientific work at Hveen had caused him to be in debt to the amount of 6000 daler. This sum was at once ordered to be paid by the Government, so that Tycho might reasonably hope, even after the death of his royal patron, to be able to continue the work so munificently supported by the late king.

Tycho to Niels Bilde, who doubtless then was lieutenant at Bergen, asking him to assist Christopher Pepler, formerly Tycho's steward at Nordfjord, to get payment for some money still due to him.

[1] Friis, p. 180; *Danske Magazin*, ii. p. 280 (Weistritz, ii. p. 228).

[2] Tycho in 1598 estimated the total cost of all his buildings and instruments at 75,000 daler (about £17,000). See below, Chapter X.

CHAPTER VI.

TYCHO'S LIFE AT HVEEN UNTIL THE DEATH OF KING FREDERICK II.

AT Uraniborg Tycho spent more than twenty years, from the end of 1576 to the spring of 1597, the happiest and most active years of his life. Surrounded by his family and numerous pupils, many of whom came from great distances to seek knowledge in the house of the renowned astronomer and assist him in his labours, frequently honoured by visits from men of distinction both from Denmark and abroad, Tycho during these years steadily kept the object in view of accumulating a mass of observations by means of which it would be possible to effect that reform of astronomy which was so imperatively demanded, and for which the labours of Copernicus had merely paved the way. But though the scientific work was never neglected, the pleasant little island afforded many means of recreation. The map in Braun's *Theatrum Urbium* shows that provision was made for games of various kinds in the orchards which surrounded Uraniborg, and in the south and east of the island there were places arranged for entrapping birds. There were plenty of hares and other small game, and Tycho caused a great number of fishponds to be made. Most of them lay in the south-western part of the island, connected by sluices into two rows which met in a lake, the second largest of all, from which a small river made its way through the cliff to the sea. On this spot Tycho afterwards built a paper-mill.

None of these fishponds are seen on Braun's map, and they would therefore seem to have been constructed after 1585, as the map bears the date 1586. Thus Tycho contrived to add to the comfort and convenience of his surroundings.

In addition to these means of recreation, Tycho Brahe possessed others of a higher kind. In 1584, the same year in which the Stjerneborg was built, he put up a printing-press in the building at the south angle of the enclosure surrounding Uraniborg. It was originally intended for the printing of his own works, but when not required for this purpose he occasionally employed it to print poems in memory of departed friends, and similar poetical effusions. Thus we have already mentioned that in 1584 he printed an epitaph of his friend Pratensis,[1] and in the same year he printed a poem addressed to a Danish nobleman, Jacob Ulfeld, to give the printer something to do, as he informs us.[2] Of greater interest is a longer poem of 288 lines, dated the 1st January 1585, and addressed to the Chancellor, Niels Kaas.[3] In this Tycho complains of the neglected state of astronomy in most countries, and contrasts this with its present flourishing state in Denmark, where buildings have been erected and instruments constructed such as the world never saw. But envy and malice attempt to speak slightingly of this great work, and he might almost be inclined to regret having undertaken it and look for another home elsewhere,[4] if he

[1] A poem in memory of another friend, Joh. Francisci Ripensis, given in Gassendi's book, p. 261, was possibly also printed at Uraniborg.

[2] Printed in *Danske Magazin*, ii. pp. 223-224 (Weistritz, ii. p. 130 *et seq.*).

[3] Printed ibid., pp. 226-234 (Weistritz, ii. 135 *et seq.*).

[4] "Undique Terra infra, cœlum patet undique supra,
    Omne solum patria est, cui mea sacra placent."

The first of these lines and part of the second occur in *Astr. Instauraiæ Mechania*, fol. D., where he mentions that one of his armillæ could be taken asunder and transported to any place where it might be wanted. It is remarkable how strongly imbued he always was with the cosmopolitan character of his science, even when Fortune smiled most on him.

did not remember that the Chancellor was interested in it for the sake of the honour thus conferred on his country, and would therefore continue to protect it. Another but shorter poem was soon afterwards printed at Uraniborg, addressed to the learned Heinrich Rantzov, governor of the Duchy of Holstein. In this poem, which is dated the 1st March 1585, Tycho complains that Rantzov, in a book on astrology which he had just published, had used the word *specula* when speaking of Uraniborg, which magnificent building did not merit so mean an appellation.[1]

The considerable building operations in which Tycho engaged at Hveen obliged him to require a great deal of work from his tenants there, and when we remember his naturally hot temper, and his habit of exacting without scruple what was due to him (and even more, as in the case of Holk's widow), it is not to be wondered at that complaints were more than once made by the tenants at Hveen of his arbitrary treatment of them. Already on the 10th April 1578 an order was issued by the king to the peasants at Hveen, that they were not to leave the island because Tycho Brahe required more labour than had formerly been demanded from them.[2] But Tycho, who was perhaps not worse (and certainly not better) than his fellow-nobles were generally in the treatment of their inferiors, continued in the following years to make such great demands on the peasantry at Hveen to get his buildings, plantations, fish-ponds, &c., finished, that fresh complaints were made. The king therefore sent two noblemen, the governor of Helsingborg Castle and the governor of Landskrona Castle, to Hveen to investigate matters. When these two officials had presented their report, the king, on the 8th January 1581 (the same day on which he ordered his lieutenant at

---

[1] Printed in *Danske Magazin*, ii. pp. 235-238 (Weistritz, ii. p. 148 *et seq.*).
[2] Friis, *Tyge Brahe*, p. 89.

Bergen to restore the Norwegian estate to Tycho), issued an " Arrangement and rule for Tycho Brahe and the inhabitants at Hveen," which both parties were ordered to obey and follow.[1] In this document the amount of labour to be furnished by each farm was fixed at two days a week, from sunrise to sunset, and rules were laid down about various other matters; thus a tenant who did not keep his dikes and fences in order was to pay a fine in money to the landlord and a barrel of beer to the townsmen; nobody was to gather nuts or cut wood without leave from Tycho Brahe or his steward; a petty sessions court was to be held every second Wednesday,[2] and appeals were to be heard in Scania in future, instead of in Seeland.[3] The peasants were not to consider their holdings as their own property, as they had no legal authority for doing so, but in future, when any farmer died, his holding was to be treated as any other farm on a Crown estate.

If the buildings and other works at Hveen required much manual labour, the scientific researches for the sake of which they were erected required a great deal of work to be done by practised observers and computers, and these Tycho readily found in the young men who soon, began to flock to Hveen in order to enjoy the privilege of studying under his guidance. The first to arrive seems to have been Peder Jakobsen Flemlöse, born about 1554 in a village called Flemlöse, in the island of Fyen (Funen). He had already, in 1574, published a Latin poem on the solar eclipse of that year, in which he showed that though eclipses have a perfectly natural cause, they are signs of the anger of God; but the eclipse of 1574 he believed to mean that the second coming of Christ was soon to take place. This

---

[1] Printed in *Danske Magazin*, ii. pp. 213–217 (Weistritz, ii. pp. 110–116).
[2] The court was held on a hill close to Uraniborg (*I* on the map).
[3] This does not seem to have been carried out. See above, p. 88.

little book he dedicated to Tycho Brahe.[1] He seems to have studied medicine in his youth, for his second publication, in 1575, was a translation of Simon Musæus' book against melancholy. He must have entered Tycho's service in the beginning of 1578, and did so (according to Longomontanus) on account of the supposed intimate connection between medicine and astronomy.[2] That Tycho had great confidence in him may be seen from the fact that he sent him to Cassel in 1586 to deliver a letter to the Landgrave and report to Tycho on the new instruments lately mounted there. In June 1579 he received by royal letter a promise of the first vacant canonry in Roskilde Cathedral, on condition that "he shall be bound to let himself be used *in studiis mathematicis* at Tyge Brahe's." He had, however, to wait a long time for this reward of his services at Hveen, as he did not obtain the canonry till 1590, when he had left Tycho, after more than ten years' service in the observatory, and had become physician to Axel Gyldenstjern, one of the two noblemen whom the king had sent to Hveen to report on the affairs of the tenants, and who had since been made Governor-General of Norway. Flemlöse died suddenly in 1599, just when about to proceed to Basle to obtain the degree of Doctor of Medicine.[3] Whether his medical studies had derived much benefit from his astronomical labours is not known, but while at Uraniborg, he not only spent his time on "pyronomic" (*i.e.*, chemical) and astronomical matters, but also compiled a little book which was printed there in 1591, some years after his

[1] "Æcloga de eclipsi solari anno 1574 mense Novembri futura et tempore plenilunii ecliptici anno 1573 conspecti, Succularum ortu obiter descripto, breuique Meliboei pastoris querela. . . Autore Petro Jacobo Flemlossio." Hafniæ, 1574, 4to.

[2] He made observations with the sextant on the 15th March 1578, and the distance measures on and after January 21 are possibly also by him.

[3] See N. M. Petersen, *Den Danske Literaturs Historie*, iii. pp. 176–179, and the preface by Friis to the reprint of Flemlöse's book (1865).

departure, containing 399 short rules by which to foretell changes in the weather by the appearance of the sky, the sun, moon, and stars, or by the behaviour of animals.[1] In the absence of the author, the introduction was written in his name by his fellow-student, Longomontanus, at the dictation of Tycho. In this it is stated that King Frederick took a great interest in weather prognostications, and had desired Tycho Brahe, from books and his own experience, to compile a treatise on the subject, but as Tycho had other and more important work to look after, he had requested Flemlöse to do so. It is not said whether the author had collected his materials at Hveen, but most of the rules contained in the book are chiefly such as farmers and similar observers might imagine they had deduced from their experience, and here and there it affords curious reading, at least to a modern student.[2]

Another of the early assistants of Tycho was a German, Paul Wittich, from Breslau, whose name, but for his early death, would probably be much better known in the history of astronomy than it is. He had been recommended by

[1] " En Elementisch or Jordisch *Astrologia* Om Lufftens forendring. . . . Til-sammen dragen aff Peder Jacobsön Flemlös paa Hueen. Prentit paa Vrani-borg Aff Hans Gaschitz, Anno 1591," xvi. + 143 pp., 12mo. Reprinted at Copenhagen in 1644 (by Longomontanus), 1745, and 1865. According to Friis, *Tyge Brahe*, p. 362, a German translation was printed at Hveen in September 1591, of which there is a copy in the library of the Polytechnic Institute at Vienna (see also *Kepleri Opera*, viii. p. 705, first line). Of the Danish original, only two copies are known to exist, both at Copenhagen.

[2] I shall give a few examples :—Flies and fleas announce rain when they are more than usually troublesome to men, horses, and cattle (ccv.). When goats are so very greedy that you can neither by words nor blows drive them away from small shrubs, which they bite off though they are not very hungry, then it is a sure sign of rain or storm (ccix.). When pigs with their snouts are throwing sheaves of corn or bundles of straw round about as if they were mad, you need not doubt that there will soon be rain (ccxxii.). All kinds of unusual fire in the air, appearing like an army or like stars running to and fro or against each other, or falling down to the earth, are forewarnings of comets (ccclx.). [This looks like an unconscious anticipation of modern ideas about the nature of comets.] Earthquakes generally follow after great and long-continuing comets (ccclxiii.).

Hagecius, and arrived at Hveen in the summer of 1580, where he took part in the observations of the comet of that year from the 21st to the 26th October.[1] He showed himself a very able mathematician, according to Tycho's own testimony,[2] and declared it to be his wish to stay at Uraniborg and be a "fidus Achates" to Tycho. But when he had been about three months at Hveen, he announced that he had to go home to Breslau, as a rich uncle of his was dead and he wanted to secure the inheritance, but he would return to Hveen in seven or eight weeks. He took with him a letter from Tycho to Hagecius (dated 4th November 1580), and Tycho became very uneasy when he neither heard anything from Wittich (who never returned to Hveen) nor received an answer from Hagecius for more than a year. He learned at last, in 1582, that the letter had been duly delivered.[3] A few years after he heard that Wittich had, about 1584, turned up at Cassel, where his descriptions of Tycho's improvements in instruments, particularly of the sights and the transversal divisions, as well as of Tycho's sextants for distance measures, created so great a sensation that the Landgrave immediately had his instruments improved and altered by his mechanician, Joost Bürgi, in accordance with Wittich's descriptions.[4] When Tycho

---

[1] In the observations (*Tychonis Brahe Observationes Septem Cometarum*, Hafniæ, 1867, p. 30) there is a note written in October 1600, and signed Jacob Monaw, certifying that the observations of October 21st to 26th were written in Wittich's hand. I find in Jöcher's *Gelehrten Lexicon* that this Monaw was a Jesuit from Breslau (1546–1603), where he had evidently known Wittich.

[2] In Tycho's *Mechanica*, fol. I. 3, he is mentioned as "quidam insignis mathematicus," and in *Progymn.*, ii. p. 464, he is called "quidam Vratislauiensis non vulgaris Mathematicus." In a letter to Rothmann (*Epist. Astr.*, p. 61) Tycho says that Wittich ingratiated himself with him "quod hominem ob ingeniosam in Mathematicis, præsertim quo ad Geometriam attinet, solertiam magnifacerem." We shall see farther on that Tycho and Wittich together deduced convenient formulæ whereby multiplication and division of trigonometrical quantities were avoided. See also *Epist.*, p. 296.

[3] *T. Brahe et Doct. Vir. Epistolæ*, pp. 54, 58, 64.

[4] *Epist. Astron.*, p. 3.

learned this he was extremely annoyed, and seemed to think
that Wittich had pretended to be the inventor of all he had
described to the Landgrave (although the latter had not
said so), and in his first letter he took care to tell the
Landgrave that Wittich had seen all these things at Hveen,
as might already be seen from the word "sextant."[1] How
long Wittich remained at Cassel is not known; he was
there in November 1584, when he observed a lunar eclipse,
and the Landgrave's astronomer, Rothmann, mentions him
in a letter of April 1586 as having left a good while
previously.   He died on the 9th January 1587,[2] and
Tycho seems on learning this to have regretted that he had
suspected Wittich of robbing him of his fame, for he wrote
in August 1588 that he would have written more mode-
rately about him had he known he was dead.[3] Though
Wittich spent but a short time at Uraniborg, his name
deserves to be remembered by astronomers, as he was
apparently the ablest of all Tycho's pupils.[4]

Most of these pupils spent a much longer time at Uraniborg than Wittich had done.   Thus Gellius Sascerides
stayed about six years there.   He was born at Copenhagen
in 1562, and was a son of Johannes Sascerides of Alkmaar,
in Holland, professor of Hebrew in the University of
Copenhagen.   Gellius had studied at Copenhagen and at
Wittenberg, and came to Hveen early in 1582, where he

---

[1] Ibid., p. 7.

[2] According to a MS. in the library at Breslau, quoted by Rud. Wolf in
the *Vierteljahrsschrift der Astron. Gesellschaft*, xvii. p. 129.

[3] *Epist. Astron.*, p. 113. Tycho here again praises his cleverness "in
Geometricis et Triangulorum ac numerorum tractatione."   In the letter of
20th January 1587 (to which he refers) he had, after all, only said : "Si mea
inventa . . . pro suis venditat, nec fatetur per quem ea habuerit, rem a viro
bono et grato, ac sinceritate integritateque Mathematica alienam committit."

[4] In Chalmers' *General Biogr. Dictionary*, London, 1815, vol. xx. p. 243, it
is stated, on the authority of a Life of the Scotch mathematician Duncan
Liddel by Prof. Stuart (1790), that Liddel studied mathematics at Breslau,
1582–84, "under Paul Wittichius, an eminent professor."

remained until 1588, when he went abroad to continue his
medical studies in Italy.   Tycho gave him a letter for
Rothmann, to whom he recommended Gellius as having
assisted him both in astronomical and in chemical work.[1]
We shall afterwards hear how he and Tycho got on
together after his return.

We know much less about another assistant who observed
at Hveen about the same time as Gellius, called Elias Olsen
Cimber (or Morsing, i.e., from the Isle of Mors, in the
Limfjord), although he must have spent a number of years
with Tycho.   When he first came to Hveen is not known,
but he seems to have been there in April 1583, when his
handwriting is believed to occur in the meteorological
diary.   This diary (of which the original is now in the
Hofbibliothek at Vienna) was regularly kept from the 1st
October 1582 up to the 22nd April 1597, about the time
when Tycho left Hveen for ever.[2]   It contains for every day
short notes about the weather, stating whether it was clear
or cloudy, hot or cold, rainy or dry, &c.   These notes are
always written in Danish, except where halos, auroras, or
similar phenomena are described, which is generally done
in Latin.   But the principal interest attached to this diary
arises from the numerous very short notes about the arrival
or departure of Tycho, his pupils or visitors, which occur
frequently from April 1585.   These historical notes are
always written in Latin ; they are often very much abbre-
viated and difficult to decipher.   This diary, which forms a
most interesting record of the life at Hveen, was kept now

[1] *Epist. Astron.*, p. 104.
[2] It was published at Copenhagen in 1876 : *Tyge Brahe's meteorologiske
Dagbog holdt paa Uraniborg for Aarene 1582–1597. Appendice aux Collectanea Meteorologica publiés sous les auspices de l'Académie Royale des Sciences
et des Lettres à Copenhague.*   The value of the diary (263 pp. 8vo) is greatly
increased by an index to the historical names by a Danish historian, H. F.
Rördam.   There is also a discussion of the meteorological results by P. la
Cour (with a French resumé).

by one, now by another assistant (though their names are
not given), and a great deal of it was written by the above-
mentioned Elias Olsen, whose writing appears in it for the
last time in April 1589. Probably he left Tycho's service
at that time, as he is mentioned in the diary as having
arrived and departed several times after that date.[1]

In 1584 Elias Olsen was sent by Tycho on an astronomical expedition of some importance. At Hveen the
inclination of the ecliptic had been found equal to 23° 31′.5,
while Copernicus had found 23° 28′. Tycho correctly explained this by pointing out that Copernicus had measured
the meridian altitudes of the sun at the summer and winter
solstices without taking refraction into account, and for the
latitude of Frauenburg in Prussia this would at the winter
solstice cause an error of over 4′ in the altitude. Tycho,
however, believed the solar refraction at the altitude of 12°
to be equal to 9′; but, on the other hand, he assumed with
Copernicus, that the solar parallax was 3′, so that one mistake is somewhat compensated by the other. He had also
found that the solar theory of Copernicus often deviated
considerably from the observed places of the sun, and he
suspected that Copernicus had reduced his solar observations
with an erroneous value of the latitude. He, therefore,
gladly took an opportunity of verifying this latitude when,
early in 1584, an embassy from George Frederic, Margrave
of Ansbach,[2] headed by a nobleman of the name of Levin

---

[1] He was at Hveen June 9 to 11, and July 1 to 3, 1589, November 5 to
March 11, 1590. Under the last date the printed edition has "Elias obiit H.
11½ noct.," but doubtless the original has *abiit* and not *obiit*, for the words
"Elias Olai" occur again on the 8th May 1596, so he cannot have died in
1590. In 1589 he went with Vedel on a tour through Denmark to observe
latitudes and azimuths for Vedel's topographic survey of the country. See
E. O. Morsing og hans Observationer, af F. R. Friis, Copenhagen, 1889, 28 pp.
8vo.

[2] Regent of the Duchy of Prussia (for his cousin, Duke Albrecht Frederic,
who was insane). The house of Hohenzollern is descended from him.

Bülow, returned to Germany after having carried out its
mission to the Danish Court. As the embassy was sent to
Dantzig in some royal ships, it was easy for Tycho Brahe
to obtain permission for Elias Olsen to make the voyage on
one of these. He happened to be keeping the meteorological diary at that time, and continued on the journey to
record in it the state of the weather. We learn thus that
he started from Copenhagen on the 1st May, reached
Dantzig the 10th, and Frauenburg on the 13th. In this
quiet little cathedral town Copernicus had lived many
years, engaged solely in building up his great astronomical
work, and only now and then turning aside from this to
assist with his clear mind in the government of the little
diocese-principality of Ermland or in the affairs of the
chapter of Frauenburg. Elias Olsen remained on this classical spot from the 13th May till the 6th June, and, with
a sextant which he had brought with him, he found by
meridian altitudes of the sun and stars the latitude to be
$54° 22\frac{1}{4}'$, while Copernicus made it $54° 19\frac{1}{2}'$ (the modern
value is $54° 21' 34''$). Tycho remarks that the solar declinations of Copernicus are consequently $2\frac{3}{4}'$ in error, which,
together with his omission of refraction, was sufficient to
explain the shortcomings of his solar theory. We shall
afterwards examine this question again when discussing
Tycho's labours on the solar theory. While Elias Olsen
was at Frauenburg he was requested to determine the latitude of Königsberg, and went there on the 8th June. He
found $54° 43'$, greatly different from $54° 17'$, which Erasmus
Reinhold had assumed in the Prutenic tables on the authority of Apianus.[1] On the 28th June Elias left Königsberg

---

[1] *Progymnasmata*, pp. 34–35 ; *Epist. Astr.*, p. 74. The latitude of the
Königsberg observatory is $54° 42' 51''$. Most of the observations made at
Frauenburg are given in Baretti *Historia Cœlestis*, p. 104, and are correctly
reproduced, except that the date of the observations of May 11 should be
May 17. In the *Hist. Cœl.* are not given the "Observationes factæ in

for Frauenburg, spent five days there, departed for Dantzig
on the 4th July, started from thence on the 7th, and was
back at Hveen on the 23rd.[1]

Valuable as these results of the journey were, Elias
brought something else home with him which was perhaps
even more valued by Tycho.   One of the canons at Frauenburg, Johannes Hannov, sent him the instrument used by
Copernicus and made by his own hands.   It was a triquetrum eight feet long, made of pine-wood, and divided by
ink-marks, the two equal arms into 1000 parts, the long
arm into 1414 parts.   Tycho placed this scientific relic in
the northern observatory at Uraniborg, and the very day he
received it (the 23rd July) he composed a Latin poem expressing his enthusiastic delight at possessing an instrument
which had belonged to this great man, whose name he never
mentioned without some expression of admiration.[2]   This
feeling he also gave vent to in the poem which he a few
months later wrote and placed under the portrait of Copernicus in his library.   Possibly he had received this portrait
on the same occasion as the instrument.[3]

The name of Elias Olsen is also connected with the first
book printed at Uraniborg, an astrological and meteorological diary for the year 1586, somewhat similar to the one
drawn up by Tycho for the year 1573.   It also contains an
account of the comet of 1585, which had been observed at
Hveen from the 18th October to the 15th November.   The
little book is dated the 1st January 1586, and is dedicated

Ædibus Hortensibus illustrissimi Marchionis ducis Borussiæ Regiomonti;"
they are similar to those made at Frauenburg, and extend from June 11 to 26
(MS. volume of Obs.).

[1] The dates are from the meteorological diary.   Friis (*T. Brahe*, p. 133)
tells his readers that Elias went to Regensburg (Regiomontum !!) without
remarking the wonderful speed with which he would have had to travel to
reach Regensburg from the shore of the Baltic in less than two days.

[2] *Epist. Astr.*, p. 235; Gassendi, p. 57.

[3] *Epist.*, p. 240.

to the Crown Prince, who was then between eight and nine years of age.[1]

Of Tycho's other pupils, Longomontanus is the best known. Christen Sörensen Longberg was born on the 4th October 1562, at the village of Longberg or Lomborg, in the north-west of Jutland, where his father was a poor farmer.[2] When his father died in 1570, his uncle took charge of him for some time, but as the means of the family were too small to allow the boy to follow his inclinations and go to school, the uncle sent him home to his mother to help her on the farm. The boy persuaded the mother to allow him to get some lessons during the winter-time from the clergyman of the parish, but during the summer he had to lay aside his books and take to farming again. At last he got tired of this, and in the spring of 1577 he took his books, and, without telling any one, walked off to the town of Viborg, some fifty miles from his home. He attended the grammar-school of Viborg for eleven years, and in addition to the ordinary school course of those days he learned the rudiments of mathematics. At the age of twenty-six he left the school for the University of Copenhagen, and the following year (1589) he was, on the recommendation of some of the professors, received as an assistant at Uraniborg, where he remained till 1597, when he left it together with Tycho.[3]

Of most of the other young men who for a longer or

---

[1] " Diarium astrologicum et metheorologicum anni a nato Christo 1586. Et de Cometa qvodam rotundo omniqve cavda destituto qui anno proxime elapso, mensibus Octobri et Nouembri conspiciebatur, ex observationibus certis desumta consideratio Astrologica : Per Eliam Olai Cimbrum, Nobili viro Tychoni Brahe in Astronomicis exercitiis inservientem. Ad Loci Longitudinem 37 Gr. Latitudinem 56 Gr. Excusum in Officina Vranibvrgica." See Weidler, *Hist. Astr.*, p. 623 ; Petersen, *Danske Literaturs Historie*, iii. p. 180.

[2] West of the town of Lemvig, about four miles from the west coast. In Latin, Longberg called himself Christianus Severini Longomontanus.

[3] Petersen, *Danske Literaturs Historie*, iii. p. 177.

shorter time assisted Tycho Brahe, we know little but the names. A certain Hans Crol, or Johannes Aurifaber, who had charge of the workshop, must have been with him a long time, as he is mentioned as observing in 1585, and he died at Hveen in 1591.[1] Many details as to the life at Hveen were communicated to Gassendi by Willem Janszoon Blaev, the celebrated printer at Amsterdam, who in his youth (he was born at Alkmaar in 1571) had spent a few years at Hveen, and to whom we also owe the large map of the island in his son's Grand Atlas.[2]

Two other inmates of Tycho's house may also be mentioned here. One was a maid of the name of Live (or Liuva) Lauridsdatter, who afterwards lived with Tycho's sister, Sophia, and later was a sort of quack-doctor at Copenhagen, where she also practised astrology, &c. She died unmarried in 1693, when she is said to have reached the ripe age of 124.[3] The other was his fool or jester,

---

[1] *Observationes Septem Cometarum* (1867), pp. 63–64; *Baretti Historia Cœlestis*, p. 429; *Diary*, 30th November 1591.

[2] The map was made "cum sub Tychone Astronomiæ operam daret." Blaev must have been at Hveen during the last few years of Tycho's residence there. He is mentioned in the *Observations of Comets*, p. 41, as being there in 1596. For a list of Tycho's other disciples and assistants, as far as their names are known, see Note B. at end of this volume. In 1589 Rothmann inquired, on behalf of Professor Victor Schönfeld of Marburg, whether Tycho would receive a son of Schönfeld among his pupils, adding that the young man had just been made a Master of Arts; to which Tycho answered that he might come, but whether he was a master or not did not make much difference, that it was better to *be* a master than to be called one, and it would be sufficient if he was a student of the free arts (*Epist.*, pp. 154, 168). In Wolf's *Encomion Regni Daniæ*, 1654, p. 526, it is stated that there were small bells in the rooms of the students, which could be rung by touching hidden buttons in the observatories or sitting-rooms, by which Tycho, to the surprise of his guests, could make any of the students come to him, apparently merely by calling their name in a low voice. Wolf also tells how Tycho could lie in bed and observe the stars through a hole in the wall, with some mechanism which could be turned round. Probably this refers to the mural quadrant, which had a "hole in the wall."

[3] Kästner, *Gesch. der Math.*, ii. p. 408, quoting *Nova Literaria Maris Balthici*, August 1698, p. 142. There is a portrait of this woman in the National Historical Museum at Frederiksborg Castle.

a dwarf called Jeppe or Jep, who sat at Tycho's feet when he was at table, and got a morsel now and then from his hand. He chattered incessantly, and, according to Longomontanus, was supposed to be gifted with second-sight, and his utterances were therefore listened to with some attention. Once Tycho had sent two of his assistants to Copenhagen, and on the day on which they were expected back the dwarf suddenly said during the meal, " See how your people are laving themselves in the sea." On hearing this, Tycho, who feared that the assistants had been shipwrecked, sent a man to the top of the building to look out for them. The man came back soon after and said that he had seen a boat bottom upwards on the shore, and two men near it, dripping wet. Whenever Tycho was away from home, and the pupils relaxed their diligence a little, they set Jeppe to watch for him, and when the dwarf saw Tycho approach he would call out to them, " Junker paa Landet," i.e., the squire [is] on land.[1] When any one was ill at Hveen, and the dwarf gave an opinion as to his chance of recovery or death, he always turned out to be right.

There was plenty to do for all the young men at Uraniborg. Of course the astronomical work was always their principal occupation, but the laboratory was also in constant use. We have no knowledge of the particular direction of Tycho's chemical researches, but that he always took a very deep interest in chemistry is evident from more than one allusion to this subject in his writings. In several of his books are found a pair of vignettes, which illustrate the view of Nature as a whole, representing one idea under various

---

[1] Gassendi (p. 197), who had these details from letters written to him and Peyresc by the Danish physician and historian Ole Worm, has misspelt the exclamation of the dwarf as "Juncher xaa laudit." See also *O. Wormii et doct. vir. ad eum Epistolæ*, Hafniæ, 1751, and Gassendi, *Epistolæ* (*Opera*, vol. vi.), p. 527, where the name is misspelt Leppe. The word "Junker" (esquire), which always is used of T. Brahe, shows that he was not a knight.

aspects, with which not only Tycho, but most thinkers of the Middle Ages were imbued.[1] On both these vignettes is seen a man in a reclining posture, with a boy at his side; but in the one case the man is leaning on a globe and holds a pair of compasses in his hand, while his face is turned upward; in the other case he has at his side some chemical apparatus, and holds in his hand a bunch of herbs, while the snake of Æsculapius is coiled round his arm, and he is looking downwards. At the sides of the former picture is the motto, " Suspiciendo despicio; " round the latter, " Despiciendo suspicio," expressing beautifully the mystical reciprocal action and sympathy between the " æthereal and elementary worlds." In a letter to Rothmann, Tycho enters at some length on this subject, but his remarks contain nothing which may not be read in any book of the time in which the " occult philosophy " is taught, and we have already sufficiently alluded to these matters in previous chapters. He mentions the principal authors whom he has followed,[2] but adds that Paracelsus has truly said that nobody knows more in this art than what he has experienced himself *per ignem*, for which reason he cultivates the " terrestrial astronomy " with the same assiduity as the celestial. In the laboratory Tycho also occupied himself with the preparation of medicine, and as he distributed his remedies without payment, it is not

---

[1] These vignettes seem first to have been used for a poem to a friend of Tycho's, Falk Gjöe, printed at Uraniborg between 1584 and 1587, and of which I am not aware that any copy now exists. Rothmann came across a copy at Frankfurt, and asked Tycho to explain the vignettes. *Epist. Astron.*, p. 89; Tycho's reply, ibid., p. 115–117.

[2] Among these are Hermes Trismegistus, Geber, Arnoldus de Villa Nova, Raymundus Lullius, Thomas Aquinas, Roger Bacon, Albertus Magnus, &c. He does not allude to the fact that the idea expressed in the two vignettes occurs already in the second of the thirteen sentences of the so-called *Hermes Trismegistus* : " What is below is like what is above, and what is above is like what is below, to accomplish the miracles of one thing " (see *Nature*, vii. p. 90). There is, however, an allusion to this sentence in *Epist. Astron.*, p. 164.

strange that numbers of people are said to have flocked to
Hveen to obtain them.[1]   In the official Danish *Pharmacopœa*
of 1658 several of Tycho's elixirs are given, and in 1599 he
provided the Emperor Rudolph with one against epidemic
diseases, of which the principal ingredient was *theriaca
Andromachi*, or Venice treacle, mixed with spirits of wine,
and submitted to a variety of chemical operations and ad-
mixtures with sulphur, aloes, myrrh, saffron, &c.   This
medicine he considered more valuable than gold, and if the
Emperor should wish to improve it still more, he might add
a single scruple of either tincture of coral or of sapphire, of
garnet, or of dissolved pearls, or of liquid gold if free from
corrosive matter.   If combined with antimony, this elixir
would cure all diseases which can be cured by perspiration,
and which form a third part of those which afflict the human
body.[2]   This prescription Tycho begged the Emperor to
keep as a great secret, and he had evidently as much confidence in the powers of his elixir as the ingenious Hidalgo
of La Mancha had in the efficacy of his celebrated balsam.

We can form some slight idea as to the principles which
guided Tycho in his medical practice from a remark in one
of his letters to Rothmann, where he speaks of the Aurora
Borealis.   This he takes to be sulphurous vapour, indicating that the air is apt to engender infectious diseases, "for
such illness has a good deal in common with the nature of
sulphur, and it can therefore be cured by perfectly pure
earthly sulphur, particularly if this is made into a pleasant
fluid, as like cures like (*tanquam simile suo simili*), for the

---

[1] Tycho seems to have had an apothecary in his service, as Paulus Phar-
macopola is often alluded to in the diary ; *e.g.*, 22nd July 1596 : "Elisabetha,
filia Pauli pharmacopolæ, Joachimus et Theodoricus propter seditionem
dimittuntur."

[2] The prescription is printed by Gassendi, p. 242 *et seq.* ; he had it from
Worm, who in 1653 informed Gassendi that the elixir was still much used in
Denmark, frequently by the writer himself, who found it to be most powerful
in causing perspiration (*Opera*, vi. p. 526).

principle of the Gallenians, *contraria contrariis curari*, is not always true." [1]

We have repeatedly had occasion to quote from Tycho's letters. Both before and after he had become settled at Hveen, to all appearance for life, he kept up a correspondence with friends at home and with scientific colleagues abroad. Of the former, only Vedel and Dancey were left, and with these he occasionally exchanged friendly letters, [2] but between him and the acquaintances he had made on his foreign travels very lengthy epistles passed as often as an opportunity offered of sending these by a carrier, merchant, or by some casual traveller. Among Tycho's principal foreign correspondents were Paul Hainzel and Johannes Major at Augsburg, Scultetus at Leipzig, the Emperor's physician, Hagecius, at Prague, and Brucæus at Rostock. Being always anxious to increase his library, Tycho in many of his letters inquires about new books, or asks his friends to procure them for him, especially such as were about the new star or the recent comets. These comets had also been observed by Hagecius, and Tycho pointed out the erroneous result his correspondent had come to in giving the comet of 1577 a parallax of five degrees, which would place it far within the sphere of the moon, whereas the observations made at Hveen showed that the horizontal parallax was less

---

[1] *Epist.*, p. 162. See also an article, "T. Brahe als Homöopath," by Olbers, in Schumacher's *Jahrbuch für* 1836, p. 98. Olbers remarks that of course Tycho Brahe had too much common sense to believe in infinitesimal doses.

[2] It is characteristic that while Tycho in his letters to Vedel generally sends his regards to Vedel's wife, neither of them ever alludes to the mother of Tycho's children. Dancey died in 1589; he had first been sent to Denmark by Henry II., and came afterwards again when King Frederick II. was negotiating to recover the Orkney Isles from Scotland. Owing to the disturbed state of France, his salary was often considerably in arrear, which placed him in a very humiliating position both to the Danish king and to private people who had lent him money. Notwithstanding his troubles, Dancey was greatly liked and respected in Denmark.

than a third of a degree.[1]   Tycho also told Hagecius of the
corrections to the elements of the solar orbit of Copernicus,
which his own observations indicated; but neither to the
Bohemian physician nor to his other correspondents did he
allude to the new system of the world which he had constructed, possibly because (as he wrote to Hagecius) Wittich's conduct had given him a lesson which he should not
forget.[2]   As Tycho had understood from Wittich that Hagecius had lost his post in the Emperor's household, he invited
him to come to Denmark, where he might be sure of being
well remunerated by the king and the nobility for his services as a physician; but Hagecius declined to leave Prague,
as he had not lost his post, and found it too risky for a man
who was no longer young and had a family to settle abroad.[3]
With Johannes Major, Tycho corresponded about the Gre-
gorian reform of the calendar, which was promulgated in
1582, and ordered to be adopted by the Catholic world under
threat of excommunication.   In consequence of this, Protestants refused to make any alteration in the calendar.   At
Augsburg several members of the civic council had voted
against the adoption of the new calendar for theological
reasons, and when the mayor, in consequence, tried to arrest
and carry off the principal theologian of Augsburg, the popu-
lation rose in arms and set him free.   When asked for his
opinion, Tycho very sensibly remarks that if the Pope at the
time of Regiomontanus (*i.e.*, before the Reformation) had
improved the calendar, Luther would most assuredly not
have wished to interfere with it, as this matter had nothing
to do with religious doctrines; and why should not the
new calendar, approved of by the Emperor, be accepted, as
the Nicean calendar-rules were still accepted even by Protestants?   Of Tycho's letters to his old fellow-student at

---

[1] *T. B. et ad eum Doct. Vir. Epist.*, pp. 55, 60, and 62.
[2] Ibid., p. 59.   [3] Ibid., pp. 56, 65, and 68.

Leipzig, Scultetus, five are preserved, although of these but three are printed in accessible places;[1] one of these (of 1581) deals chiefly with the comet of 1577, for which Scultetus also imagined that he had found a parallax; another (of 1592) is written in a jovial manner, Tycho promising to drink his friend's health that evening, and expecting him to return the compliment. Another former University acquaintance with whom Tycho occasionally exchanged letters was Professor Brucæus, who had been appointed to a chair of medicine in the University of Rostock in 1567 while Tycho was studying there. He was one of the comparatively few learned men of the time who would have nothing to do with astrology, and it is therefore not to be wondered at that he expressed his disapproval on hearing about the intended printing of an astrological calendar by Elias Olsen at Hveen. He wrote, for instance, that weather predictions reminded him of Cato's saying of the Roman haruspices, that he wondered if they could keep from laughing whenever they met each other.[2] But though adverse to astrology, Brucæus had no objection to an astronomer dabbling in medicine, and in one of his letters he asked Tycho to let him know if he was in possession of any remedy against epilepsy. They also corresponded on astronomical matters, and Tycho pointed out to him the difficulty in accepting the theory of Copernicus, and commented on the errors of the Alphonsine and Prutenic tables.[3]

Of far greater importance than the above correspondence

[1] The first one in *T. B. et Doct. Vir. Epist.*, p. 57; the second in Kästner's *Geschichte der Mathematik*, ii. p. 409. The source of both is *Singularia Historico-literaria Lusatica oder historische und gelehrte Merckwürdigkeiten von Ober- und Nieder-Lausitz*, 27te Sammlung, 1743, pp. 178 *et seq.*, where there are two more letters printed. Scultetus (Schultz) died in 1614 as burgomaster at Görlitz. A letter dated January 1600 is printed in *Aus T. Brahes Briefwechsel*, von F. Burckhardt. Basel, 1887.

[2] *T. B. et Doct. Vir. Epist.*, p. 93.

[3] Ibid., p. 75 *et seq.*

were the letters exchanged between Tycho and Landgrave
Wilhelm of Hesse, and his astronomer Christopher Rothmann.   We have seen how Tycho's visit to Cassel in the
year 1575 seems to have given a fresh impetus to the
scientific tastes of the Landgrave, who in 1577 engaged
Christopher Rothmann of Anhalt as his *mathematicus*, a man
not without some knowledge of astronomy and mathematics,
though not possessing the genius of the man who, two years
later, was engaged as his assistant.[1]   Joost Bürgi was born
in 1552, at Lichtensteig, in the county of Toggenburg, in
Switzerland, and seems to have been a watchmaker in his
youth, but nothing is known of his life until Landgrave
Wilhelm, in 1579, appointed him court-watchmaker at
Cassel.   The methods of observing adopted in the observatory at Cassel rendered good clocks indispensable, and both
these and the constantly improved instruments made the
services of the ingenious mechanician most valuable to the
Landgrave, who, indeed, was well aware what a treasure he
had found, as he in one of his letters to Tycho calls Bürgi
a second Archimedes.   Observations were regularly made
at Cassel by Rothmann and Bürgi, especially of the fixed
stars, with the object of constructing a new star-catalogue,
but other celestial phenomena were not altogether neglected,
and the comet of 1585 gave rise to a correspondence
between Tycho and the Landgrave.   They had lost sight
of each other since 1575, but the Landgrave was well
aware that a magnificent observatory had been erected at
Hveen, and that work was steadily carried on there, particularly since the visit of Wittich had put him in possession
of the important improvements which Tycho had introduced
in the construction of instruments.   He was therefore

---

[1] Neither the year of Rothmann's birth nor that of his death are known.
About him and Bürgi, see in particular Rudolph Wolf's *Geschichte der Astronomie* and his *Astronomische Mittheilungen*, Nos. 31, 32, and 45.

anxious to learn what observations Tycho had made of the comet of 1585, as it was not a very conspicuous one, and probably would not be observed by many astronomers. With this view the Landgrave wrote a letter to the learned Heinrich Rantzov, governor of Holstein,[1] asking that his compliments might be sent to Tycho, with a hint that he would be glad to hear something of the observations of this comet made at Hveen. Tycho was very happy to renew his acquaintance with the Landgrave, to whom he wrote a long letter on the 1st of March 1586, in which he enclosed an abstract of his observations of the comet. In the letter he suggested an exchange of observations of the star of 1572 and of the recent comets, claimed the instrumental improvements already announced to the Landgrave as his own, and gave an account of various instruments he had designed, such as a bifurcated sextant, to be used by two observers, and the equatorial armillæ. He pointed out the great convenience of the latter instrument, which directly gave the right ascension and declination of an object, from which the longitude and latitude could be found either by calculation, or by a specially prepared table, or by a large globe. Tycho also sent the Landgrave a solar ephemeris for the current year, and asked him to compare his observations with it. This letter and its appendices were sent to Cassel by Tycho's assistant, Flemlöse, who was bound for the book-mart at Frankfurt, and could take Cassel on his way, where Tycho doubtless also wished him carefully to inspect the improved instruments.

To this letter the Landgrave at once replied, and Rothmann also took the opportunity of entering into correspondence with Tycho. During the next six years letters

---

[1] Rantzov (1526–1599) was celebrated as a collector, not only of books (of which he possessed about 7000 volumes), but also of works of art. He also wrote on astrology.

continued to be sent backwards and forwards between
Cassel and Uraniborg, in which were discussed the methods
of observing, the instruments in use, and, after the publication of Tycho's system of the world, also the question
whether this system or that of Copernicus was the true one.
We shall in the sequel have many opportunities of quoting these letters, or rather astronomical essays, of which
Tycho recognised the interest to the scientific world by
sending copies of some of them to several other correspondents, and finally by publishing them all in a volume
printed at Uraniborg, which forms an excellent supplement to his other writings, and completes the picture of
his scientific activity.[1]

All the details about Tycho's observatory which the
Landgrave had learned from Tycho's letters to himself and
Rothmann had naturally made him anxious to see it for
himself, and an opportunity of doing so seemed to offer itself
in 1588, as there was to be a meeting of North German
princes at Hamburg, which the Landgrave was going to
attend. King Frederick had already given orders to have
ships ready to carry the Landgrave over to Seeland, when
the king's death prevented the meeting at Hamburg, and
with it a second meeting of Tycho and the Landgrave.[2]

But though Wilhelm IV. never came to Hveen, Tycho
had from time to time the pleasure of welcoming other
distinguished guests at Uraniborg. Among these we shall
here mention Johan Seccerwitz, professor in Greifswalde,
who is known as a Latin poet. He came to Denmark in
1580 with the Duke of Pomerania to attend the christening

---

[1] A short chronological summary of the principal points of interest in this
correspondence is given by Delambre in his *Histoire de l'Astronomie Moderne*,
tom. i. pp. 232 *et. seq.* See also Gassendi, p. 65 *et seq.*

[2] *Epist. Astron.*, in the dedication to the Landgrave's son, and also p. 104.
The "Comitia Hamburgensia" was, I suppose, a meeting of the princes of
the Nether-Saxon circle, to which King Frederick belonged as Duke of
Holstein.

of a new-born princess, and met Tycho in the house of the Bishop of Lund. He has left a versified description of his journey, in which he expresses his joy at having made the acquaintance of Tycho. In 1584 the French historian Jacques Bongars was at Uraniborg.[1] Another learned visitor was Duncan Liddel, who was born at Aberdeen in 1561, and had studied at Frankfurt-on-the-Oder and at Breslau. In 1587 he went to Rostock, and while studying there paid a visit to Hveen on the 24th June.[2] He was professor at Helmstadt from 1591 to 1607, and is said to have been the first person in Germany who explained the motions of the heavenly bodies according to the three systems of Ptolemy, Copernicus, and Tycho.[3] Some travellers who were not of a scientific turn of mind were nevertheless attracted to Hveen by the wonderful things to be seen there. Thus, in 1582 Lord Willoughby d'Eresby, who had been sent by Queen Elizabeth to invest King Frederick with the Order of the Garter, paid a visit to Uraniborg, and brought with him a physician, Thomas Muffet, in whom Tycho was pleased to find an acquaintance of his friend Hagecius.[4] Daniel Rogers, who was on several occasions employed by Queen Elizabeth on missions to the Netherlands and Denmark, was also acquainted with Tycho, and in 1588, when he came to condole on the king's death, he went to Hveen, where he promised Tycho to obtain for him the copyright of his books in England.[5] Below we shall see that Tycho

---

[1] *Danske Magazin*, ii. pp. 210 and 220 (Weistritz, ii. pp. 105 and 126).

[2] *Meteorol. Diary.*

[3] About Liddel, see above p. 121, footnote. He died at Aberdeen in 1617.

[4] Letter to Hagecius, *T. B. et Doct. Vir. Epist.*, p. 70. Tycho does not mention that Lord Willoughby had landed at Elsinore on the 22nd July, but that the king's installation as a K.G. did not take place till the 19th August, because the king for a long time refused to be dressed in the full costume, &c., in public. Dancey had to assist in settling the matter.

[5] Tycho tells this in a letter to Peucer, and adds that he had already secured the copyright in France and Germany (Weistritz, i. p. 264). Rogers (1540–1590) was a man of considerable learning, particularly in British antiquities;

was to receive even more exalted visitors from abroad during
the last years of his residence at Hveen.

It is needless to say that Danish visitors frequently
crossed over the Sound to the little island which had so
suddenly become famous. Both learned and unlearned men
were ready to pay court to the great astronomer who had
raised a beautiful building full of curious apparatus on the
lonely island. Though this spot had expressly been selected
for his residence in order that Tycho might undisturbedly
devote himself to the studies he loved, he had probably no
objection now and then to receive as his guests even some
of those who had in former days sneered at his scientific
tastes,[1] and not a few among the Danish visitors were men
of learning. Among those who paid repeated visits was
Tycho's former tutor and his friend through life, Anders
Sörensen Vedel, who was now royal historiographer, and
lived at Ribe in Jutland, as a canon of the cathedral there.
He was on a tour through Denmark to collect topographical
and other information for his Danish history, when he
arrived at Hveen on the 13th June 1586. He must have
stayed there some weeks, as he was still with Tycho when
a stately little fleet on the 27th June approached the island
from Seeland with Queen Sophia on board. The queen was
a daughter of Duke Ulrich of Mecklenburg-Güstrow, and was
an able and accomplished lady. Tycho's mother, Beate Bille,
acted as her Mistress of the Robes (to which post she was
regularly appointed in 1592 after the death of his aunt
and foster-mother, Inger Oxe), and the queen was therefore
interested beforehand in Tycho and his work. She was

he had studied at Wittenberg during the persecution of the Protestants by
Queen Mary. Perhaps he is alluded to in the *Meteorol. Diary*, 9th July 1588,
"Angli aderant."

[1] Fortunately for him, Tycho lived before the age of telescopes, so he was
not annoyed by constant requests "to see the moon" or to "take an observation through the big telescope."

detained on the island by a storm till the 29th, so that she had time enough to see everything of interest, and to converse with Tycho and Vedel on the various topics which the scenery of the island and the curiosities of the observatory and laboratory suggested. At table Tycho called the queen's attention to Vedel's historical researches and his collections of ancient ballads and other folk-lore, a subject in which she took a great interest. She asked Vedel for a copy of these ballads or *Kjæmpeviser*, which he promised to send as soon as he could, and this incident gave rise to Vedel's collection of ancient ballads being printed five years later.[1] The queen must have enjoyed herself well (she is said to have had a taste for chemistry), and two months afterwards, on the 23rd August, she brought her father and mother[2] and a cousin to see Uraniborg, and was on this occasion attended by a large suite. The Duke was also fond of chemistry, which in those days was a fashionable occupation, owing to the prevailing opinion that it would sooner or later lead to the discovery of the art of making gold.[3]

King Frederick did not accompany the queen on either of these occasions, and it is not certain that he ever was at Hveen.[4] In contemporary documents and in Tycho's own writings there is no allusion to the king's having visited

---

[1] Wegener's *Life of Vedel*, p. 148. The reader may recollect that Vedel's edition of the *Kjæmpeviser* is referred to in Note K. to "The Lady of the Lake." The queen wrote somewhere at Uraniborg her motto "Gott verlest die Seinen nicht."

[2] Elizabeth, daughter of King Frederick I. of Denmark.

[3] Tycho told the Landgrave about this visit in a letter dated 18th January 1587 (*Epist. Astron.*, p. 36). In March 1592 the queen wrote to T. Brahe requesting him to send her father a small barrel of emery, as the Duke had heard that some had lately been found at Hveen, and for herself she wanted some "burnt antimony," such as she had got from him before. Friis, *Breve og Aktstykker angaaende T. Brahe.* Copenhagen, 1875, p. 5.

[4] The king was in Jutland and North Slesvig during the last days of June (Wegener's *Life of Vedel*, p. 149), and that he did not accompany the queen in August is evident from Tycho's letter to the Landgrave just quoted.

him, and if he had done so during the last three years
of his life, it would certainly have been mentioned in the
meteorological diary, in which during this period all events
of that character were noted.[1]   But this does not exclude
a visit of the king to Hveen before 1585, and it would
indeed be strange if he had never during the years he was
building the castle of Kronborg at Elsinore, with the island
before his eyes, crossed over the narrow strip of water to
see the buildings of which he must have heard so much,
and to whose owner he continued to show favour on every
occasion.   The king might also have taken the opportunity
of seeing Uraniborg in the year 1584, when his eldest son,
Prince Christian, was elected his successor.   On the 20th
July the nobility of Scania swore fealty to the prince at
Lund, where Tycho Brahe also appeared among the other
nobles of the province, and the king was apparently in
Scania at that time.   A remarkable document, which is
still in existence, and is printed among the many important
letters in the *Danske Magazin*,[2] seems to show that the king
was expected at Hveen at that time.   It is a draught of an
act, written in Latin and in the king's name, dated " Huenæ
in Avtopoli Vranopyrgensi," the 1st July 1584.   In this
document the king, in recognition of Tycho Brahe's scientific
work, and following the memorable examples of former ages,
grants to him and his heirs male for ever the island of
Hveen in fief, with all privileges and honours, provided that
they do nothing to injure the king or kingdom, and keep
the buildings of the island solely for the furtherance of
mathematical studies.   But this document was never en-
grossed and signed by the king, and even if Tycho could
have persuaded the king to grant him so great a favour, it

[1] It is curious that the very first note of an historical character is under
the 27th April 1585: "Nuncium de adventu Regis," but in the following
there is nothing about him.

[2] ii. pp. 220–221 (Weistritz, ii. p. 124).

would have been very hard for the king to obtain the consent of the Privy Council, although its principal members were at that time very friendly disposed to Tycho; and in particular these great nobles would have protested against so monstrous a proceeding as the transmission of a valuable fief to the children of a " bondwoman." Probably the act was only drawn up in an idle moment, while the writer [1] was thinking about the chance of a visit from the king, but it shows at any rate that Tycho's wishes went in the direction indicated by the draught, and that he felt the insecure position in which all his creations at Hveen were placed. All his endowments were only enjoyed by him during the king's pleasure, and even the island was only granted for his own lifetime. Were then the beautiful buildings and wonderful instruments some day to vanish again, as the observatories of Alexandria, Cairo, Meragah, Cordova, and Nürnberg had vanished? This thought was doubtless a painful one to Tycho, who, the more he studied the stars in the heavens and the elements in the earth, could not but feel that life was short and art was long.

While his royal protector lived, Tycho and his observatory were, however, safe enough; that much he knew, not only by the readiness with which one pecuniary grant after another was made to him, but also by many more private acts of kindness and good feeling which emanated from the king, and of which we have ample proofs in various letters still extant. The king evidently looked on Tycho not only as a great man, whose achievements conferred honour on the country and on the monarch who supported him, but also as a confidential servant to whom he could turn for advice on matters within his province, and whom he in return delighted to honour and befriend. All the existing

[1] According to Friis (*Elias Olsen Morsing*, Copenhagen, 1889, p. 6), the handwriting seems to be Vedel's.

portraits of Tycho Brahe represent him as wearing round
his neck a double gold chain, by which is suspended an
elephant. It is not known on what occasion the king presented him with this mark of favour, but the source whence
it came is evident from the king's initials, motto, or miniature, which on different portraits are shown on the elephant.[1]
But in addition to this more ornamental than useful present,
the king frequently bestowed others of a more practical
nature on Tycho. Thus he sent in June 1581 an order to
the treasury to pay the cost of a bell which had just been
cast at Copenhagen for Tycho, and which was to be used at
Hveen. Perhaps it was this bell which was suspended in
the cupola at Uraniborg. Again, in November 1583 the
king ordered the treasury to hand over to Tycho " a good
new ship or pilot-boat," with all necessary tackle, &c.[3]

From some letters of the king's it appears that Tycho
entertained plans of some work of a geographical and historical character, for in September 1585 the king instructed
his librarian to lend to Tycho Brahe " as many *chartas
cosmographicas* or maps as are to be found in our library at
our castle of Copenhagen, and which are of our kingdom,
Denmark, or Norway, or any other of our dominions, for
information in some undertaking of which he has told us." [4]
A few weeks later the governor of Kronborg Castle was
informed that whereas Tycho Brahe had stated his intention
of publishing something about Danish kings, and had re-

---

[1] On the contemporary painting, which was destroyed in the burning of
Frederiksborg Castle in 1859, (of which there is a copy in Friis's book), a small
miniature is seen on the middle of the elephant. The engraving, which occurs
in several of Tycho's printed works (by Geyn, dated 1586), shows on the
elephant the letters F. S. (Fredericus Secundus), while the portrait of 1597
(copied in this book) has the miniature, and underneath the elephant the
letters M. H. Z. G. A. (Meine Hoffnung zu Gott allein), the king's motto.

[2] *Danske Magazin*, ii. p. 217 (Weistritz, ii. p. 118).

[3] Ibid., ii. p. 219 (Weistritz, ii. p. 121).

[4] Werlauff, *Historiske Efterretninger om det Store Kongelige Bibliothek.*
Copenhagen, 1844, p. 9.

quested that he might get their portraits as shown on the new tapestries at Kronborg, the king's painter was to be ordered to copy all the portraits and Danish and German rhymes on the tapestries.[1] Possibly Tycho may have wished to find some work for his newly-acquired printing-office, but if he really intended preparing a work on the geography and history of Denmark, he never carried out this plan. It seems, however, more probable that he had intended to assist his friend Vedel, who just at that time was collecting materials of this kind in connection with the work on Danish history on which he was engaged.

In return for all the kindness shown by the king, Tycho from time to time rendered such service to his patron as he was able to offer. Thus his name is associated with the castle of Kronborg by a couple of Latin poems with which he ornamented this favourite building of the king. On one of the gables was placed a lengthy versified inscription praying for a long life and success to the builder and his work; on the dial of a clock in one of the towers he put these lines :

" Transvolat hora levis neque scit fugitiva reverti,
　Nostra simul properans vita caduca fugit." [2]

Tycho was scarcely settled at Uraniborg before the king wished to consult him. In September 1578 he wrote to the astronomer from Skanderborg, in Jutland, that it was said by the common people about that place that a new star

[1] Visitors to Copenhagen may still see some of these tapestries in the upper storey of the Museum of Northern Antiquities. They were made between 1581 and 1584 from designs by Hans Knieper of Antwerp, whom we have mentioned above as having painted part of the picture on Tycho's mural quadrant. The tapestries (which originally numbered 111) represent each a Danish king in full figure, with the name and a short account of his reign in German rhymes above.

[2] Pontoppidan's *Danske Atlas*, tom. ii. (1764), p. 272 *et seq.* A poem with which Tycho ornamented one of the clocks in the study at Stjerneborg is printed in *Epist. Astron.*, p. 245.

had again appeared in the heavens, and he therefore asked what planet or other star might have been mistaken for a new star.[1]   Again, in December 1584 the king turned to Tycho for help, writing that he was under the impression that he had returned to Tycho a compass made by the latter, as there was something wrong with it.   If this was the case, Tycho was to send back the compass ; but if not, he was to make two new ones similar to the old one.[2]

But the most important service (according to the ideas of the time) which Tycho had to render to the king was by astrological predictions.   The first occasion on which he was ordered to show his skill in such matters was probably in 1577, when the king's eldest son, Prince Christian, was born.   The king and queen had been married since 1572, and two daughters had been born of the marriage, when at last a son was born at Frederiksborg Castle at half-past four o'clock in the afternoon on the 12th April 1577.   Popular tradition has preserved several strange circumstances in connection with the birth of this prince, who afterwards became one of the most popular kings of Denmark.   An old peasant announced to the king in the previous autumn that a mermaid had appeared to him and commanded him to tell the king that the queen was to be delivered of a son who should be counted among the most renowned princes in the northern countries.   The infant prince was christened on Trinity Sunday, the 2nd June, at Copenhagen, with the solemnities and festivities usual on such occasions, and among those who attended the ceremony and had an opportunity on the two following days of being edified by the stories of the virtuous Susanna and David and Goliath, which the students of the university acted in the courtyard of the castle, was Tycho Brahe, to whom doubtless more

---

[1] *Danske Magazin*, ii. p. 204 (Weistritz, ii. p. 93).
[2] Friis, *Tyge Brahe*, p. 147.

than one eye was directed when hopes and wishes were
uttered for the future of the little prince. In those days,
when most people of note had their nativities worked out
for them, it must have been a comfort to the king that he
could get this done for the infant by so great an authority
as his renowned star-gazer was already considered. Tycho
Brahe was accordingly directed to prepare the horoscope of
the prince, and on the 1st July following he handed in a
detailed report of his investigations. The original document
does not appear to have been preserved, but there are two
copies (apparently of a somewhat later date) in the Royal
Library at Copenhagen.[1] The report contains, first, a dedication to the young prince, after which follow the calculation
of the requisite astronomical data and the discussion of the
astrological signification of these, all written in Latin, but
followed at the end by a German translation of the astrological predictions, probably prepared for the convenience
of the queen. The dedication alludes shortly to the origin
and importance of astrology, and uses the same arguments
as we have met with in Tycho's oration on this subject.
The positions of the planets are next calculated for the date
of the prince's birth by the Prutenic tables (the successive
steps being given for each planet), while those resulting from
the Alphonsine tables are also given, but merely for the
sake of comparison. Being a practical astronomer, the
writer was not content with this, but corrected by means
of his own observations the tabular places of Jupiter, Mars,
Venus, and the sun, adopting the positions of the other
planets as given in the Prutenic tables because he had no

---

[1] " Horoscopus Sr. Regis Christiani IVti., ad Mandatum Sr. Regis Fride-
rici IIdi., a Tychone Brahe Ottonide conscript. in Insula Hvena Cal. Julij
Ao. 1577." The two copies must be of slightly later date, as the prince did
not become king till 1588. The dedication is to " Inclyto et Illustri Infanti
Christiano, Opt. et Potentiss. Principis Friderici IIdi. Daniæ et Norvegiæ
Regis, Domini Clementissimi Filio primogenito."

recent observations of them. The *figura natalis* is not of
the square shape generally used by astrologers,[1] but circular,
in accordance with the plan already followed by Tycho in
the case of the new star.

Before giving a short account of the further contents of
Tycho's report on the horoscope of Prince Christian, it may
not be useless to say a few words about the general principles followed by astrologers in preparing horoscopes; referring for further particulars to works in which this subject
is treated in detail.[2]

The point of the heavens of greatest importance for the
fate of man was the point of the ecliptic which was rising at
the precise moment of his birth (*punctum ascendens*). The
next step for the astrologer was to see how the planets
and the signs of the zodiac, as well as a few of the most
important fixed stars, were at the same moment situated in
the twelve " houses " into which the heavens were divided.[3]
The first house, *ascendens* or *horoscopus*, was considered the
foundation of fate, and if Mercury or a favourable star was
found in this house, it would announce a happy and pros-
perous life, while, on the other hand, an unfavourable planet
(Saturn or Mars) would indicate a short and unhappy life.
The second house (north of the first one) gave information
about riches and possessions; it was an unlucky house, be-

---

[1] See, *e.g.*, Wallenstein's and Kepler's horoscopes in Kepler's *Opera Omnia*
vol. i. p. 293, and vol. v. p. 476.

[2] See, in particular, Origani *Novæ Cœlestium Motuum Ephemerides*, Frankfurt, 1609, vol. i. ; or of modern books, Max Uhlemann, *Grundzüge der Astronomie und Astrologie der Alten, besonders der Ægypter*, Leipzig, 1857 ; Kepler's
Opera Omnia, ed. Frisch, i. p. 293 ; Delambre, *Hist. de l'Astr. Anc.*, ii. p. 546 ;
Moyen Age, p. 290 and p. 496 *et seq.*

[3] As already remarked, different astrologers divided the heavens in different
ways (Delambre, *M. A.*, p. 496 *et seq.*), by dividing the zodiac or the equator
by circles through their poles, or (as Tycho did) by circles through the north,
south, east, and west points of the horizon. About the Babylonian origin of
these " houses," see Mr. G. Bertin's lectures on Babylonian Astronomy in
Nature, vol. xl. p. 237.

cause it was not in favourable aspect to the first one, and while a favourable star (Jupiter or Venus) would here point to great riches, a questionable character like Mercury might make a thief and a vagabond of the new-born infant. Similarly the other houses had each a separate significa- tion; the third refers to brothers, friends, or journeys; the fourth, or most northern house (*imum cœlum*), refers to parents, because it is in quadrature with the first house, and therefore closely allied to it; the fifth (*bona fortuna*) tells about children, and is a very favourable house, because it is in *aspectus trigonus* with the first one, and Venus placed here would have great effect. The sixth house is a bad one (*mala fortuna*), because it has no aspect to the first, and, perhaps on this account, is allotted to servants, health, women, &c. The seventh and easternmost house, opposite the first, refers to marriage; the eighth is a bad one (no aspect) and refers to death, and here only the moon is favourable. The ninth house is intimately connected with the first (*aspectus trigonus*), and the sun is here of particular value; this house deals with religion and journeys. The tenth house (*medium cœli*) gives information about life, deeds, country, residence, &c. The eleventh (*bonus dæmon*), is in *aspectus sextilis* with the first, and is generally speaking a favourable house; but at a birth in the night, Saturn would here cause cowardice and poverty, and for a person born in the daytime, Mars would here induce loss of property. The twelfth house is, like the second, a bad one (*malus dæmon*), and tells of enemies and illnesses. Having drawn all these "houses" on a diagram and inserted the planets in them, the astrologer proceeded to examine the aspects of the latter (conjunction, opposition, quadrature, &c.[1]), and make out the *prognosticum* by means of rules, as

---

[1] Of these conjunction, *aspectus trigonus* and *sextilis* were favourable, opposition and quadrature unfavourable.

to which much difference of opinion existed. Some of the
most important things, however, were the *directions*. So-
called circles of position were drawn through the north and
south points of the horizon and any two points of the zodiac,
called the *significator* and the *promissor* (the sun, moon,
or planets, according as they had to be considered), and
the arc of the equator included between these circles was
their *directio*.[1] Thus Tycho computes the direction of the
ascendant to the planets (remarking that an error of four
minutes in the stated time of birth will alter these directions by one degree, which corresponds to an error of one
year in the time of any event foretold by a direction), and
also the directions of sun, moon, and Venus to the other
planets. There were various methods of "directing" or
referring the effects of the planets, as they might be placed
at any subsequent time, to their positions at the moment
of birth. Thus Kepler says that if the sun at this moment
be in a certain place in the zodiac, and a planet afterwards
comes to an important place, it should be computed how
many days after the birth the sun took to reach that place,
and the number of days corresponds to the number of
years which will elapse from the birth before the power of
that configuration will be felt.[2]

The action of each planet was very different according to
the house and sign of the zodiac which it occupied. The
sun and moon had each a sign (by some also called house)
specially belonging to it (Leo and Cancer), and the other
planets had each two, and a planet exercised the greatest
power when it was in its own house. The sun and moon
are the most powerful, while the others have the greater
effect the nearer they are to one of those. If a planet is

---

[1] Directions might also be taken along the ecliptic. See, *e.g.*, some remarks
on this matter in a letter from Tycho Brahe to Ludolf Riddershusen of
Bremen, of April 1600, in *Breve og Aktstykker* (1875), p. 121.

[2] *Kepleri Opera*, i. p. 295.

not in its own sign, but in that of another planet, the two
bodies act together, either with increased effect if they are
of the same nature (*e.g.*, both favourable), or neutralising
each other more or less if of opposite nature.

After this necessarily very crude outline of the principles
of judicial astrology, we return to Tycho's forecast of the
fate of the new-born prince. It would, however, lead us
too far if we were to follow him through the various proofs
which he adduces for his statements, and we can only
mention some of the more important ones. The years of
infancy will pass without danger, as Venus is favourably
placed in the ninth house, and though in the second year
the opposition of Mercury to the ascending point indicates
some small illness, it will be nothing serious. The years of
the prince's life are then enumerated in which he will be
afflicted with illness. For instance, in his twelfth year the
ascendant will be in quadrature with Saturn, which indicates
some serious illness "arising from black bile," but it will
not be mortal. In his twenty-ninth year he will have to
be very careful both about his health and his dignity,
because the sun will be in quadrature with Saturn at the
same time as Venus and the latter are in opposition. A
very critical time will be about the fifty-sixth year, when
the sun and Mars are most unfavourable, and even Venus
cannot help, as she is in the eighth house. The methods of
the Arabians do not show any life beyond fifty-six years,
and Ptolemy's rule gives the same result. As the sun's direction to its setting gives $41\frac{1}{4}$ years,[1] the moon, Venus, and
Jupiter add together twenty-six years, and Saturn in quad-
rature subtracts $10\frac{2}{3}$, so that the result is about $56\frac{1}{2}$ years.
As there are so many concurring signs, the prince will

[1] This is easy enough to understand. On the 12th April the sun would
set about 7h. 17m. or 2h. 47m. after the prince's birth. As four minutes or
one degree corresponds to a year, 2h. 47m. is not quite forty-two years.

hardly survive that age, unless God, who alone has power
over human destiny, specially prolongs his life; and if the
prince gets over the critical period, he will have a happy
old age.   Passing to the question as to what planets are the
ruling ones, it appears that Venus is ruler of the nativity
(*dominus genituræ*), being close to the tenth house or *sum-
mum cæli;* but Mars is in conjunction with Venus, and in
the sign belonging to Mercury (Gemini), so that these two
also have great influence.   Venus will make him pleasant,
comely, and voluptuous, fond of music and the fine arts;
Mars makes him brave and warlike, while Mercury adds
cleverness and acuteness to his other faculties.   He will be
of a sanguine temperament, because nearly all the planets
which indicate the temperament are in sanguine signs, but
at the same time he will not be without some saturnine
gravity.   Venus, as the ruler, determines his character, and
as Mars is joined to her, the prince will indulge too much
in sensual enjoyment, but Mars in the sign of Mercury will
make him generous and ambitious.   He will be healthy
and not subject to illness, but in various years (which are
enumerated) he must be careful, as the ascendant will be
influenced by the malevolent rays of Saturn.   His mental
abilities will be very good, because Mercury is favourably
situated; and as this planet is in a good aspect with Venus
and Mars, the prince will be fond of warlike occupations
and field sports, and take an interest in surgery and other
sciences.   He will have good luck in his undertakings, as
Saturn is in the fourth house and in his own sign of Capri-
corn, while Mercury and Mars occupy each other's signs;
but as Jupiter is badly placed, the prince will be less successful in ecclesiastical matters.[1]   As regards honours and

---

[1] Probably because Jupiter had an oracle at Dodona, and therefore was of
a clerical turn of mind.   Here he was in the twelfth house and in the sign of
Mercury (Virgo), both circumstances bad.

dignities, it is an excellent circumstance that the most brilliant of all stars, Alhabor, in the mouth of Sirius, is in the
corner of *medium cœli*, and there are also other fixed stars
of importance in favourable positions, such as the Twins in
the tenth house, Spica in the first, with Corona borealis a
little above, and the Southern Crown exactly in the corner
of the fourth house. Among the planets, the sun has most
influence on honours and dignities and is well placed, and
only Saturn in opposition to *medium cœli* shows that the
prince will meet with some serious adversities, which, however, will be overcome as everything else is so favourable.
The years are mentioned in which he will be specially
fortunate or unfortunate; and here again it appears that
after his fifty-fifth year, " when the direction of the sun
overtakes Mars," there will be serious adversities awaiting
him. As to riches, it is especially of importance that *pars
fortunæ* [1] is well situated in the eleventh house, and the
sun is in the seventh, and the prince will therefore become
rich; but as the sign of Mars (*Scorpio*) is in the second house
(*domus divitiarum*), his riches will principally be acquired
by war. At great length it is set forth in which years of
his life the position of *pars fortunæ* with regard to the
planets portends the acquisition of riches. The prospects
with regard to marriage are not altogether favourable, as
the moon is in the sixth house, and the position of Venus
with regard to Mars and Saturn signifies some adversity
in matrimony; but, on the other hand, Mercury is in the
seventh house (*domus conjugii*), which promises some happiness. Tycho here adds the remark, that in his opinion
the prince will be more inclined to other amours than to
matrimony (which turned out true enough). The time when
he will be inclined to marry will be about the age of twenty

---

[1] *Pars fortunæ* is the difference of longitude between sun and moon added
to the longitude of the *punctum ascendens;* it is indicated by the sign ⊕.

or twenty-one, when Venus comes in sextile aspect with Jupiter about the *medium cœli*, or in his thirty-fourth or thirty-fifth year, or, if not married before, in his forty-seventh year, when the moon reaches the seventh house. But all this depends more on man's free will than on the stars. It does not seem that he will have many children, as Saturn is master of the fifth house, and is in a sterile sign, but if he has any, they will be healthy and long-lived. His friends will be "solar people," such as kings and princes, because the sun is ruler of the eleventh house, where *pars fortunœ* is placed. His enemies will be "jovial and mercurial people," because Jupiter is unluckily placed in the twelfth house, and Mercury ruling the twelfth is in the seventh, but the latter planet assumes the nature of Mars, which is in its sign. His enemies will, therefore, be ecclesiastics and warriors, but he will defeat them, because Venus, the ruling planet, is much higher in the sky than Mars, and is in the apogee of its excentric; but he must beware of captivity or exile on account of the position of Mercury, which is also injured by being in quadrature with Saturn. There is nothing to indicate a violent death, and the prince will die from natural causes, but Venus shows that he will cause his own death by immoderate sensuality.

Finally, Tycho ends this dissertation by saying that all this is not irrevocably settled, but may be modified by many causes. God is, besides, the origin of all, and the giver of life and all good things, and He disposes freely of everything according to His own judgment. He alone is therefore to be implored that He may rule our life, grant us prosperity, and avert evil.[1]

The reader will pardon this long digression, but judicial astrology has played so important a part in the history of

---

[1] "Ille potest Solis currus inhibere volantes,
Ille augere potest, tollere fata potest."

the world, and been so beneficial in furthering the study of astronomy, that it cannot be left out of consideration if we wish to get a full view of the scientific life and doings of former ages. Having devoted so much space to the horoscope of the first-born son of the king, we shall not review those of the younger sons, which Tycho was afterwards called on to prepare, although in these cases the originals (and not merely copies) have been preserved in the Royal Library at Copenhagen. The second son, Prince Ulrich, was born on the 30th December 1578, and Tycho worked out his *Genethliaca* by royal command, and presented it in May 1579. It is a handsome volume in small 4to, bound in pale green velvet with gilt edges, containing about 300 pages, all written in Tycho Brahe's own hand. The arrangement of the contents is like that of the previous prognostication, the results being, as before, given first in Latin and afterwards in German. Mars is the ruler, as he is in his own sign, and in every way most favourably situated, but the sun is *dominus ascendentis*, and the solar eclipse of the 21st July 1590 in the eighth degree of Leo, and "in the very degree of the ascendant," will be of great importance, and may injure the prince. It is again repeatedly pointed out how uncertain the whole thing is.[1] In 1583 the king's third and last son, Hans, was born on the 26th July, and Tycho had again to attack the twelve houses, aspects, &c. He sent in a volume like the last one, bound in the same manner, and containing about the same number of pages, but the Latin part is neatly written by one of Tycho's assistants, and only the German part by himself. To show his readiness to please the king, he has, in addition to the circular figure, divided into "houses" in the same way as on the two previous occasions, drawn two square figures, divided

---

[1] Round the four sides of the central part of the *figura natalis* Tycho has written : "Potest—fata augere—Deus—tollere fata."

by distributing the houses evenly round the equator and round the ecliptic. In the preface he talks about the possibility of averting the inclinations of the stars in the same strain as before, and throughout the whole dissertation he seems more doubtful about the results to be expected than he was in 1577. He has again corrected the places of the planets by his own observations. Mercury is here the strongest planet, free from the rays of the sun, though somewhat weakened by being retrograde and moving slowly, but particularly by being in the sixth house. The prince seems only to have "mediocre" luck in store, but Tycho remarks that everybody shapes his own fortune.[1]

That Tycho did not take much interest in nor attach any importance to these astrological prognostications will be evident to anybody who has read the foregoing pages. Whatever he had thought about these matters in his youth, the great work of his life now stood so clearly before him, that he did not care to waste his time on work of so very doubtful value as astrological forecasts.[2] We possess even stronger testimony to this effect than any we have yet quoted, in a letter which he wrote on the 7th December 1587 to Heinrich von Below, a nobleman from Mecklenburg, who in 1579, through the queen's influence, had received an estate in Jutland in fief, and who was married to a first cousin of Tycho's.[3] Duke Ulrich of Mecklenburg-Güstrow,

---

[1] "Quisque suæ fortunæ faber : tamen non est dubium astra in his plurimum posse, ut non immerito dixerit Poeta ille :

'Esse igitur sapiens et felix nemo potest qui
Nascitur adverso cœlo stellisque sinistris.'"

[2] All the same, he was naturally looked upon by the common people in Denmark as nothing but an astrologer, and thirty-two unlucky days are attributed to his authority (Hofman, *Portraits historiques des hommes illustres de Denmark*, vi. Partie, p. 23), though nothing could be more opposed to the principles of astrology than the fixing upon certain *dates* as lucky or unlucky.

[3] C. G. F. Lisch, *Tycho Brahe und seine Verhältnisse zu Meklenburg*, in the *Jahrbücher des Vereins für Meklenburgische Geschichte*, vol. xxxiv. (1869). I quote from a reprint, 20 pp. 8vo. See also Note C.

the queen's father, had procured two prognostica for the
year 1588, the one by Tobias Möller, the other by Andreas
Rosa; and as they were so far from agreeing, that one let
the year be governed by the two beneficent planets, while the
other put it under the dominion of the two malevolent ones,
the Duke requested Below to inquire from his kinsman Brahe
which of them was correct. In his answer Tycho remarked that
he did not care to mix in astrological matters, but for some
years had endeavoured " to put astronomy into proper order,"
because only in this way, by reliable instruments and mathematical methods and certainty, could the truth be arrived
at. He shows that the two prognostics differ so much
because one is built on the Prutenic, the other on the
Alphonsine tables, which differ nineteen hours as to the time
of the vernal equinox. It is therefore not surprising that
the two astrologers find different rulers for the year, as
these are found from the *figura cœli* for the time of vernal
equinox. These astrological predictions are like a cothurnus,
which may be put on any foot, large or small; and when he
every year sends his Majesty a prognosticon, he only does
it by the king's express command, although he does not
like to have anything to do with such doubtful predictions,
in which one cannot come to the truth, as in geometry and
arithmetic, on which astronomy is founded, by means ot
diligent observations. As to the two prognostications about
which the Duke inquires, neither the Prutenic nor the
Alphonsine tables are correct, as he had found by his own
observations, and he had as usual sent the king a prog-
nostic for the coming year, but had not kept a copy of it,
and if the Duke wanted to see it, he might apply to the
king about it.

This letter shows with all desirable distinctness what
Tycho thought of judicial astrology, with which philosophical
speculations on the unity of the kosmos and the analogy

between its celestial and terrestrial parts must by no means
be confounded.   He was not, like Kepler, obliged to waste
his time on work of that kind in order to get daily bread
for himself and his family; but he was highly paid, and
his scientific researches were most liberally supported by
the king, who could not be expected to appreciate their
real value; and it was only natural that he should annually
send the king an offering of a kind that the latter could
understand, and which by the king was considered an
acceptable gift.   Tycho showed clearly enough in the horoscopes which he drew up for the royal children that he
was inclined to agree with Horace when he said—

> " Tu ne quæsieris, scire nefas, quem mihi, quem tibi
>   Finem di dederint, Leuconoe, nec Babylonios
>   Tentaris numeros."

None of the almanacs which Tycho prepared for the king
have been preserved, but a letter from the king is extant,
dated 24th September 1587, in which he reminded Tycho
about sending him the usual almanac for the ensuing year
by the bearer of the letter, or, if it was not ready, as soon as
possible.[1]

Tycho doubtless obeyed the king's command, and it
turned out to be the last time he had to do so.   King
Frederick II. died on the 4th April 1588, in his fifty-fourth
year, to the great regret of Tycho, who owed him so much,
as well as of the country at large.   His character was open and
chivalrous, and he was sincerely religious, while he at the
same time tried to keep himself free from the intolerance
prevailing everywhere in those days.   He was less free
from another weakness of his time, and, with characteristic
frankness, Vedel said in a funeral oration, that " if His
Grace could have kept from that injurious drink which is
much too prevalent all over the world among princes and

---

[1] *Danske Magazin*, ii. p. 247 (Weistritz, ii. p. 171).

nobles and common people, then it would seem to human eyes and understanding that he might have lived for many years to come." But if he was not better than his contemporaries in this respect, he was at any rate far superior to most of them by honouring and protecting the peaceful student of science, and in the history of astronomy his name will always be gratefully remembered as long as that of Tycho Brahe continues to be reckoned among the heroes of science.

CHAPTER VII.

TYCHO'S BOOK ON THE COMET OF 1577, AND HIS SYSTEM OF THE WORLD.

THE year 1588 is one of great importance in the life of Tycho Brahe, not only because his firm friend and benefactor died in that year, but also because he then published a volume containing some of the results of his work at Uraniborg, and embodying his views on the construction of the universe. The subject specially dealt with in this volume was the great comet of 1577, the most conspicuous of the seven comets observed in his time.

This comet was first noticed by Tycho on the 13th November 1577, but it had already been seen in Peru on the 1st, and in London on the 2nd November.[1] On the evening of the 13th, a little before sunset, Tycho was engaged at one of his fishponds, trying to catch some fish for supper, when he remarked a very brilliant star in the west, which he would have taken for Venus if he had not

---

[1] According to Tycho, it had been seen by mariners on the 9th. In a copy of *Cometæ anno humanitatis 1577 a 10 VIIIIbris . . . adparentis descriptio*, by Bart. Scultetus (Gorlicii, 1578), which I picked up at Copenhagen some years ago, and which now belongs to the Royal Observatory, Edinburgh, there is written in a neat hand the following on the last blank page :— "Ego Londini in Anglia cometam hoc libro descriptum, et 2 die Nouembris visum, tertio obseruare coepi ut potui radio nautico necdum sesquipedali, ita ut triangulum faceret cometa cum stellis subnotatis, caudæ arcu comprehendente gradus 6m.30 et amplius." [Then follow distance measures on November 3, 9, 13, 15, 24, and 25, but without indication of time.] "Tanto lumine corruscabat hic cometes primo meo aspectu idque per nubes obuersantes, ut antequam integram ejus formam vidissem, Lunam esse suspicarer, quam tamen eo tum loci et temporis lucere non potuisse statim, idque in tanto maiore admiratione, colligebam."

158

known that this planet was at that time west of the sun.
Soon after sunset a splendid tail, 22° in length, revealed
itself, and showed that a new comet had appeared. It was
situated just above the head of Sagittarius, with the slightly
curved tail pointing towards the horns of Capricornus, and
it moved towards Pegasus, in which constellation it was
last seen on the 26th January 1578. During the time it
was visible Tycho observed it diligently, measuring with a
radius and a sextant the distance of the head from various
fixed stars, and occasionally also with a quadrant furnished
with an azimuth circle (four feet in diameter), the altitude
and azimuth of the comet. The sextant, which afterwards
was placed in the large northern observing room at Uraniborg,[1] was constructed on the same principle as the one
which Tycho had made at Augsburg in 1569, and was
mounted on a convenient stand, which enabled the observer
to place it in any plane he liked; the arms were about four
feet long. The quadrant was about 32 inches in semi-
diameter, and the arc was graduated both by the transversals nearly always employed by Tycho, and by the concentric
circles on the plan proposed by Nunez; and on the back of
the quadrant was a table, by means of which the readings
of the latter could be converted into minutes without calculation.[2] When observing this comet, Tycho had not yet
at his disposal as many instruments and observers as in after
years, nor had he as yet perceived the necessity of accurate
daily time determinations by observing altitudes of stars,
but merely corrected his clocks by sunset.[3] The observa-

[1] Figured in *De Mundi Æth. Rec. Phenom.*, p. 460, and *Astr. Inst. Mech.*,
ol. D. 6 *verso*.

[2] The quadrant is figured in *De Mundi Æth. Rec. Phen.*, p. 463, and *Mech.*,
fol. A. 2.

[3] The orbit of the comet of 1577 was computed from Tycho's sextant
observations by F. Woldstedt, "De Gradu Præcisionis Positionum Cometæ
Anni 1577 a celeberrimo T. B. . . . determinatarum et de fide elementorum
orbitæ," &c. Helsingsfors, 1844, 15 pp. 4to.

tions of this comet cannot therefore compare in accuracy with
his later ones, but still they were immeasurably superior to
those made by other observers, and they demonstrated most
decisively that the comet had no perceptible parallax, and
was consequently very far above the "elementary sphere" to
which the Aristotelean philosophy had consigned all comets
as mere atmospherical phenomena. By showing that the
star of 1572 was situated among the stars, Tycho had
already dealt the Aristoteleans a heavy blow, as it was now
clear that new bodies could appear in the æthereal regions.
But still that star was not a comet, and Tycho, who had
formerly believed in the atmospherical origin of comets,
now took the opportunity of testing this matter, and found
that the comet had no appreciable daily parallax. Though
he was not the only observer who placed the comet beyond
the moon, his observations were known by his contemporaries to be of very superior accuracy, and his authority
was so great that this question was decided once for
all.[1]

Before proceeding to pass in review the book which
Tycho prepared on this comet, we shall shortly allude to
the other comets observed at Hveen. On the 10th October
1580 Tycho found a comet in the constellation Pisces. It
was observed at Hveen till the 25th November, and again
after the perihelium passage on the morning of the 13th
December. The observations are more numerous and better
than those of the previous comet, and time determinations
with a quadrant were made nearly every night, while
there are very few quadrant observations of the comet.
Moestlin had seen it already on the 2nd October, and both
he and Hagecius observed it assiduously, but their observa-

---

[1] Except that Scipione Chiaramonte and an obscure Scotchman, Craig,
vainly endeavoured to deduce the very opposite result from Tycho's observations, but they were easily reduced *in absurdum*.

tions are worthless compared with Tycho's.[1] The next
comet was visible in May 1582, and was observed by Tycho
on three nights only, the 12th, 17th, and 18th, after which
date the strong twilight prevented further observations;
but in Germany it was still seen on the 23rd, and in China
it was seen for twenty days after the 20th May.[2] Of
greater interest are the observations of the comet of 1585,
which appeared at a time when Tycho's collection of instruments was complete, and when he was surrounded by a
staff of assistants. The comet was first seen by Tycho on
the 18th October after a week of cloudy weather, but at
Cassel it had already been seen on the 8th (*st. v.*).[3] Tycho
compares its appearance when it was first seen with the
cluster (or nebula, as it was then called) Præsepe Cancri,
without any tail. The observations are very numerous, and
were made partly with a sextant, partly with the large
armillæ at Stjerneborg, with which newly-acquired instrument the declinations of the comet and the difference of
right ascension with various bright stars were observed at
short intervals on every clear night up to the 12th November. The excellence of the observations and the care with
which the instruments were treated are fully demonstrated
by the most valuable memoir on this comet by C. A. F.
Peters.[4] We have already mentioned that this comet gave
rise to the correspondence between Tycho and the Landgrave and Rothmann. The next comet appeared in 1590,

---

[1] The orbit was determined by Schjellerup from a complete discussion of
Tycho's sextant observations (*Det kgl. danske Videnskabernes Selskabs Skrifter*,
math. *Afdeling*, 5te Række, 4de Bind, 1854).

[2] The orbit is very uncertain. D'Arrest, *Astr. Nachr.*, xxxviii. p. 35.

[3] Tycho returned home from Copenhagen on October 18th. Elias Olsen
Morsing had seen it on the 10th, as he wrote in the meteorological diary,
"*Stellam ignotam vidi.*" See also Introduction to the Observations.

[4] *Astr. Nachr.*, vol. xxix. The observations had been published by Schumacher in 1845 (*Observationes cometæ anni* 1585 *Uraniburgi habitæ a Tychone
Brahe.* Altona, 4to).

and was observed from the 23rd February to the 6th March
inclusive, the declination with the armillæ, altitudes and
azimuths with a quadrant, and distances with a sextant.
The time determinations are numerous.[1] In July and
August 1593 a comet appeared near the northern horizon.
It was not observed at Hveen, but only by a former pupil of
Tycho's, Christen Hansen, from Ribe in Jutland, who at
that time was staying at Zerbst in Anhalt. He had only
a radius with him, and his observations were therefore not
better than those made by the generality of observers in
those days.[2] The last comet observed at Hveen was that
of 1596, which was first seen by Tycho at Copenhagen
on the 14th July, south of the Great Bear. It was not
properly observed till after his return home on the 17th,
and then only on three nights. It was last seen on the
27th July.[3]

The star of 1572 and the comets observed at Hveen had
cleared the way for the restoration of astronomy by helping
to destroy old prejudices, and Tycho therefore resolved to
write a great work on these recent phenomena which should
embody all results of his observations in any way bearing
on them. The first volume he devoted to the new star, but
as the corrected star places which were necessary for the
reduction of the observations of 1572–73 involved researches
on the motion of the sun, on refraction, precession, &c., the
volume gradually assumed greater proportions than was originally contemplated, and was never quite finished in Tycho's

---

[1] Orbit computed by Hind, *Astr. Nachr.*, xxv. p. 111.

[2] Orbit by Lacaille, in Pingré's *Cométographie*, i. p. 560.

[3] Orbits by Hind and Valz, *Astr. Nachr.*, xxiii. pp. 229 and 383 ; the observations are published ibid., p. 371 *et seq.* Pingré gives the results of most
of the observations of the seven comets from a copy of them which is still
preserved at the Paris Observatory. A complete edition of all the observations was published in 1867 at Copenhagen, under the supervision of D'Arrest,
*Tychonis Brahe Dani Observationes Septem Cometarum. Nunc primum edidit
F. R. Friis.* 4to.

lifetime. On account of the wider scope ot its contents he gave it the title *Astronomiæ Instauratæ Progymnasmata*, or Introduction to the New Astronomy, a title which marks the work as paving the way for the new planetary theory and tables which Tycho had hoped to prepare, but which it fell to Kepler's lot to work out in a very different manner from that contemplated by Tycho. The second volume was devoted to the comet of 1577, and as the subject did not lead to the introduction of extraneous matter, this volume was finished long before the first one. The third volume was in a similar manner to treat of the comets of 1580 and following years, but it was never published, nor even written, though a great deal of material about the comet of 1585 was put together and first published in 1845 with the observations of this comet.[1]

The two volumes about the new star and the comet of 1577 were printed in Tycho's own printing-office at Urani-borg, and after some delay caused by want of paper, the second volume was completed in 1588.[2] The title is "Tychonis Brahe Dani, De Mundi ætherei recentioribus phænomenis Liber secundus, qui est de illustri stella cau-data ab elapso fere triente Nouembris anno MDLXXVII usque in finem Januarii sequentis conspecta. Vraniburgi cum Privilegio." The book is in demy 4to, 465 pp., and the

---

[1] The third volume is alluded to in several places in Tycho's writings, *e.g.* *Progym.*, i. pp. 513 and 714 ; *Epist.*, pp. 12, 20, 104, &c.

[2] In the above-mentioned letter to Below, Tycho wrote in December 1587 that he should soon be in want of paper for a book which was being printed in his office, and had applied to the managers of two paper-mills in Mecklen-burg without getting an answer. He therefore asked Below to write to the managers of these mills, and to ask some friend at the Duke's court to intercede for him ; that he would willingly pay for the paper, which might be sent through his friend Brucæus at Rostock. Below wrote at once (28th December 1587) to Duke Ulrich, and asked him to do Tycho this favour, " der löblichen Kunst der Astronomie zur Beförderung " (Lisch, *l. c.*, p. 6). To avoid a repetition of this inconvenience the paper-mill at Hveen was built a few years later.

colophon is the vignette "Svspiciendo Despicio," with the
words underneath "Uranibvrgi In Insula Hellesponti Danici
Hvenna imprimebat Authoris Typographus Christophorus
Weida. Anno Domini MDLXXXVIII."

The book is divided into ten chapters. The first contains
most of the observations of the comet; the second deduces
new positions for the twelve fixed stars from which the distance of the comet had been measured. Tycho mentions
that while the comet was visible he had not yet any armillæ,
and he therefore carefully placed a quadrant in the meridian,
and thus determined the declination of the star, and by the
time of transit (through the medium of the moon and the
tabular place of the sun) also the right ascension. He does
not give any particulars about the observations and method,
but he goes through the computations of the latitude and
longitude of each star from the right ascension, declination
and the point of the ecliptic culminating with the star. In
a note at the end of the chapter he gives improved star
places from the later observations with better instruments
and methods, and, as might be expected, these later results
are really much better than those found in 1578.[1] In the
third chapter the longitude and latitude of the comet for
each day of observation are deduced from the observed distances from stars; but though he gives diagrams of all the
triangles, and gives all the numerical data, the trigonometrical process is not shown. In the fourth chapter the right
ascensions and declinations of the comet are computed from
the longitudes and latitudes.[2] The fifth deals with the deter-

---

[1] In the above-mentioned paper Woldstedt compares the two sets of positions with modern star places (Åbo or Pond with proper motions from Bessel's
Bradley or Åbo). The means of the errors of Tycho's places, irrespective of
sign, are in longitude and latitude, for the older positions, 4′.8 and 1′.1, for
the later ones, 1′.4 and 1′.5. About the methods by which these positions were
found, see Chapter XII.

[2] By the method of Al Battani, which employs the point of the equator having
the same longitude as the comet. Delambre, *Astr. du Moyen Age*, p. 21.

mination of the inclination and node of the apparent path
of the comet with regard to the ecliptic, which Tycho found
from two latitudes and the arc of the ecliptic between them ;
seven different combinations give results which only differ a
few minutes *inter se*. The sixth chapter is a more lengthy one,
and treats of the distance of the comet from the earth ; and
as this was of paramount importance as a test of the Aristotelean doctrine, he endeavours to determine the parallax in
several different ways.   First, he shows that the comet had
moved in a great circle, and though not with a uniform velocity throughout, yet with a very gradually decreasing one ;
and if it had been a mere "meteor" in our atmosphere, it
would have moved by fits and starts, and not in a great circle.
The velocity never reached half that of the moon, the nearest
celestial body.   He next discusses two distance measures
from ϵ Pegasi, made on the 23rd November, with an interval of three hours, and finds that if the comet had been at
the same distance from the earth as the moon,[1] the parallax
would have had the effect of making the second angular
distance from the star equal to the first, even after allowing
for the motion of the comet in the interval, while the second
observed distance was 12' smaller than the first one.   At
least the comet must have been at a distance six times as
great as that of the moon, and all that can be concluded
from the distance measures is that the comet was far beyond
the moon, and at such a distance that its parallax could not
be determined accurately.   The same appears from comparisons between distance measures from stars made at Hveen
and those made at Prague by Hagecius, which should differ
six or seven minutes if the comet was as near as the moon,
whereas they only differed one or two minutes.   The observations of Cornelius Gemma at Louvain, when compared with
those at Hveen, point in the same direction, but are much

---

[1] Which he, with Copernicus, assumes = 52 semi-diameters of the earth.

too inaccurate to build on.   Again, Tycho takes two observations of altitude and azimuth ; from the first he computes
the declination, corrects this for the motion of the comet in
the interval, and with this and the second azimuth computes
the altitude for the time of the second observation.   For a
body as near as the moon there would be a considerable
difference, while several examples show none. Finally, Tycho
employs the method of Regiomontanus for finding the actual
amount of parallax from two altitudes and azimuths, but
several combinations gave the same result, that no parallax
whatever could be detected in this way.   Tycho was well
aware that this was a bad method, and evidently only tried
it as a duty.[1]   (The comet of 1585 was chiefly observed
with the large armillæ, and the want of parallax was
demonstrated by comparing the right ascension and declination observed with an interval of some hours with the daily
motion of the comet.[2])

In the seventh chapter the position of the comet's tail
is examined.   The increased attention which had been paid
to comets during the sixteenth century had led to the discovery of the fact that their tails are turned away from the
sun, and not only Peter Apianus, who is generally credited
with the discovery, but also Fracastoro, and after them
Gemma and Cardan, had pointed out this remarkable fact
from observations of different comets.   Tycho, who took
nothing on trust, examined the matter, and computed from
twelve observations of the direction of the tail of the comet
of 1577 the position of the tail with regard to a great
circle passing through the sun.   He found that the direc-

---

[1] See his remarks about the method, *De mundi æth. rec. phen.*, p. 156, and
in a letter to Hagecius (who had found a parallax of five degrees by the
method), *T. B. et doct. vir. Epist.*, p. 60. Delambre sets forth the method
with his usual prolixity in *Hist. de l'Astr. du Moyen Age*, p. 341 ; *Astr. Moderne*,
i. p. 212 *et seq.*

[2] *Epist. Astron.*, pp. 16–17.

tion of the tail never passed exactly through the sun, but seemed to pass much nearer to the planet Venus; he adds, that though the statement of Apianus was only approximately true, the opinion of Aristotle was far more erroneous, for according to him, the tails, as lighter than the head, should be turned straight away from the centre of the earth. The curvature he considers merely an illusion, caused by the head and the end of the tail being at different distances from the earth.

The eighth chapter is the most important in the whole book, as the consideration of the comet's orbit in space leads Tycho to explain his ideas about the construction of the universe. The " æthereal world," he says, is of wonderfully large extent; the greatest distance of the farthest planet, Saturn, is two hundred and thirty-five times as great as the semi-diameter of the " elementary world " as bordered by the orbit of the moon. The moon's distance he assumes equal to fifty-two times the semi-diameter of the earth, which latter he takes to be 860 German miles.[1] The distance of the sun he believes to be about twenty times that of the moon. In this vast space the comet has moved, and it therefore becomes necessary to explain shortly the system of the world, which he had worked out " four years ago," i.e., in 1583.[2] The Ptolemean system was too complicated, and the new one which that great man Copernicus had proposed, following in the footsteps of Aristarchus of Samos, though there was nothing in it contrary to mathematical principles, was in opposition to those of physics, as the heavy and sluggish earth is unfit to move,

[1] The value for the earth's semi-diameter was probably taken from Fernels well-known *Cosmotheoria*, Paris, 1528. We shall see in the next chapter what ideas Tycho had formed as to the distance of the outer planets and the fixed stars (*Progym.*, i. p. 465 *et seq.*).

[2] The book was written in 1587, as appears from several allusions to time in it.

and the system is even opposed to the authority of Scripture. The vast space which would have to be assumed between the orbit of Saturn and the fixed stars (to account for the want of annual parallax of these), was another difficulty in the Copernican system, and Tycho had therefore tried to find a hypothesis which was in accordance with mathematical and physical principles, and at the same time would not incur the censure of theologians. At last he had, " as if by inspiration," been led to the following idea on the planetary motions.

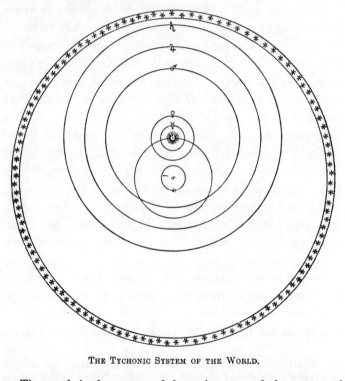

THE TYCHONIC SYSTEM OF THE WORLD.

The earth is the centre of the universe, and the centre of the orbits of the sun and moon, as well as of the sphere of the fixed stars, which latter revolves round it in twenty-four hours, carrying all the planets with it. The sun is the centre of the orbits of the five planets, of which Mercury and Venus move in orbits whose radii are smaller than that of the solar orbit, while the orbits of Mars, Jupiter, and Saturn encircle the earth. This system accounts for the irregularities in the planetary motions which the ancients explained by epicycles and Copernicus by the annual motion of the earth, and it shows why the solar motion is mixed up in all the planetary theories.[1] The remaining inequalities, which formerly were explained by the excentric circle and the deferent, and by Copernicus by epicycles moving on excentric circles, could also, in the new hypothesis, be explained in a similar way. As the planets are not attached to any solid spheres, there is no absurdity in letting the orbits of Mars and the sun intersect each other, as the orbits are nothing real, but only geometrical representations.

This is all which Tycho considered it necessary to set forth about his system in the book on the comet, but he stated his intention of giving a fuller account of it on a future occasion, which never came. We shall finish our account of his labours connected with the comet of 1577 before we consider his system a little more closely.

The comet was by Tycho supposed to move round the sun in an orbit outside that of Venus, and in the direction opposite to that of the planets, the greatest elongation from the sun being 60°. He was unable to represent the observed places of the comet by a uniform motion in this orbit, and was obliged to assume an irregular motion, slowest when in inferior conjunction, increasing when the comet was first discovered, and afterwards again decreasing. Tycho remarks

---

[1] This alludes to the circumstance, which had appeared so strange to the ancients, that the period of the motion of each upper planet in its epicycle was precisely equal to the synodical period of the planet, while in the case of the two inferior planets the period in the deferent in the Ptolemean system was equal to the sun's period of revolution.

that an epicycle might be introduced to account for this, but as the inequality was only 5′, he did not deem it necessary to go so far in refining the theory of a transient body like a comet; and besides, it is probable that comets, which only last a short time, do not move with the same regularity as the planets do. He finds the inclination of the orbit to the ecliptic equal to 29° 15′, and shows how to compute the place of the comet for any given time by means of the table of its orbital motion with which he concludes the first part of the book. The ninth chapter is a very short one, and treats of the actual size of the comet; as the apparent diameter of the head on the 13th November was 7′, the diameter was 368 miles, or $\frac{3}{14}$ of the diameter of the earth. Similarly he calculates the length of the tail, and finds it equal to 96 semi-diameters of the earth. This is on the assumption that the tail is really turned away from Venus, and though he adds that he had also found this to be the case with the comet of 1582, he suspects that some optical illusion must be the cause of this, as it would be more natural that the tail should be turned from the sun than from Venus. In a letter to Rothmann in 1589, he expresses the opinion that the tail is not a mere prolongation of the head, for in 1577 head and tail were of a different colour, and stars could be seen through the tail. He apparently thought that the tail was merely an effect of the light from the sun or Venus shining through the head, and referred to the opinion of Benedict of Venice that the illumination of the dark side of the moon was due to Venus, about which he, however, does not express any decided opinion.[1]

The only part of the tables of the comet's motion which requires notice is that relating to the horizontal parallax. This he makes out from his theory to have been nearly 20′

[1] *Epist. astron.*, p. 142.

in the beginning of November, and then rapidly to have decreased; and, as an excuse for this considerable quantity not having been detected, he adds his belief that refraction would counteract the parallax near the horizon where the comet was observed.

The remainder of Tycho's book is devoted to a detailed examination of the writings and observations of other astronomers on the comet. This was the first comet which gave rise to a perfect deluge of pamphlets, in which the supposed significance of the terrible hairy star was set forth, and for more than a century afterwards every comet was followed by a flood of effusions from numberless scribblers. The astrological significance of the comet Tycho does not trouble himself about, though he takes the opportunity of stating that he does not consider astrology a delusive science, when it is kept within bounds and not abused by ignorant people. For the sun, moon, and fixed stars would have sufficed for dividing time and adorning the heavens, and the planets must have been created for some purpose, which is that of forecasting the future.[1] But he goes through the observations or speculations of eighteen of his contemporaries, taking first those who had acknowledged the comet to be beyond the lunar orbit (Wilhelm IV., Moestlin, Cornelius Gemma, and Helisæus Roeslin), and afterwards the great herd of those who believed it to move in the "elementary world." Among these there are no generally known names except those of Hagecius and Scultetus. A theory very like that of Tycho was proposed by Moestlin, who also let the comet move in a circle round the sun outside the orbit of Venus, and accounted for the irregular motion by a small circle of libration perpendicular to the plane of the orbit, along the diameter of which the comet moved to and fro. This idea was borrowed from

---

[1] *De mundi æth. rec. phen.*, p. 287. Compare above, Chapter IV. p. 75.

Copernicus, whose lead Moestlin also followed with regard to the motion of the earth.

That the great Danish astronomer did not become convinced of the truth of the Copernican system, but, on the contrary, set up a system of his own founded on the immovability of the earth, may appear strange to many who are unacquainted with the state of astronomy in the sixteenth century, and it may to them appear to show that he cannot have been such a great reformer of astronomical science, as is generally supposed. But it is not necessary to concoct an apology for Tycho; we shall only endeavour to give an intelligible and correct picture of the state of science at that time with regard to the construction of the universe.

That Copernicus had precursors among the ancients who taught that the earth was in motion, is well known, and he was well aware of this fact himself. But none of those precursors had done more than throw out their ideas for the consideration of philosophers; they had not drawn the scientific conclusions from those ideas, and had not worked them into a complete system by which the complicated motions of the planets could be accounted for and made subject to calculation. Neither had this been done by the philosophers who made the earth the centre of the universe, and let it be surrounded by numerous solid crystal spheres to which the heavenly bodies were attached. All this was only philosophical speculation, and was not founded on accurate observations; but the only two great astronomers of antiquity, Hipparchus and Ptolemy, have handed down to posterity a complete astronomical system, by which the intricate celestial motions could be explained and the positions of the planets calculated. But this "Ptolemean system," in which a planet moved on an epicycle, whose centre moved on another circle (the deferent), with a velocity which was uniform with regard to the centre of a third circle, the equant,[1] was only a most ingenious mathematical representation of the phenomena—a working hypothesis; it did not pretend to give a physically true description of the actual state of things in the universe.[2] No doubt there were many smaller minds to which this did not become clear, but both by the great mathematician who completed it, and by astronomers of succeeding ages the Ptolemean system was merely considered a mathematical means of computing the positions of the planets.

When astronomy towards the end of the fifteenth century again began to be cultivated in Europe, the inconvenience of the extremely complicated system became felt, and soon the great astronomer of Frauenburg conceived how a different system might be devised on the basis of the earth's motion round the sun. But Copernicus did a great deal more than merely suggest that the earth went round the sun. He worked out the idea into a perfect system, and developed the geometrical theory for each planet so as to make it possible to construct new tables for their motion. And though he had but few and poor instruments, and did not observe systematically, he took from 1497 to 1529 occasional observations in order to get materials for finding the variations of the elements of the orbits since the time

[1] The earth, the centre of the deferent, and the centre of the equant were in a straight line and equidistant; only in the case of Mercury the centre of the equant was midway between the earth and the centre of the deferent.

[2] Perhaps we may illustrate this by an example from modern science. When the deflection of a magnetic needle in the neighbourhood of an electric current was first discovered, some difficulty was felt in giving a rule for the direction in which either pole of a needle is deflected by a current, whatever their relative positions may be, until Ampère suggested that if we imagine a human figure lying in the current facing the needle, so that the current comes in at his feet and out at his head, then the deflection of the north-seeking pole will be to his left. Nobody ever suspected Ampère of believing that there really was a little man lying in the current, but to many people in the Middle Ages the epicycles were doubtless really existing.

of Ptolemy. He was therefore able to produce a complete new system of astronomy, the first since the days of the Alexandrian school, and the first of all which gave the means of determining the relative distances of the planets. And it was in this way that he showed himself as the great master, and was valued as such by Tycho Brahe, who was better able than any one else to appreciate Copernicus, since his own activity left no part of astronomy untouched. But unfortunately the edifice which Copernicus had constructed was not very far from being as artificial and unnatural as that of Ptolemy. The expedient of letting the earth move in a circular orbit round the sun could explain those irregularities in the planetary motions (stations and retrogradations) of which the synodic revolution was the period (the second inequalities, as the ancients had called them), because they were caused by the observer being carried round by the moving earth. But this could not account for the variable distance and velocity (the first inequality) of which the orbital revolution was the period, and of which Kepler gave the explanation when he found that the planets move in ellipses, and detected the law which regulates the velocities in these. Until Kepler had discovered the laws which bear his name, there was no way of accounting for these variations, except by having recourse to the same epicycles and excentrics which Ptolemy had used so liberally ; and the planetary theory of Copernicus was therefore nothing but an adaptation of the Ptolemean system to the heliocentric idea.[1] And the motions were not referred to the real place of the sun, but to the middle sun, *i.e.*, to the centre of the earth's orbit, while the orbit of Mercury required a combination of seven circles, Venus of

[1] The chief claim of the system of Copernicus to be considered simpler than the Ptolemean was that it dispensed with the equant (which really violated the principle of uniform motion, so much thought of), and let the motion on the deferent be uniform with regard to its centre.

five, the earth of three, the moon of four, and each of the
three outer planets of five circles; and even with this
complicated machinery the new system did not represent
the actual motions in the heavens any better than the
Ptolemean did. Copernicus himself said that he would be
as delighted as Pythagoras was when he had discovered his
theorem, if he could make his planetary theory agree with
the observed positions of the planets within $10'$.[1] But the
accuracy was very far indeed from reaching even that limit.[2]
Doubtless the Prutenic tables were better than the Alphonsine ones, but that was simply because Copernicus had been
able to apply empiric corrections to the elements of the
orbits, and because Reinhold did his work better than the
numerous computers at Toledo had done theirs. The
Copernican system as set forth by Copernicus, therefore, did
not advance astronomy in the least; it merely showed that
it was possible to calculate the motions of the planets without having the origin of co-ordinates in the centre of the
earth. But of proofs of the physical truth of his system
Copernicus had given none, and could give none; and though
there can hardly be any doubt that he himself believed in
the reality of the earth's motion, it is extremely difficult to
say of most of his so-called followers whether they had any
faith in that motion, or merely preferred it for geometrical
reasons.[3]

It is always difficult to avoid judging the ideas of former
ages by our own, instead of viewing them in their connection
with those which went before them and from which they

---

[1] *Rhetici Ephemerides novæ*, 1550, p. 6.

[2] Möbius has shown that the use of the mean place of the sun (*i.e.*, the
centre of the earth's orbit) instead of the true place might, in the Copernican
theory of Mars, lead to errors of $2°$. See a note in Apelt's *Die Reformation
der Sternkunde*, Jena, 1852, p. 261.

[3] The contemporaries of Copernicus were not aware that the introduction
to his book, in which the system is spoken of as a mere hypothesis, was
written without the knowledge of the author by Osiander of Nürnberg.

were developed. The physical objections to the earth's motion, which to us seem so easy to refute, were in the sixteenth century most serious difficulties, and the merits of Galileo in conceiving the principles of elementary mechanics and fixing them by experiments must not be underrated. Neither should the advantage be forgotten which the seventeenth century had over the sixteenth from the invention of the telescope, which revealed the shape of the planets, the satellites of Jupiter, and the phases of Venus, and thus placed the planets on an equal footing with the earth, to which the unassisted vision could never have seen any similarity in them.

Tycho Brahe evidently was not content with a mere geometrical representation of the planetary system, but wanted to know how the universe was actually constructed. He felt the " physical absurdity " of letting the earth move, but, on the other hand, the clearness of mind which made him so determined an opponent of the scholastic philosophy enabled him to see how unfounded some of the objections to the earth's motion were. In a letter to Rothmann in 1587 Tycho remarks that the apparent absurdity is not so great as that of the Ptolemean idea of letting a point move on one circle with a velocity which is uniform with regard to the centre of another circle. He adds that the objections which Buchanan had made to the revolution of the earth in his poem on the sphere are futile, since the sea and the air would revolve with the earth without any violent commotion being caused in them.[1] But all the same he thought that a stone falling from a high tower ought to fall very far from the foot of the tower if the earth really turned on its axis. This remark is made in another letter to Rothmann in 1589, in which he made several objections to the annual motion of the earth.[2] The immense space between Saturn

[1] *Epist. astron.*, p. 74.   [2] Ibid., p. 167.

and the fixed stars would be wasted. And if the annual
parallax of a star of the third magnitude was as great
as one minute, such a star, which he believes to have an
angular diameter of one minute, would be as large as the
annual orbit of the earth. And how big would the brightest
stars have to be, which he believes to have diameters of
two or three minutes? And how enormously large would
they be if the annual parallax was still smaller?[1] It was
also very difficult to conceive the so-called "third motion"
of the earth, which Copernicus (so needlessly) had introduced
to account for the immovable direction of the earth's axis.

Tycho alludes in several places to the difficulty of recon-
ciling the motion of the earth with certain passages of
Scripture.[2] He was far from being the only one who
believed this difficulty to be a very serious one against
accepting the new doctrine. The Roman Church had not
yet taken any official notice of the Copernican system, but
in Protestant countries the tendency of the age was de-
cidedly against the adoption of so stupendous a change in
cosmological ideas. Nobody cared to study anything but
theology, and theology meant a petrified dogmatism which
would not allow the smallest iota in the Bible to be taken
in anything but a strictly literal sense. Luther had in his
usual pithy manner declared what he thought of Copernicus,[3]
and even Melanchthon, who was better able to take a dis-
passionate view of the matter, had declared that the authority

---

[1] Tycho had in vain tried to find an annual parallax of the pole star and
other stars. Letter to Kepler, December 1599, *Kepleri Opera Omnia*, viii.
p. 717.

[2] *Epist.*, p. 148. He says here that Moses must have known astronomy,
since he calls the moon the lesser light, though sun and moon are apparently
of equal size. Therefore the prophets must also be assumed to have known
more about astronomy than other people of their time did.

[3] "Der Narr will die ganze Kunst Astronomiä umkehren! Aber wie die
heilige Schrift anzeigt, so hiess Josua die Sonne still stehen und nicht das
Erdreich."—*Luther's Tischreden*, p. 2260.

of Scripture was against accepting the theory of the earth's
motion.[1] This may have had some weight with Tycho, at least
it might at first have made him indisposed openly to advocate
the Copernican system, as the most narrow-minded intolerance was rampant in Denmark (as in most other countries),
notwithstanding the king's more liberal disposition.   But
the king did not wish to be considered unorthodox, and had
yielded to the importunity of his brother-in-law, the Elector
of Saxony, by dismissing the distinguished theologian Niels
Hemmingsen from his professorship at the University, as
suspected of leaning to Calvinism.   It would certainly not
have been prudent for the highly-salaried and highly-envied
pensioner of the king, to declare himself an open adherent
of a system of the world which was supposed not to be
orthodox.

How far this consideration influenced Tycho it is not
easy to decide, but the supposed physical difficulties of the
Copernican system and a disinclination to adopt a mere
geometrical representation, in the reality of which he could
not believe, led him to attempt the planning of a system
which possessed the advantages of the Copernican system
without its supposed defects.   In a letter to Rothmann in
1589 [2] Tycho states that he was induced to give up the
Ptolemean system by finding from morning and evening
observations of Mars at opposition (between November 1582
and April 1583) that this planet was nearer to the earth
than the sun was, while according to the Ptolemean system
the orbit of the sun intervened between that of Mars and
the earth.   To the modern reader who knows that the
horizontal parallax of Mars can at most reach about $23''$,

---

[1] Melanchthon's *Initia doctrinæ physicæ*, in the chapter "Quis est motus
mundi."

[2] *Epist. astr.*, p. 148; see also ibid., p. 42, and letter to Peucer of 1588,
Weistritz, i. p. 243.

a quantity which Tycho's instruments could not possibly measure, this looks a surprising statement, particularly when it is remembered that Tycho, like his predecessors, assumed the solar parallax equal to 3'. This mystery was believed to have been solved by Kepler, who states that he examined the observations of 1582–83, and found little or no parallax from them; but, to his surprise, he found among Tycho's manuscripts one written by one of his disciples, in which the observed places were compared with the orbit of Mars according to the planetary theory and numerical data of Copernicus, and a most laborious calculation of triangles ended in the result that the parallax of Mars was greater than that of the sun. Kepler suggests that Tycho meant his pupil to calculate the parallax from the observations, but that the pupil, by a misunderstanding, worked out the distance of Mars from the diameters of the excentrics and epicycles of Copernicus.[1] The subject of the parallax of Mars is alluded to by Tycho in a letter to Brucæus, written in 1584. Here he does not hint at having already constructed a new system himself, but merely tries to disprove that of Copernicus, and among his arguments is, that, according to Copernicus, Mars should in 1582 have been at a distance equal to two-thirds of that of the sun, and consequently have had a greater parallax, whereas he found by very frequent and most exquisite observations that Mars had a far smaller parallax, and therefore was much farther from us than the sun.[2] In other words, Tycho could not find any parallax of Mars from his observations, but somehow he afterwards imagined that he had found Mars to be nearer

---

[1] Kepler, *De motibus stellœ Martis*, ch. xi., *Opera omnia*, iii. p. 219; see also p. 474. In his *Progymnasmata*, i. p. 414, Tycho says that the outer planets have scarcely perceptible parallaxes, but that he had found by an exquisite instrument that Mars at opposition was nearer than the sun. On p. 661 he alludes to it again.

[2] *T. Brahei et doct. vir. Epistolœ*, p. 76.

the earth at opposition than the sun was, and this decided
him to reject the Ptolemean system. He adds in his letter
to Rothmann, that the comets when in opposition did not
move in a retrograde direction like the planets, for which
reason he had to reject the Copernican system also. It did
not strike him that comets might move in orbits greatly
differing from those of the planets. Having rejected the
two existing systems, there was nothing to do but to design
a new one.

The Tychonic system could explain the apparent motions
of the planets (including their various latitudes), and it
might have been completed in detail by being furnished
with excentrics and epicycles like its rival. Copernicus
had referred the planetary motions, not to the sun, but to
the centre of the earth's orbit, from which the excentrici-
ties were counted, and through which the lines of nodes
passed, so that the earth still seemed to hold an exceptional
position. The Copernican system, so long as it was not
purged of the artificial appendage of epicycles by the laws
of Kepler, was not very much simpler than the Tychonic,
and, mathematically speaking, the only difference between
them was, that the one placed the origin of co-ordinates in
the sun (or rather in the centre of the earth's orbit), the
other in the earth.[1] Tycho's early death prevented the
turther development of the theory of the planets by his
system, which he intended to do in a work to be called
Theatrum astronomicum. He only gives a sketch of the
theory of Saturn in the first volume of his book, in which
the planet moves in a small epicycle in retrograde direction,

---

[1] Might Tycho have got the idea of his system by reading the remark of
Copernicus (*De revol.*, iii. 15) when talking about the earth's orbit : " Estque
prorsus eadem demonstratio, si terra quiesceret atque Sol in circumcurrente
moveretur, ut apud Ptolemæum et alios " ? According to Prowe (*Nic. Coppernicus*, Bd. i. Part 2, p. 509), this is one of the sentences struck out in the
original MS., but reinserted by the editor of the first edition.

making two revolutions while the centre of the small epicycle moves once round the circumference of a larger one
in the same direction in which the centre of the latter
moves along the orbit of Saturn.[1]

The Tychonic system did not retard the adoption of the
Copernican one, but acted as a stepping-stone to the latter
from the Ptolemean. By his destruction of the solid spheres
of the ancients and by the thorough discomfiture of the scho-
lastics caused by this and other results of his observations
of comets, he helped the Copernican principle onward far
more effectually than he could have done by merely acqui-
escing in the imperfectly formed system, which the results
of his own observations were to mould into the beautiful and simple system which is the foundation of modern
astronomy.

The book on the comet of 1577 was ready from the press
in 1588, and though not regularly published as yet, copies
were sent to friends and correspondents whenever an opportunity offered.[2]   Thus Tycho's pupil, Gellius Sascerides, who
in the summer of 1588 started on a journey to Germany,
Switzerland, and Italy, brought copies to Rothmann and
Maestlin, to whom he was also the bearer of letters.[3]   The
Landgrave did not receive a copy, but studied Rothmann's
copy with great interest, and thought that it must have
been meant for himself, until Rothmann suggested that it
was only part of an unfinished work, and that he would get
one later on, which of course he did as soon as Tycho heard
of this incident.   In the following year, while he was at
the fair of Frankfurt, Gellius received another copy of the
book, which he was to bring to Bologna to Magini, and this

---

[1] *Progymn.*, i. p. 477, where he also alludes to the "Commentariolus" of
Copernicus, see above, p. 83).

[2] The book was not for sale till 1603.   There are three copies in the Royal
Library at Copenhagen with the original title-page of 1588.

[3] About Maestlin see *Kepleri Opera*, i. p. 190.

he forwarded from Padua in 1590, together with a letter
in which he gave an account of the unfinished first volume
of Tycho's work.[1]   A copy was sent to Tycho's old friend
Scultetus, who let Monavius of Breslau partake of his joy
over it.   To Thomas Savelle of Oxford, a younger brother
of the celebrated founder of the two Savillian professorships,
who was then travelling on the Continent, Tycho sent two
copies of the book, together with a letter in which he,
among other things, asked him to remind Daniel Rogers
about the copyright which he had promised to procure
Tycho for his books in England.[2]   To Caspar Peucer, who
had already heard of the book from Rantzov, Tycho sent a
copy, and added a very long letter in which he entered
fully into his reasons for rejecting the Copernican system,
and discussed some passages of Scripture which had been
made use of to prove the solidity of the celestial spheres.
In this letter he also gives an interesting sketch of the plan
of the great work to which the three volumes on the new
star and comets were to be introductory.   It was to consist
of seven books; the first was to describe his instruments,
the second the trigonometrical formulæ required in astronomy, the third the new positions of fixed stars from his
observations, the fourth was to deal with the theories of the
sun and moon, the fifth and sixth with the theories of
the planets, the seventh with the latitudes of the planets.[3]
With the exception of the first chapter (which he made
into a separate book), the contents of this projected work

---

[1] *Carteggio inedito di Ticone Brahe, G. Keplero, &c.*, con G. A. Magini.   Ed.
Ant. Favaro, Bologna, 1886, p. 193.

[2] A Collection of letters illustrative of the progress of science in England.
Edited by J. O. Halliwell.   London, 1841, p. 32.   Tycho also sent Savelle four
copies of his portrait engraved at Amsterdam (by Geyn, 1586), and inquired
whether there were any good poets in England who would write an epigram
on this portrait or in praise of his works.   He added that Rogers might also
show his friendship by helping him in this matter.

[3] Weistritz, i. pp. 239-264, reprinted from Resen's *Inscriptiones Hafnienses.*

(or at least the outlines of them) were afterwards incorporated in Tycho's first volume of *Progymnasmata*.

When Rothmann had received the book he wrote to Tycho to thank him for it, and remarked that the new system of the world seemed to be the same as one which the Landgrave a few years previously had got his instrument-maker to represent by a planetarium.[1] Tycho, who had kept his system a deep secret until the book was ready, was at first unable to understand from whom the Landgrave could have got a description of it,[2] but he soon after received from a correspondent in Germany a recently published book which solved the riddle. The title of the book was *Nicolai Raymari Ursi Dithmarsi Fundamentum astronomicum*, printed at Strassburg in 1588. The author, Nicolai Reymers Bär, was a native of Ditmarschen, in the west of Holstein, and a son of very poor parents. He is even said to have earned his bread as a swineherd, but possessing great natural abilities, he rapidly acquired considerable knowledge both in science and in classics. In 1580 he published a Latin Grammar, and in 1583, at Leipzig, a *Geodaesia Ranzoviana*, dedicated to his patron, Heinrich Rantzov, Governor of Holstein.[3] Having for some time worked as a surveyor, he seems to have entered the service of a Danish nobleman, Erik Lange of Engelholm, in Jutland, who was a devoted student of alchemy. Lange went on a visit to Tycho in September 1584, and brought Reymers with him, but this probably somewhat uncouth self-taught man seems to have been treated with but scant civility

---

[1] *Epist. astron.*, pp. 128, 129.

[2] So Tycho says in his reply to Rothmann (*Epist.*, p. 149), but before Rothmann's letter was written Tycho had in his letter to Peucer (dated 13th September 1588) mentioned that a German mathematician had two years previously heard of the system "per quendam meum fugitivum ministrum " (Weistritz, i. p. 255), and this he also mentions in the letter to Rothmann.

[3] Kästner, *Geschichte der Mathematik*, i. p. 669; *Kepleri Opera* ed. Frisch, i. p. 218.

at Uraniborg. After having spent a winter as tutor in Pomerania, Reymers went to Cassel in the spring of 1586, where he informed the Landgrave that he had the previous winter, while living on the outskirts of Pomerania, designed a system of the world. This was exactly like Tycho's, except that it admitted the rotation of the earth. The Landgrave was so pleased with the idea, that he got Bürgi to make a model of the new system; but though he had been well received at Cassel, Reymers was not long in favour there, as he fell out with Rothmann, to whom he abused Tycho. Rothmann mentioned this in a letter to Tycho in September 1586,[1] but did not mention Reymers' system, which first became known in 1588 by the above-mentioned book.[2] This contains some chapters on trigonometry and some on astronomy, and in the last chapter the new system is explained and illustrated by a large diagram on about twice as large a scale as that in Tycho's book. The only important difference is, that the orbit of Mars does not intersect that of the sun, but lies quite outside it.

Tycho was apparently very proud of his system, and (as in the case of Wittich) he immediately jumped to the conclusion that Reymers Bär had robbed him of his glory.[3]

---

[1] *Epist. astron.*, p. 33, where Rothmann (who thought that Reymers had been employed in Tycho's printing-office) calls him a dirty blackguard ("plura scriberem, præsertim de impuro illo nebulone"), which expression Tycho now found very suitable (ibid., p. 149).

[2] For accounts of this book see Kästner, i. p. 631 ; Delambre, *Astron. moderne*, i. p. 287; and Rudolf Wolf's *Astronomische Mittheilungen*, No. lxviii.

[3] Already in 1589 or 1590 Duncan Liddel lectured at Rostock on the Tychonic system, calling it by this name. A report afterwards reached Tycho to the effect that Liddel privately took the credit of the new system to himself, and that he later on did so openly at Helmstadt (see letter from Cramer, a clergyman of Rostock, to Holger Rosenkrands, in *Epistolæ ad J. Kepplerum*, ed. Hanschius, p. 114 *et seq.*). It appears, however, that Liddel indignantly denied the charge, though he claimed to have deduced the system himself, and to owe Tycho nothing except the incitation to speculate on the matter, for which reason he had mentioned the system as the "Tychonic" (*Kepleri Opera omnia*, i. pp. 227, 228).

He wrote at once to Rothmann (in February 1589) that Reymers must have seen a drawing of the new system during his stay at Uraniborg in 1584, and as a proof of this he refers to the orbit of Mars, which in a drawing made before that time, by a mistake, had been made to surround the solar orbit instead of intersecting it. This cancelled drawing had got in among a number of maps in a portfolio, where Reymers must have seen it, as he copied the erroneous orbit of Mars in the diagram of his book. He therefore expressed his concurrence in the not very flattering expression which Rothmann had applied to Reymers Bär in a former letter.[1]

It must, however, be said that this accusation of plagiarism is founded on very slight evidence, and the verdict of posterity can only be "not proved." In his writings Reymers has shown himself an able mathematician, and there is no reason whatever why he should not independently have arrived at a conclusion similar to the idea which Tycho conceived on the planetary motions. We shall afterwards see what a curious end this affair got, and how Tycho and Rothmann may have regretted that they had not let the *bear* alone.

[1] *Epist. astr.*, pp. 149, 150.

CHAPTER VIII.

FURTHER WORK ON THE STAR OF 1572.

AFTER the publication, or rather the completion, of the second volume of his book Tycho pushed on the preparation of the first volume on the new star, of which the printing began long before the manuscript was approaching completion. From many direct or indirect allusions to time in various places in the book, it appears that it was written in the years 1588 to 1592,[1] and as Tycho had several times been inconvenienced by want of paper, he resolved to build a paper-mill on the south-western coast of the island, which could be driven by the water from the fish-ponds. This mill was finished in 1589 or 1590, and the same water-wheel which turned it could also be connected with a corn-mill and machinery for preparing skins.[2] It was, however, by no means only the want of paper which delayed the completion of Tycho's book; it had come to embrace

---

[1] See, *e.g.*, *Progym.*, i. pp. 34, 52, 102, 335, 559, 710, 721, 745. In the appendix written by Kepler it is stated that the book was written between 1582 and 1592, but the printing cannot have commenced before 1588, and that the first chapter was written in 1588 is evident.

[2] In a letter to Rothmann (24th November 1589) Tycho mentions the mill as having been for some time at work. The inscription on a slab in the wall of the mill is given slightly different by Resen., *Inscript. Hafn.*, 1668, p. 335 (Weistritz, i. p. 69), and in the *Danske Magazin*, ii. p. 265 (Weistritz, ii. p. 198); and according to the former it was begun in 1589 and completed in 1590; according to the latter, commenced 1590, finished 1592. But the former dates must be correct, as they agree with Tycho's statement that the mill was at work in 1589; and in the meteorological diary we read under 22nd July 1590: "Abiit Valentinus opere aggeris apud molendinum confecto." In March 1590 the widow of Steen Bille was ordered to allow Tycho to cut down an oak in the wood at Heridsvad for use in the mill (*D. Magazin*, ii. 264).

many branches of astronomy, and as the current observations continued to reveal imperfections in the values of astronomical constants handed down from antiquity, Tycho was unwilling to finish the book and deprive himself of the power of inserting in it further results of his work. The book was never issued in a complete state in his lifetime; only a very few friends or correspondents received incomplete copies or portions of the book; and after Tycho's death an important section (32 pp.), separately paged, was inserted at the end of the first chapter. When completed, the book numbered more than 900 pages, divided into three parts and a "conclusion;" and it bears many traces of having been both written and printed in the course of many years, succeeding sheets frequently before preceding ones. The first chapter deals with the apparent motion of the sun, the length of the year, the elements of the solar orbit, refraction, and gives tables for the motion of the sun. As there were a few pages to spare (the second chapter having been printed and paged first), Tycho determined to devote them to the lunar theory, though this had nothing to do with the determination of star places, and was not even mentioned in the title of the chapter; and as this subject grew in importance and difficulty, it eventually delayed the publication of the volume considerably. The second chapter describes the methods of determining the places of stars, investigates the amount of precession, and contains Tycho's own catalogue of star places.[1] This finishes the first part of the book, and as we shall examine in our last chapter the various subjects dealt with, we may pass to the second part of Tycho's book, which is devoted to his own observations of the star of 1572.

---

[1] The Catalogue was printed after the succeeding sheets, and the sheets KK. and LL. are therefore double ones, as the Catalogue filled more space than anticipated.

The third chapter describes the appearance of the star, the gradual fading of its light, the variation of colour; how it was seen by carriers, sailors, and similar people, a good while before the astronomers in their chimney-corners heard of it; then branches off into a mythological account of Queen Cassiopea, and gives a map of the constellation with the star in it. Tycho refers to the Aristotelean idea of the unchangeable nature of the heavens, and to the star of Hipparchus, which he believes to have been similar to his own star, and then to that of the Magi, which he says could not have been a star in the heavens, since it showed the way to a particular town, and even to a house, and was only seen by the wise men. He therefore summarily dismisses the idea that the star in Cassiopea should signify the return of Christ. Lastly, he mentions the stars said by Cyprianus Leovitius to have appeared in the years 945 and 1264. All these subjects we have dealt with in sufficient detail in the chapter on the new star, where the book of which we are now summarising the contents is frequently quoted.

In the fourth chapter are given descriptions and illustrations of the sextant with which he observed the star, and of the great quadrant at Augsburg. This chapter also contains the measured distance of the new star from twelve stars in Cassiopea, the distances *inter se* of most of these stars (from observations made at Hveen in 1578 and 1583), and a number of observations made with the Augsburg quadrant. As this instrument was designed and constructed by Tycho, he naturally wished to prove its excellence, and inserted a number of observed declinations of circumpolar stars (which give values for the latitude of Göggingen agreeing *inter se* within a minute), and the declinations of six zodiacal stars in equally good accordance with the results obtained at Hveen.

In the fifth chapter the co-ordinates of the star both

with regard to the ecliptic and the equator are computed
from its distance from the other stars in Cassiopea and the
places of these stars as observed at Hveen.   Seven different
combinations give results of which the extremes differ only
about half a minute.[1]   Tycho also gives the places of the
twelve comparison stars according to Alphonso and Copernicus (i.e., Ptolemy), which differ in many cases upwards of
a degree from his own.   He then turns in the sixth chapter
to the question as to where the star was situated in space,
and proves in four ways that it was far beyond the planets,
"in the eighth sphere."   First, the shape, light, continual
twinkling, immovability, daily revolution like the fixed stars,
and its having lasted more than a year, prove that it was
not a comet.   Secondly, it had no parallax, as the distance
from the pole and from neighbouring stars remained unaltered during the daily revolution, while the polar distance
would have varied 1° 5′ if the star had been as near as the
moon, 2′ 52″ if as near as the sun, and 16″ if at the distance
of Saturn, with smaller variations in the distances from the
other stars.[2]   Here he not only gives this indication of his
idea of the distance of the planets, but also shortly alludes
to his system of the world.   He remarks that if the star
was situated in the sphere of Saturn, and if we adopt the
annual motion of the earth according to Copernicus, the
star would in a year appear to move backwards and forwards
(i.e., have an annual parallax) to the extent of about ten
degrees, so that even followers of Copernicus must admit
that the star was far beyond Saturn.   The third proof of
the great distance of the star is, that the meridian alti-

---

[1] The result adopted by Tycho is for 1573, AR 0° 26′ 24″, Decl. 61° 46′ 45″,
while Argelander found 0° 28′ 6″ and 61° 46′ 23″ from a recomputation of the
distance measures, using Bessel's and Bradley's star places. *Astr. Nachr.*,
lxii. No. 1482.

[2] On p. 414 he refers to the difficulty of finding the parallaxes of the outer
planets, and how Mars was nearer than the sun. See above, p. 179.

tudes gave the same latitude (for Heridsvad and Göggingen)
as other stars gave ; and the fourth is, that observations at
far-distant places gave results in good accordance *inter se*,
as, for instance, his own and those of Munosius at Valencia.
As Tycho has so often referred to the parallax of the moon,
he verifies at the end of this chapter the value of Copernicus by computing the lunar parallax from six observations,
three on the meridian and three at the nonagesimal point,
where there is no parallax in longitude.

In the seventh and last chapter of the second part of the
book Tycho attempts to calculate the diameter of the new
star. He first recounts the crude ideas of his predecessors
as to the diameters of the planets and fixed stars, on which
he did not improve very much himself. He did not, however,
place all the fixed stars at the same distance just beyond
the orbit of Saturn, and he suggested that the fainter stars
are probably at a far greater distance than the brighter
ones, though even if they were at the same distance it
would not follow that all the stars which we consider as
belonging to one magnitude were equal in size, as Sirius
and Vega are much larger than Aldebaran, which again is
larger than Regulus.[1] The apparent diameter of the sun
Tycho had, in 1591, measured "through a canal 32 feet
long," and in this way he found that at the apogee the
diameter was barely 30′, and at the perigee slightly above
32′.[2] The instrument was, according to Kepler, a screen
on which the image of the sun fell through a small opening,
and the " canal " must have been added merely to exclude
stray light.[3] The diameter of the moon Tycho generally

---

[1] *Progym.*, p. 470.

[2] Ibid., p. 471. *Historia Cœlestis*, p. 475 *et seq.* Tycho's mean diameter
31′ is exactly 1′ too small, and the difference between apogee and perigee is
only 1′, as Kepler already found.

[3] *Ad Vitell. Paral.*, cap. xi. ; *Opera omnia*, ii. pp. 343-44, where Kepler quotes

determined by observing the difference of declination of
the upper and lower limb; he adopts the mean diameter
33'.   With these data he now calculates the real diameters
of the sun and moon, making use of the old value of the
solar parallax of 3', which neither Copernicus nor he thought
of discarding.   The distance of the sun being 1150 semi-
diameters of the earth, the semi-diameter of the sun will be
5.2 times that of the earth, and similarly the distance of the
moon is 60 and its semi-diameter 0.29.   For the planets
he assumes apparent diameters from 2' to 3', and calculates
from these their diameters and volumes in parts of those
of the earth.[1]   For the fixed stars Tycho assumes smaller
apparent diameters than other astronomers did before the
invention of the telescope.[2]   With regard to the distance
of the stars, he believes the greatest distance of Saturn to
be 12,300 semidiameters of the earth (to arrive at which he
sketches the theory of Saturn as mentioned above[3]); and
as he does not believe that there is a great void between
the orbit of Saturn and the fixed stars, he places these at

some observations made at Hveen in March and June 1578, giving 30' 35"
and 29' 53".   About the diopters of Hipparchus, see Halma's preface to the
Almegist, vol. ii. p. lviii.

[1] The diameters of the planets are measured by pointing with the armillæ
or a quadrant alternately to the upper and lower edge of the planet.   See,
e.g., *Historia Cœlestis*, p. 429, for a number of measures of Saturn.   The
diameters assumed are (*Prog.*, pp. 475–76) :—

Mercury	2' 10"	at mean distance,		1,150
Venus	3' 15"	„	„	1,150
Mars	1' 40"	„	„	1,745
Jupiter	2' 45"	„	„	3,990
Saturn	1' 50"	„	„	10,550

[2] First mag. diameter 120", second 90", third 65", fourth 45", fifth 30",
sixth 20" (ibid., pp. 481–82).   Magini took the stars of the first mag. to
be 10' in diameter ; Kepler made the diameter of Sirius 4' (*Opera*, ii. p. 676) ;
the Persian author of the *Ayeen Akbery* put the diameter of stars of the first
mag. = 7' (Delambre, *Moyen Age*, p. 238), so that Tycho's estimates were more
reasonable than any of these.

[3] The ratio of the semidiameters of the deferent of Saturn and of the solar
orbit he borrows from Copernicus.   Compare above, p. 181, footnote [1].

a distance of about 14,000, and the new star at least at 13,000 semidiameters. The apparent diameter of the new star at its first appearance he estimated at $3\frac{1}{2}$, and its real diameter must therefore have been $7\frac{1}{8}$ times that of the earth, or somewhat greater than that of the sun. He does not think that the diminution of light was caused by the star having moved away from us in a straight line, partly because no celestial body moves in a straight line, partly because it would, when about to disappear, have been at the incredible distance of 300,000 semidiameters of the earth. The star must actually have decreased in size, so that it at the end of the year 1573 was about equal to the earth in size.

This finishes what Tycho has himself found by observation and speculation concerning the star of Cassiopea, and he next devotes the third part of his book, 300 pages, to an examination of the writings of other astronomers or authors about the star. First he discusses in Chapter VIII. the observations of those who could not find any parallax (the last book considered being his own little book of 1573, of which he reprints the greater part, omitting the astrological predictions); next he deals in Chapter IX. with those authors who thought they had found some parallax, but who did not place the star within the lunar orbit; and lastly, he deals with the writers "who have not brought out anything solid or important, and either maintained that the star was not new or that it was a comet or a sublunary meteor." His remarks are often written in a sarcastic style, with puns or play upon words, by which he perhaps meant to relieve the dulness of this far too lengthy part of the book. We have above, in our third chapter, given the reader some idea of these various classes of writers, and need not, therefore, here enter into further details about these chapters of Tycho's book.

Finally, the "Conclusion" of the volume (pp. 787–816) gives first a rapid summary of the contents of the book, and then deals with two questions not yet touched upon, the physical nature of the new star and the astrological effect and signification, which the author did not wish to enter on in the body of the work, "as these matters are not subject to the senses, nor to any geometrical demonstration, but can only be speculated on." As to the nature of the star, Tycho considers that it was formed of "celestial matter," not differing from that of which the other stars are composed, except that it was not of such perfection or solid composition as in the stars of permanent duration. It was therefore gradually dissolved and dwindled away. It became visible to us because it was illuminated by the sun, and the matter of which it was formed was taken from the Milky Way, close to the edge of which the star was situated, and in which Tycho believed he could now see a gap or hole which had not been there before. This idea may to the modern reader seem absurd, but it should be remembered that the telescope had not yet revealed the true nature of the Milky Way, and Tycho's ideas about the latter were at all events a great advance from those of Aristotle (which he sharply attacks), according to which the Milky Way was merely an atmospheric agglomeration of stellar matter. With regard to the other question, the astrological signification of the star, Tycho had evidently considered it a good deal since he wrote his little book in 1573, and he does not on this occasion merely express himself in very general terms, but gives his opinion with more decision. As his prediction attracted a good deal of attention, particularly later when it seemed to have been fulfilled, it is worth while to give a short summary of it.

As the star of Hipparchus announced the extinction of the Greek ascendency and the rise of the Roman empire,

so the star of 1572 is the forerunner of vast changes, not
only in politics, but also in religious affairs, for the star was
situated close to the equinoctial colure, which by astrologers
is supposed to have something to do with religious matters.[1]
And as the star first shone with Jovial and clear light, and
afterwards with Martial and ruddy light, the effect will first
be peaceful and favourable, but afterwards become violent
and tumultuous.   And the religions which are full of
" Jovial " splendour and pomp, after having for a long time
dazzled ignorant people by their external magnificence and
more than Pharisaic formalism, will, like that pseudo-star,
fade and disappear. Though the star was so near the
equinoctial colure that it nearly touched it with its rays, it
was quite within the vernal quadrant, and it announces that
some great light is at hand, just as the sun when past the
vernal equinox conquers the darkness of night.   And as
the star was visible over most of the earth, so the effects of
it will be felt over the greater part of the globe, though
the northern hemisphere will be especially affected.   With
regard to the time when the influence of the star will begin
to be felt, this will be nine years after the great conjunction
of planets in April 1583, in the 21st degree of Pisces (because the *direction* of this and the star along the equator
is 9°), or in other words, in 1592, and those who were born
when the star appeared will about that time enter man's
estate, and be ready for the great enterprise for which they
are ordained.   But if we take the *direction* along the
zodiac, we find forty-eight years, after the lapse of which
period the effect of the star will become strongest, and will
last for some years, until 1632 or about that time, when
the effect of the fiery trigon (which the star announced)
will also be felt. The conjunction of 1583 concluded a
cycle of planetary conjunctions, the seventh since the

[1] Pisces was supposed to be the sign of Palestine.

creation of the world.[1] The first cycle ended at the time
of Enoch, the second at the deluge, the third at the exodus
from Egypt, the fourth at the time of the kings of Israel,
the fifth at the time of Christ when the Roman empire was
at its height, the sixth when the empire arose in the
western world under Charlemagne, and the seventh and
sabbath-like one was now coming. And as the first, third,
and fifth "restitution" had been salutary to the world, the
seventh, which had a particularly uneven number, will in-
augurate a very happy state of things, a peaceful and quiet
age such as that foretold by the prophets Isaiah (ch. xi.)
and Micah (ch. iv.), when the lion shall eat straw like the
ox, and the sucking child shall play on the hole of the asp,
&c. As to the place on the earth from which this change
will arise, it will be the one in the Zenith of which the star
was at its first appearance, which Tycho assumes to have
been at the time of the New Moon previous to the 11th
November, when he noticed the star.[2] The star was then
on the meridian of places about 16° east of Uraniborg and
in the Zenith of a place with north latitude $61\frac{3}{4}°$. This
fixes the ominous spot "in Russia or Moschovia where it
joins the north-east part of Finland." Having devoted so
much space to this matter, I must pass over the way in
which Tycho finds Moschovia pointed out in the Prophets,
the Revelation, and a certain ancient prophecy of Sibylla
Tiburtina, found in 1520 in Switzerland.

That Tycho when writing of the religion distinguished
by pomp and splendour which was soon to disappear was
thinking of the Roman Catholic persuasion is beyond a

---

[1] "Septima hæc est trigonorum in integrum ab Orbe condito restitutio."
About the trigoni see above, p. 49, footnote. The conjunction of Jupiter
and Saturn in Sagittarius in December 1603 commenced a new cycle with a
fiery trigon.

[2] See above, p. 50. Tycho now (*Progym.*, p. 809) gives the time of New
Moon as 7h. $31\frac{2}{3}$m. P.M.

doubt, and it is curious that the book in which we read this, though printed in Denmark, should eventually come to be published at Prague (where the religious war which he foretold raged furiously less than twenty years after his death) and was dedicated to the Roman Emperor! But it is more curious still that some of his other predictions seem to be fulfilled in the person of Gustavus Adolphus, the greatest champion of Protestantism in the seventeenth century. He was born in 1594 (only two years after the influence of the star should begin to be felt), and his glory was greatest in the year in which he fell, 1632, the very year mentioned by Tycho. He certainly was not born in Finland (for it is Finland and not the adjoining part of Russia which is indicated by 16° east of Uraniborg and 62° Latitude), but in Stockholm; but Finland was still a province of Sweden, and the yellow Finnish regiments were conspicuous for their bravery on many a blood-stained battlefield in Germany. No wonder that many contemporaries of Gustavus Adolphus were startled by these coincidences, and that the concluding part of Tycho's book was translated into several languages.[1] But the star had a truer mission than that of announcing the arrival of an impossible golden age. It roused to unwearied exertions a great astronomer, it caused him to renew astronomy in all its branches by showing the world how little it knew about the heavens; his work became the foundation on which Kepler and Newton

---

[1] I possess an English translation which seems to be very scarce : "Learned Tico Brahæ his astronomicall Coniectur of the new and much Admired ✲ which Appered in the year 1572," London, 1632, 26 pp. text, 5 pp. dedication ("To the High and Mighty Emperour Rvdolphvs the II. The Preface of the Heyres to Tycho Brahe"), and 2 pp. of epigrams by the translator and James VI. I have seen another copy in which there was a portrait of Gustavus Adolphus. Lalande has a German translation also printed in 1632, and there is a Dutch one printed at Goude in 1648 ("Generale Prognosticatie van het jaer 1572 tot desen tegenwoordigen Jare, alles in Latijn beschreven van Ticho Brahe") in the library of Trin. Coll., Dublin.

built their glorious edifice, and the star of Cassiopea started astronomical science on the brilliant career which it has pursued ever since, and swept away the mist that obscured the true system of the world.   As Kepler truly said, " If that star did nothing else, at least it announced and produced a great astronomer." [1]

[1] "Certe si nihil aliud stella illa, magnum equidem astronomum significavit et progenuit."   The last words in Kepler's Appendix to the *Progymnasmata*.

CHAPTER IX.

THE LAST YEARS AT HVEEN, 1588-1597.

AT the death of King Frederick II., in 1588, his eldest son, Prince Christian, who had been elected his successor, was only eleven years of age. The Queen claimed the right of governing the country during his minority, and asked the Privy Council if she had not in her husband's lifetime shown her love to the two kingdoms, and whether they could show cause why she should have forfeited the right of Dowager-Queens. But the powerful nobles were determined to take the reins into their own hands, and elected four Protectors from among the Privy Councillors —the Chancellor, Niels Kaas; the Chief of the Exchequer, Christopher Valkendorf; the Admiral Peder Munk; and the Governor of Jutland, Jörgen Rosenkrands. In order to keep their power as long as possible, it was decided that the minority should last till the Prince was twenty years of age. The quiet and careful education of the young King-Elect, which his father had planned, was continued, and the Government of the Protectors was on the whole well and ably conducted. To Tycho Brahe it was of great importance that the Chancellor Kaas was a member of the Government, as the precarious tenure by which he held all his endowments and pensions made it a vital matter for him to have firm friends among those at the head of affairs. Probably with a view to ascertain how far the new Government was friendly to him, Tycho in the spring of 1588 addressed a memorial to the young King in which he showed that he not only had spent his various pensions and the income from his hereditary estate on his buildings and works at Hveen, but had incurred a debt of 6000 Daler (£1360), and he begged that the King would indemnify him for this loss, as he had spent the money according to the desire of the late King and for the honour of the country. On the 8th July, Kaas and Rosenkrands paid Tycho a visit at Hveen, and probably by their advice the young King, on the 12th July, with the sanction of the Privy Council, granted the said sum of 6000 Daler to Tycho, 2000 to be paid by the Treasury and 4000 from the Crown revenue of the former Dueholm monastery in Jutland. The money was paid on the 14th December following, 2000 from the Sound dues and 4000 from Dueholm.[1] In addition to this proof of the continued favour of the Government, the Protectors on the 23rd August issued a declaration promising to keep the buildings at Hveen in repair at the public expense, and on the expiration of the King's minority to advise him to fix a certain annual endowment for the continuance of the astronomical work there by some fit person of Tycho's own family, or, failing such, by some suitable person of Danish nobility or by some other native.[2] In the following year, on the 17th July 1589, a new declaration to the same effect and very much in the same words was drawn up and signed and sealed, not only by the four Protectors, but by all the members of the Council, fourteen in number.[3]

Though these declarations were very reassuring to Tycho, he seems to have thought that the young King might possibly in future years not consider himself bound by them, for in March 1590 he procured a letter from the

---

[1] *Danske Magazin*, ii. p. 249, 253 (Weistritz, ii. 175 and 182).

[2] Ibid., p. 250 (W., p. 177).

[3] Ibid., p. 260 (W., p. 192).

Queen stating that she perfectly remembered to have heard
King Frederick, some time before his death, express his
intention of appointing one of Tycho's children to succeed
him at Hveen, if one should be found skilful in the astronomical art.[1] Tycho does not appear to have made any
use of this letter in after years, perhaps because neither
of his two sons showed any taste for astronomy.

For the present, at any rate, Tycho's position was secured
by the new Government, and we have already seen that the
grant of the Norwegian estate was renewed in June 1589.
In this year he received another mark of the friendly
feelings of the Government, as a letter from the young king
to the burgomasters and Corporation of Copenhagen (dated
Copenhagen the 13th March 1589) ordered them to lend
Tycho Brahe a stone tower next the rampart, "and a small
piece of the rampart up to his paling," as he intended to
erect a building on the tower for astronomical use, where
he wanted to keep some instruments for the use of some
people who might reside there and practise with them. He
was, however, to give up the tower and rampart whenever
they might be required for the defence of the city.[2] This
part of the rampart was doubtless close to a house which he
is known to have owned in the Farvergade, in the south-
west part of Copenhagen, perhaps at the corner where the
street (until a few years ago) adjoined the rampart, as the
latter is said to "reach to his paling." On the 25th of
March following the king furthermore gave to Tycho and
his heirs for ever two empty houses next his own, on condition that he should build another house for the dyer who

---

[1] *Breve og Aktstykker angaaende Tyge Brahe og hans Slægtninge.* Samlede af F. R. Friis, Kjöbenhavn, 1875, p. 1. From the Queen's letter-book
in the Royal archives. It appears from two other letters that the Queen had
lent Tycho 1000 Daler, which he was to pay back at Michaelmas 1590.

[2] G. F. Lassen, *Documenter og Actstykker til Kjöbenhavn's Befæstnings
Historie,* Copenhagen, 1855, p. 111.

had carried on business in one of those which Tycho got.[1] Though he frequently went to Copenhagen (as may be seen by the Meteorological Diary), nothing is known about his domestic arrangements there, nor do we know whether he really kept students or pupils. No observations were made at Copenhagen.

At Hveen the life and work continued as in previous years, and Tycho was still honoured and fêted by both compatriots and foreigners. Among his nearer friends, Vedel and Erik Lange paid him occasional visits, and early in 1590 the latter became engaged to Tycho's youngest sister, Sophia, who was a very frequent guest at Uraniborg, and who after Steen Bille's death (1586) was the only one of Tycho's relations who was capable of appreciating the work carried on there.[2] At the age of nineteen or twenty she had been married to Otto Thott of Eriksholm, in Scania, who died in 1588, leaving an only child, Tage Thott, during whose minority the young widow managed the property of Eriksholm. Here she devoted her leisure hours partly to horticulture (in which she must have excelled, since Rothmann, who paid her a visit during his stay in Denmark, thought her garden worthy of special praise to the Landgrave), partly to chemistry and medicine (which latter she made use of to relieve the poor), and especially to judicial astrology, to which she was greatly devoted, so that she is said to have always carried a book about with her in which the horoscopes of her friends were entered. She had several times met Erik Lange at Uraniborg, where he probably came to consult Tycho on matters relating to alchemy, on which pursuit he squandered his fortune.

---

[1] *Danske Magazin,* ii. p. 254 (Weistritz, ii. p. 183), where also is given Tycho's acknowledgment of his being bound to provide the dyer with a new house.

[2] About her observation of the lunar eclipse of 1573, see above p. 73.

The match was not a brilliant one for her, though Lange
was her equal as regards birth, but in searching for the
philosopher's stone he had become greatly indebted, and in
order to escape his creditors he left the country in 1591,
hoping perhaps abroad to be more successful in the gold-
making line than he had been at home.   It is needless to
say that he met with new disappointments, and Sophia and
he were not united till 1602, six months after Tycho's
death.[1]

Sophia Brahe and her future husband were not the
only guests at Uraniborg in the spring of 1590, at which
time Tycho received his most distinguished visitor, King
James VI. of Scotland.   This monarch had several years
before made overtures for the hand of Princess Elizabeth,
the eldest daughter of Frederick II.   His envoy, Peter
Young, had, however, produced powers so limited, or had
conducted the negotiations in so lukewarm a manner (it is
supposed at the instigation of Queen Elizabeth), that the
Danish king did not believe the wooer to be in earnest, and
promised his eldest daughter to the Duke of Brunswick.
Not discouraged by this failure, and yielding to the loudly
expressed wish of the Scotch nation to see the king married
soon, James solicited the hand of King Frederick's second
daughter, Anne.   The Earl Marshal, Lord Keith, was sent
to Denmark in 1589, and the marriage was celebrated at
Kronborg Castle by procuration.   The bride set out for
Scotland in September, escorted by a Danish fleet of fourteen vessels, but a storm obliged the fleet to seek shelter at

---

[1] *Danske Magazin*, iii. (1747), p. 12 *et seq.*   During Lange's absence abroad
Sophia Brahe sent him a long letter in Latin verses, which is printed in
Resenii Inscriptiones Hafnienses (1668), but from a very incorrect copy.
There is a more correct copy in the Hofbibliothek at Vienna, printed in *Breve
og Actstykker*, pp. 6–25.   The poem is most interesting from the numerous
allusions to alchemy and astrology in it, but Sophia Brahe cannot have written
it in Latin, to judge from what Tycho says of her attainments in a MS. note,
also in the Hofbibliothek (ibid., pp. 160–161).

Oslo in Norway, and James was informed that it was not likely to put to sea again for some months. Vexed at this new disappointment, he quickly made up his mind (not a very usual thing for him to do, but he was probably anxious to have the vexed question of the Orkney and Shetland Isles settled as soon as possible), and having, without communicating his intention to his Council, fitted out some ships, he started for Norway attended by the Chancellor, Sir John Maitland, and a numerous suite. He arrived at Oslo in November, and the marriage was solemnised at Aggershus Castle on the 24th of that month by his own chaplain, David Lyndsay. The timid monarch did not care to face the boisterous North Sea a second time in winter, and remained in Norway for some time, until he accepted the invitation of the Danish Government and set sail for Kronborg, where he arrived with his bride on the 20th January 1590. A month after he went to Copenhagen, where the usual festivities were held in his honour; but James did not neglect the opportunity of enjoying the conversation of learned men, and even went to see the theologian Hemmingsen at Roskilde. It is natural that he should wish to see the spot to which the eyes of all the learned men of Europe were directed, and on the 20th March he paid a visit to Tycho Brahe at Hveen, arriving at eight o'clock in the morning and remaining till three P.M.[1] King James was particularly pleased to see in the library at Uraniborg the portrait of his former tutor, George Buchanan, which had been presented to Tycho by Peter Young, who had once taught James to spell, and had afterwards several times been sent to Denmark on various missions.[2] The learned king and the astronomer had thus more than one interest in common, and it is easy to imagine the delight the former must have

---

[1] *Meteorological diary:* "Rex Schotiæ venit mane H. 8, abiit H. 3."
[2] He was now the king's almoner.

felt while conversing with his host. To show how gratified he was with his reception, he wrote at Uraniborg (whether it was in a "visitors' book" does not appear):

Est nobilis ira Leonis
Parcere subjectis et debellare superbos.
Jacobus Rex.[1]

Why he wrote these particular lines is not easy to understand, but perhaps he considered them emblematic of "kingcraft." [2] Maitland also tried his hand at Latin versemaking, expressing his admiration of the house of the Muses. The king is said to have discussed the Copernican system and other matters with Tycho, and was doubtless equally proud of his own exhibition of learning and pleased with the hospitable reception he had met with. He readily promised Tycho copyright in Scotland for his writings for thirty years, and sent him this three years later, expressing in the document the pleasure it had given him to converse with Tycho and learn with his own eyes and ears things which still delighted his mind. Two Latin epigrams accompanied the document and are printed with it at the beginning of Tycho's *Progymnasmata*.[3] King James is also said to have presented Tycho with two fine English mastiffs before his departure. Various members of his suite paid visits to Hveen during the time between the king's visit and his final departure from Denmark, which took place on the 21st April.[4]

---

[1] *Danske Magazin*, ii. p. 266 (Weistritz, ii. p. 200).

[2] These lines seem to have been a standing dish with King James, for according to Horace Marryat (*A Residence in Jutland, the Danish Isles and Copenhagen*, London, 1860, vol. i. p. 306) he also wrote them in a hymn-book belonging to Ramel, tutor to the young King Christian IV.

[3] Also in Gassendi, p. 106.

[4] *Epist. astron.*, p. 175: "Exinde quasi quotidianos hospites habuerim." *Meteor. diary*, April 21: "Rex Scotiæ circiter horam 7 P.M. Helsingora cum Regina sua et comitatu in regnum per mare discessit, Navali regis comitatu stipatus."

Two days before his departure King James had at Kron-borg assisted at the nuptials of the lady he had first wooed with Henry Julius, Duke of Brunswick-Wolfenbüttel. Of course the Duke had also to be taken over to Hveen to see the wonders there, which he and his suite did on the 4th May; but this visit does not appear to have been as pleasant to Tycho as that of King James. The Duke took a fancy to the little revolving statue of Mercury which stood on the roof of the central room of Stjerneborg, and thought it would be a pretty toy to take home with him. Tycho had to give him permission to take it away with him, when the Duke had promised to send him an exact copy of it, which promise he never took the trouble to fulfil, though Tycho sent him several reminders.[1] Gassendi tells a curious in-cident of this visit, which he had heard from Janszoon Blaev. At table the Duke remarked that it was getting late and he would have to take his leave, but Tycho, who perhaps was still annoyed at the loss of the statue, said in a joking way that it was his right to give the signal for breaking up. The Duke took offence at this and walked off towards the shore without taking leave, and when Tycho, who had first remained at table, after a little while followed him and offered him a stirrup-cup, the Duke turned away and continued his walk. Upon which Tycho let him go, and returned home without troubling himself more about his guest.[2] This may be only gossip, but Tycho was cer-tainly haughty and self-sufficient enough to have behaved in this manner even to the king's brother-in-law, and he probably made himself more than one powerful enemy by his overbearing manner.

A more welcome visitor arrived three months after this event in the person of the Landgrave's astronomer, Chris-

[1] Tycho tells this himself in *Epist.*, p. 256.
[2] Gassendi, p. 196.

topher Rothmann, who came to Hveen on the 1st August and stayed with Tycho till the 1st September. He seems to have been a somewhat peculiar character, and to have taken a rather unfair advantage of the great modesty and retiring disposition of his colleague, Bürgi, to push himself into the foreground, if not actually to try to shine with borrowed plumes.[1] In a letter of which he was the bearer, the Landgrave wrote that Rothmann had been in bad health for some time, and imagined that a little travelling and change of air would do him good. "But he has a head of his own, for which he every year buys a hat of his own, so we must leave him to himself; but we should be sorry if anything happened to him, for he is ingenious and a fine, learned fellow."[2] Tycho and he had now been in regular correspondence for about four years, and had in their letters entered very fully into the methods of observing used at Hveen and at Cassel, and the advantages or difficulties of the Copernican system. They had discussed the frequently observed "chasmata" or aurorae (which Tycho took to be sulphurous exhalations ignited in the air, and not clouds illuminated by the sun, as the latter was too far below the horizon in winter[3]); they had exchanged ideas about the celestial space, which Tycho did not believe to be filled with air, as this would produce a sound when the planets moved through it (which Rothmann denied); about the amount of refraction and the duration of twilight, for which Rothmann assumed a depression of the sun equal to $24°$, while Tycho found $16°$ to $17°$,[4] and on any other subject which their work might suggest. Rothmann had prepared for publica-

---

[1] R. Wolf, *Astr. Mittheilungen*, xxxii. p. 66.

[2] *Epist. astr.*, p. 182.

[3] *Epist.*, p. 162. In his *Handbuch der Mathematik, Astronomie, &c.*, ii. p. 337, R. Wolf states that Tycho and Rothmann had seen the zodiacal light; but I have not been able to find anything which looks like an observation of this.   [4] *Epist.*, pp. 139-140.

tion several treatises, none of which have ever been printed,
among which was one on trigonometry, which he had thought
of dedicating to King Frederick II., one on the Copernican
system, which he suppressed when he saw how badly the
Prutenic tables agreed with the observations,[1] and a treatise
on spherical and practical astronomy, which is still preserved
in MS. at Cassel.[2]

Tycho and his guest must have had plenty of subjects
for conversation, and the host no doubt did his best to
entertain the man with whom he had for years exchanged
ideas, though they had never yet met. It appears from
the diary that several foreign visitors came and went during
Rothmann's stay at Hveen, and these as well as the above-
mentioned trip to Scania to see Sophia Brahe and her
garden lent variety to the visit. There were even some
fine auroræ to be looked at and discussed. The many
interesting objects to be seen at Hveen, and the scenery, so
charming in summer-time, ought to have made Rothmann's
stay at Uraniborg very pleasant ; but unluckily Tycho seems
to have belaboured him with arguments against the Copernican system which must have become somewhat tiresome
in the end. In a lengthy note which Tycho has inserted
among his printed letters,[3] he states that during the weeks
he had Rothmann with him he pleaded his cause with this
generally very obstinate man so well, that Rothmann began
to waver, and finally declared himself convinced, and assured
Tycho that he had only held to his opinion for the sake of
argument ; he even added that he had not published, and
never would publish, anything in that direction. Doubtless
Rothmann was glad to end a dispute which could lead to
nothing, and both these skilful observers knew well that

---

[1] Ibid., pp. 89-90.

[2] R. Wolf, *Astr. Mitth.*, xlv., where there is a *résumé* of the contents.

[3] *Epist. astr.*, pp. 188-192.

there was a great deal of work to be done yet ere the true system of the world could be indisputably proved. In the note already alluded to Tycho sets forth the arguments which he had made use of. These refer only to the rotation of the earth, as he thought the two other motions would be untenable when that was disproved. He maintains that though the thin and subtle air might follow the rotatory motion, a heavy falling body would not, and if two projectiles were shot out with equal force, one towards the east and the other towards the west, he was sure they would go equally far, and thus prove that the earth was stationary. The enormous velocity with which the eighth sphere revolves in twenty-four hours he considers a proof of the wisdom and power of God; and as motion is more noble than rest, it is natural that the æthereal world should be in motion and the lower and coarser earth at rest, and this idea he dwells on at some length.

Rothmann left Tycho on the 1st September 1590, ostensibly to return to Cassel; but whatever the reason may have been, he went instead to his native place, Bernburg in Anhalt. After a long silence he wrote once more to Tycho in September 1594, but his letter was only a short one, complaining greatly of his health and inquiring why Tycho's book on the comet of 1577 had not yet been published. Tycho wrote him in January 1595 a very long answer, which is almost entirely taken up by a defence of his book on the comet against the attack made on it by John Craig, formerly Professor of Logic and Mathematics at Frankfurt on the Oder, and now Physician to the King of Scotland.[1] Craig had in 1588, through the intercession of a countryman (no doubt Liddel), obtained a copy of the book, and had written to Tycho trying to disprove the conclusion at which

[1] Chalmers' *Gen. Biograph. Dictionary*, London, 1815, vol. xx. p. 243. Tycho does not mention his name.

the latter had arrived, that the comet was far beyond the moon. Tycho had in reply taken the trouble to prepare a detailed "apology" for his book, and had sent it to Craig in 1589, and three years later the latter published a refutation of Tycho's book, in which he ("nec tam scotice quam scoptice") made a violent attack on all who would not follow Aristotle's doctrine about comets.[1] In this last letter to Rothmann, Tycho, in a needlessly prolix manner, defends his observations and results against this obscure writer, who, but for his attack on Tycho, would be quite unknown in the annals of science.[2]

Rothmann never returned to Cassel, and nothing further is known of him. He was still alive in 1599, when Tycho heard from him through a mutual acquaintance, and he must have died before 1608, when a theological pamphlet by him was published, which is designated as posthumous.[3] At Cassel, where the astronomical work was carried on by Bürgi, his continued absence created much surprise, and the Landgrave and Tycho, in the letters which they frequently exchanged in 1591, repeatedly expressed their wonder at his disappearance. These letters are not like the earlier ones, almost entirely devoted to astronomy, though Tycho did not omit to tell the Landgrave that the printing of the first volume of his work was approaching completion, and that he had shown Rothmann as much as was in type; it was partly want of paper which delayed the

---

[1] "Capnuraniæ restinctio seu cometarum in aethera sublimationis refutatio." Kepler began a refutation of Craig's book (*Opera*, i. p. 279), and Longomontanus also (Gassendi, p. 206), but neither were printed.

[2] *Epist.*, pp. 284–304. Tycho's first *Apologia* of 1589 was never published, though Lalande in his *Bibliographie* (and following him Delambre) mentions it as printed at Uraniborg in 1591. Tycho might have treated Craig's attack with the same contempt with which he met Christmann's attack on his solar theory, which he only answered by putting up in one of his rooms a picture of a dog barking at the moon, with the inscription "Nil moror nugas." Gassendi, p. 119.

[3] R. Wolf, *Geschichte der Astronomie*, p. 274.

finishing of the book.　The Landgrave offered to inquire at
Frankfurt whether there might be one or two papermakers
there who might be willing to go to Hveen, but Tycho
wrote back that he had already got one.[1]　In February
1591 the Landgrave wrote that he had heard of an animal
from Norway, taller than a stag, of which there were some
at Copenhagen and in the royal deer park at Frederiksborg,
and he would like a drawing of it.　Tycho answered that
he did not know of any such animal, but some time ago a
reindeer had been sent to Copenhagen from Norway, but
had died during the summer; he sent, however, a drawing
of it.　The Landgrave again wrote that he had also twice
got a number of reindeer from Sweden, and they seemed
to thrive well in winter, and could draw a sledge on the
ice, but they died as soon as warm weather came on.　In
his deer-park at Zapfenburg, the Landgrave had an elk
since the previous autumn, and it skipped about well, and
when he came driving in his little green carriage, the elk
would run alongside like a dog.　If Tycho could send him
one or two tame ones, he would be very pleased.　This
Tycho promised to do, and added that he had himself had an
elk on his estate in Scania, and had wanted to get it over
to Hveen.　In the first instance the animal had been sent
to Landskrona Castle, where Tycho's niece's husband kept
it for some days, until unluckily the elk one day walked up
the stairs into a room, where it drank so much strong beer,
that it lost its footing when going down the stairs again
and broke its leg, and died in consequence.　Tycho never
succeeded in getting an elk for the Landgrave, nor in
satisfying his curiosity concerning the gigantic animal called
Orix, about which the Landgrave had first inquired.[2]

[1] *Epist. astr.*, pp. 198, 202, 205, 215.　In 1592 Tycho was buying rags in
Seeland for his paper-mill (*Danske Magazin*, ii. 280).

[2] *Epist. astr.*, pp. 195, 200, 201, 205, 210, 212, 214.　In April 1592 the
Landgrave wrote that he had got four elks from Sweden, but three of them
had died, probably from eating too many rotten acorns (p. 269).

Among other matters, the Landgrave inquired about the state of affairs in Denmark, and Tycho gave him the required information in detail, telling him that the young king-elect was being carefully educated, and that there was every prospect of his walking in his father's footsteps; that among the four protectors, the one of greatest influence was the Chancellor Kaas, a man conspicuous not only by his illustrious descent, but also by his experience, judgment, and prudence, while he was also a very well-read man, particularly in historical and political matters. If anything of special importance occurred, it was referred to the annual assembly of nobles. The form of government was thus an aristocratic one (which was not a bad one), until the king-elect should attain his majority.[1] In return, the Landgrave sent Tycho some abstracts from newspapers about the state of France, and gave it as his opinion (in the curious mixture of German and Latin in which he always wrote), "dass es misserimus status totius Europæ ist."[2]

As Rothmann did not return to Cassel, and the Landgrave, therefore, did not see the drawings and descriptions of the instruments at Hveen which he had collected during his stay there, Tycho caused his German amanuensis to prepare a description of all the instruments, twenty-eight in number, which was sent to the Landgrave, and afterwards was inserted in the printed volume of letters, together with a Latin translation, which is somewhat longer, and furnished with woodcuts of the buildings and a map of the island, as well as with copies of the versified inscriptions on various portraits in Tycho's collection. Tycho was still anxious to have good mechanics in his service, and wrote to the Landgrave in February 1592 that his goldsmith, Hans Crol, was dead,[3] who had had charge of his instruments for many

---

[1] *Epist. astr.*, p. 199, translated in Weistritz, ii. p. 209.
[2] Ibid., p. 210.   [3] He died on the 30th November 1591 (Diary).

years, and had made several of them. He therefore inquired if Bürgi knew of an able man who might succeed
him. It so happened that Bürgi shortly afterwards had to
go to Prague to present to the Emperor a mechánical representation of the motions of the planets, and the Landgrave
promised that he should inquire about some goldsmith who
was accustomed to instruments and clocks. Whether Tycho
got such a man is not known.[1]

The Landgrave died at Cassel on the 25th August 1592,
at the age of sixty. His son and successor, Maurice, did
not share his father's taste for astronomy (though he continued to keep Bürgi in his service till 1603, when Bürgi
removed to Prague), but he was a man of literary tastes,
and at Tycho's request sent him a Latin poem for insertion
in one of his publications, though he modestly disclaimed
the poetical talent which had been attributed to him.[2]

Before Tycho lost his diligent correspondents at Cassel he
had opened a literary intercourse with Giovanni Antonio
Magini, from 1588 Professor of Mathematics at Bologna, a
man who by his extensive correspondence and his literary
activity gained a position of some importance in the history
of science.[3] We have already mentioned that Tycho's pupil,
Gellius Sascerides, during his stay at Padua, sent Magini a
copy of the volume on the comet of 1577, and in 1590
Magini wrote to Tycho thanking him for the welcome present, and expressing his approval of the new system of the
world. In this he could only have wished that the orbits
of Mars and the sun had not intersected each other, though

---

[1] *Epist.*, pp. 266 and 270.

[2] The verse is printed in *Epist.*, p. 281. Tycho wrote to Landgrave Maurice
in December 1596, when he at last sent him the two elks which the late Landgrave had wished for so much (ibid., p. 305).

[3] Carteggio inedito di Tichone Brahe, Giovanni Keplero e di altri celebri
astronomi e matematici dei secoli xvi. e xvii. con G. A. Magini, tratto dall'
Archivio Malvezzi de' Medici in Bologna, pubblicato ed illustrato da A. Favaro.
Bologna, 1886, 8vo.

this would be admissible if (as Gellius had told him) Tycho
had found Mars in opposition to be nearer than the sun.
He begged Tycho particularly to observe Mars, as he suspected its excentricity to be variable and periodical, so that
an equation to this effect should be introduced in the
theory.[1] In reply, Tycho remarked that he had found this
well-known difficulty not only in the theory of Mars, but in
a lesser degree also in the theories of the other planets, and
he wanted to observe the oppositions of Mars all round the
zodiac. He also gave a short account of the reasons why
he found it necessary to devise a new system. He would
have sent Magini a copy of his star-catalogue, but the distance was so great, and the difficulties of transit so considerable, that it might fall into wrong hands and somebody
might publish it as his own.[2] In conclusion, Tycho remarks
that he had read in Magini's *Tabulæ Secundi Mobilis* that
geographical latitudes, since the days of Ptolemy, had increased more than a degree, but he does not believe it, as
the latitude of Rome, according to Pliny, was $41° 54'$,
while Regiomontanus found $42° 7'$ and $42° 0'$; likewise
Pliny says that at 'Venice the gnomon and its shadow were
of equal length at the time of equinox, which gives the
latitude $45° 16'$, agreeing to the minute with Pitati's result ;
also Pliny gave $44° 10'$ for Ancona, which was more than
the modern value, $43° 20'$, instead of less, as had been
imagined.[3]

Gellius spent about two years at Padua, where he was
matriculated at the University in October 1589. In
1591 Magini had a sextant made from his description, and

---

[1] This letter is printed in Tycho's *Astr. inst. Mechanica*, fol. H., reprinted
in Carteggio, p. 392.

[2] In February 1591 Gellius wrote to Magini that Tycho had determined
the places of 500 stars to within a minute of arc. Carteggio, p. 202.

[3] Carteggio, p. 403. It was Domenico Maria Novara (whose lectures Copernicus had attended) who had suggested that the latitudes had increased.

they observed Mars with it together on some evenings in
June and July, as Tycho had called Magini's attention to
the singularly favourable opposition.[1] Magini wrote to
Tycho that he was going to get a large quadrant and a
radius (cross-staff) made with sights like Tycho's.   He
added that he could not procure for Tycho the copyright of
his books in the Venetian dominions, as they had not been
printed there.[2]   In the following year Magini dedicated to
Tycho a book on the extraction of square root, but the copy
sent to Denmark never reached its destination, and Tycho
did not see the book till five years later, when he came across
another copy and reprinted the dedication in his *Mechanica*.

In 1592, the year in which Tycho wrote the concluding
part of his book on the new star, an event occurred which
seemed to augur well for his future. The young king-
elect, then fifteen years of age, paid a visit to Hveen on
the 3rd July. We possess a detailed account of the way
in which this visit was brought about, through the Latin
exercise-book of the Prince, in which he was in the habit of
writing letters, sometimes fictitious, sometimes really addressed to those about him.[3] In the beginning of April
1592 he was obliged to leave Copenhagen, where the plague
had appeared, and on the way to Frederiksborg Castle he
received from Tycho's friend Kaas so lively a description of
Uraniborg, that he became very anxious to pay a visit to
Hveen.   He at once composed a Latin letter, probably
addressed to his governor, in which he requested leave to
proceed to Hveen, and as he met with a refusal, he appealed

---

[1] *Barretti Historia Cœlestis*, p. 498 ; Kepler, De *Stella Martis*, (*Opera omnia*,
iii. p. 211).

[2] Carteggio, p. 407. In a footnote Favaro quotes a statement by J. D.
Cassini, that he had seen a sextant which T. Brahe had got made for Magini
by a workman sent from Denmark, and that M. sold the sextant as soon as
the workman was gone. No doubt this "workman" was Gellius.

[3] Published in the Danish *Nyt Historisk Tidskrift*, vol. iii. ; compare T.
Lund, *Historiske Skitser efter utrykte Kilder*, Copenhagen, 1876, p. 322 *et seq.*

to the Chancellor, from whom he at once obtained the desired permission, as Kaas was only too glad to see the future king interested in Tycho and his work. Unluckily the plague had made its appearance on the island, and Tycho, who on the 29th of April had been informed of the intended visit, thought it his duty the next day to send one of his pupils over to Seeland to announce this.[1] The messenger found the Prince at the shore, just about to embark, and the youth could only console himself in his disappointment by composing a new exercise the next day, in which he expressed the hope that there might be nothing to hinder the visit in the coming month of June. The prince was evidently determined that nothing should prevent him from seeing Tycho's observatory as soon as possible, and it is much to be regretted that he did not, five years later, show an equally strong desire to keep the great astronomer in his own country, notwithstanding all the complaints brought against Tycho by his detractors. On the present occasion he got his own way. In June the Prince's governor was obliged to take his charge to the manor of Hörsholm (about fourteen miles north of Copenhagen, and only a mile and a half from the sea), and this temporary residence, which the spreading of the plague had rendered necessary, was most convenient for the visit to Hveen. When Tycho therefore arrived on the 30th of June to announce that the plague had vanished from Hveen, the Prince's governor could not find any excuse for preventing the future king from visiting the astronomer, and on the 3rd July the Prince started, attended by two of the protectors, Admiral Munk and Jörgen Rosenkrands, and his governor, Hak Ulfstand. The weather was most favourable, and the trip was no doubt thoroughly enjoyed by the Prince, whose excellent education enabled him to view with

---

[1] Cort Axelsen from Bergen ; see the Meteorol. Diary.

intelligent interest the many strange objects which Tycho had to show him. In particular, he admired a small brass globe, which, by an internal mechanism, showed the motions of the sun and moon. Tycho immediately begged him to accept it, and in return the Prince took off a massive gold chain in which his own portrait was suspended, and hung it round his host's neck.[1] The conversation between the astronomer and his youthful guest turned to fortification, navigation, shipbuilding, and other branches of applied science, in which the Prince had been instructed, and it is stated that Tycho on this occasion received a promise of an annual grant of 400 daler (£91) for instructing young men in the theory of navigation and astronomy, and an allowance of 120 daler annually for the keep of each pupil.[2] The Prince was greatly pleased with his visit, and seems to have regretted that it was but a short one, as he wrote in his Latin exercise the next day that he returned to Hörsholm "long before supper."

Historical events, whether trifling or important, are often by posterity, without any reason, connected with others or supposed to have caused them. Tradition afterwards made out that Christopher Valkendorf was in the Prince's suite on this occasion, and it has been related in detail how he and Tycho became enemies because Valkendorf kicked one of the dogs which King James had presented to Tycho, while the latter in turn abused the offender. But this story, which, according to other writers, refers to a later date, rests on a very slender foundation indeed, and at any rate

---

[1] *Astron. inst. Mechanica*, fol. B. (where for 1590 should be read 1592). Tycho had already in 1589 procured a couple of globes for the Prince from the Dutch artist Jacob Floressen (Florentius), who sent his son to Hveen to obtain correct star-places for his globes, which Tycho declined to give in writing, while he allowed him to examine the great globe in the library (*Progym.*, p. 274). In 1600 Tycho sent a star-globe to the Elector of Saxony.

[2] Lund, *Historiske Skitser*, p. 353, quoting Slange's *Christian den Fjerde's Historie* (1749). I have not seen this mentioned elsewhere.

the incident cannot have occurred on this occasion, as it is quite certain that Valkendorf did not attend the Prince on his visit to Hveen. During the summer of 1592 Tycho had a number of other visitors, among them Prince Vilhelm of Courland, a brother to the Duke.[1]

But though Tycho's position still seemed an excellent one, and he continued in the undisturbed possession of all his sources of income, he seems about this time to have become dissatisfied and annoyed by various circumstances. In the letter to the Landgrave in 1591 in which he described the state of Denmark,[2] he remarked that there were certain unpleasant obstacles which hindered him from carrying out successfully all that he had planned for the restoration of astronomy, but he hoped to get rid of these and other obstacles in some way or other, and any soil was a country to the brave, and the heavens were everywhere overhead. These last words ("omne solum forti patria est cœlum undique supra est") are very similar to some which he had used six years before in the poetical letter to Kaas, and they seem to indicate that already at this time he was not unfamiliar with the idea of seeking a home outside Denmark, if circumstances should make the stay at Hveen unpleasant to him. Among his causes of annoyance was a quarrel with one of the tenants on the estate of the Roskilde prebend, which does not place Tycho in a very favourable light, and which may, perhaps, account for the coldness shown by the governor of the Prince when the visit to Hveen was proposed. It appears that Tycho and the tenant, Rasmus Pedersen, had had some difference in the

---

[1] Tycho calls him in the diary Vilihelmus, Dux Curlandiæ et Semigalliæ (*i.e.*, Semgallen or Samogitia, the south-east part of Courland), but he was not a Duke, and never became one. He probably visited Denmark to endeavour to enlist the sympathy of that country for Courland, which had a difficult position between Russia and Sweden.

[2] *Epist.*, p. 198, last line. The letter is dated *Cal. Augusti*, which should be *Cal. Aprilis*, as may be seen from the Landgrave's answer.

year 1590, as the latter got the mayor of Roskilde and
another man to go over to Hveen on the 15th July to try
and settle the matter.[1] They cannot, however, have succeeded, and Tycho wanted a decree of eviction against the
tenant, but the court which tried the matter, and which was
composed of four noblemen, did not grant the decree. Tycho
now appealed to the king, who summoned the four nobles
and the litigants to appear before the High Court of Justice
in July 1591. From the judgment of the four Commissioners it appeared that the tenant had been disobedient
and had refused to come to Tycho when ordered, but that
Tycho, notwithstanding the lease for life which the tenant
held of his farm, had let other people plough and sow the
land, and in the previous October (six months before he
tried to have him legally evicted) had taken the farm from
him. Tycho had furthermore taken the law into his own
hands by having Rasmus put in irons at his own table,
from whence he was carried off to Hveen, where he was
detained for six weeks or more. And as Rasmus had
represented that he had feared the severity of Tycho, and
did not go to him when ordered because Tycho would not
give him a safeguard, the Commissioners had thought that
six weeks' imprisonment was a sufficient punishment for
this act of disobedience, and that Rasmus should not be
evicted from the farm, of which he had only purchased the
lease four years before, and on which he had built a house.
Although Tycho had forbidden Rasmus to work his farm,
this was not according to law, as long as the tenant had
not legally forfeited his lease. As to the house on the farm,
which Tycho complained had been sold by the tenant, it appeared that it was still standing in the garden, and formed

[1] Meteor. Diary. In *Breve og Aktstykker*, p. 3, is a letter from Tycho dated
June 1591 to the Rector of the University asking for the testimony of Dr.
Krag and Magister Kolding, who had been present at Hveen when the two
men came over from Roskilde.

part of the farm. Further Tycho alleged that a certain house had been sold and not entered on the accounts; but as Rasmus Pedersen denied all knowledge of any such matter, and Tycho had not submitted any evidence to prove his assertion, the tenant had been acquitted of this charge. He had been obliged to go over to Hveen to work for his landlord with four horses and two carts, and two of the horses had died; but the Commissioners had found that Rasmus was in no way bound to work for his landlord. Some boxes belonging to him had been sealed and carried off while he had been locked up, and these Tycho had been ordered to hand over unopened to the tenant. When asked in the Court of Appeal what objection he had to the judgment of the Commissioners, Tycho stated that he had hardly any objection to make, except that the case had been greatly delayed, and that the Commissioners had not tried the entire case, but had referred some parts of it to the local court,[1] others to the provincial court,[2] notwithstanding that the royal command had ordered them to judge in all matters between Tycho and his tenant; they had also passed over some questions, and not tried them at all. To this the Commissioners answered that the delay had merely been in having their seals affixed to the judgment; that they had been justified in referring some matters to the local court, as the case about a man who was drowned in a well was clearly a case for a jury, while other things were under the jurisdiction of the provincial court; and as to the matters which they were said to have passed over altogether, they were not aware of any such matters. The High Court of Justice concurred with the Commissioners in every respect, and ordered that their judgment should stand.[3]

---

[1] Herredsthing, *i.e.*, court of the hundred.

[2] Landsthing.

[3] The whole judgment is printed in *Danske Magazin*, ii. pp. 274-278 (Weistritz, ii. pp. 216-224).

It might have been expected that the humiliation of having had the judgment upheld, against which he so need-lessly had appealed, would have been enough for Tycho, and that he might have left Rasmus Pedersen alone in future, but this he evidently could not persuade himself to do, although we know very little about the further progress of the case. In February 1592 Tycho had to attend the pro-vincial court at Ringsted, in Seeland, as a Latin epigram has been preserved which he wrote on the way home, and in which he complains of unfair treatment by the judge.[1] The case tried on that occasion was probably one of those referred to the provincial court by the Commissioners. From the end of the same year a draught of a royal letter has been found (dated 17th November 1592) which shows that Tycho was still persecuting the tenant. The letter states that whereas Rasmus Pedersen has complained that he was still kept out of his farm, and that his brother and his servant had been imprisoned, and were still detained by Tycho, while he was most anxious to be left in peace, since he had built a house on the farm, and would be utterly ruined if this quarrel did not cease, the king desired that Tycho should remember the misery of the man, and act in a Christian, reasonable, and lawful manner towards him, so that the Crown would not be obliged to interfere and pro-tect him, particularly as he was a tenant of an estate which was merely granted to Tycho during pleasure. It was therefore the royal command that this case be finally settled and arranged by the judge of the provincial court of Seeland and some other gentlemen, who were to judge in all matters which had not already been judicially decided. Tycho was therefore desired to nominate some impartial gentlemen who might be ordered to act on this commission.[2]

---

[1] Also printed in *Danske Magazin*, ii. p. 279 (Weistritz, ii. p. 226).
[2] *Danske Magazin*, 3rd Series, vol. iv. p. 263.

Nothing further is known about Rasmus Pedersen and his disagreeable landlord, who seems to have acted in a remarkably high-handed manner in the whole affair. He certainly did not by this conduct improve his credit with the young king, who throughout his life wished to act justly by everybody, irrespective of rank and social position.

Another source of trouble to Tycho, for which he also had himself alone to blame, arose soon after out of his prebend at Roskilde. We have seen that the possession of this prebend brought with it the obligation to keep in good repair not only the residence attached to it, but also the "Chapel of the Holy Three Kings" in the Roskilde Cathedral. Though perfectly willing to enjoy the income of the prebend, Tycho seems altogether to have neglected to look after the state of the chapel, and in August 1591 the Government found it necessary to draw the attention of the Chapter to the want of repair of the chapel. Having been informed that Tycho was bound to see to this, a letter was written on the 30th of August to Tycho in the king's name about this matter. Tycho does not appear to have taken any notice of this reminder, and the king had to write to him again in August 1593, that he had himself been in the cathedral, and found the roof, the woodwork, and the vault in so dilapidated a state, that it was to be feared that it would all come down unless something was done before the winter. Tycho was therefore desired to repair the chapel at once, and if he did not do so, the king would have it done by a builder at Tycho's expense. But Tycho did nothing, and in consequence received in September 1594 a third reminder, in which he was informed that if the chapel was not repaired at latest before Christmas, the prebend would be conferred on somebody else. Now at last Tycho thought he had better do something,

and applied for leave to take down the arched roof of the
chapel and put a flat ceiling in its place, which would sim-
plify the repairs, and this he received permission to do in
November 1594; but he did not carry out his proposal, and
he must have managed to repair the chapel in some other
manner.[1]

Tycho's conduct in these various transactions could not
but undermine his position in Denmark, and there was
doubtless more than one of his fellow-nobles who took the
opportunity of fanning the flame of discontent with the
self-willed and highly-paid astronomer which gradually
sprang up among the rulers in Denmark. Among these,
Tycho had hitherto had a very powerful friend in the
Chancellor, Niels Kaas, but he died in June 1594, and
after his death Tycho must have felt himself less secure in
the enjoyment of his several endowments. Possibly Tycho
may also gradually have become tired of the continued
residence on the lonely little island, from which his very
frequent trips to Scania and to Copenhagen[2] cannot always
have been pleasant, particularly in winter, and he may by
degrees have become desirous of making a change. He
had not been outside Denmark since 1575, and must have
longed for the easy intercourse with learned men which he
had once hoped to find at Basle, and for which the occasional visits of learned foreigners to Hveen was not a
sufficient compensation. Reports must also have reached
him of the great love of astronomy and alchemy of the
Emperor Rudolph II., and the thought may easily have
arisen in his mind that he might find the same liberality
in the German monarch as he had formerly found in King
Frederick. With the Emperor's physician, Hagecius, Tycho

---

[1] *Danske Magazin*, ii. 281–283 ; Werlauff, *De hellige tre Kongers Kapel i
Roeskilde Domkirke*, Copenhagen, 1849, p. 18 *et seq.*

[2] See the Diary, *passim*.

had continued to correspond, and he had even a more in-fluential ally in the Imperial Vice-Chancellor, Jacob Curtz of Senftenau, with whom he had also exchanged letters, and who in 1590 had sent him the privilege for his writings which Hagecius had some years before promised to get for him, together with a description of a method of subdividing arcs designed by Clavius, which is based on the same principle as that afterwards, in a more practical form, proposed by Vernier.[1]   According to Gassendi, Curtz went to Denmark not long before his death (which took place in 1594), on the pretence of coming on the Emperor's business, and offered Tycho to intercede with the Emperor to procure an invitation to Bohemia in case he should wish to leave Denmark ; he is even said to have offered Tycho his own house at Prague, and to have left a plan of it with Tycho in case he might wish to have any alterations made in it.[2]   After Curtz's death, Hagecius is said to have assured Tycho that the new Vice-Chancellor, Rudolph Corraduc, would not fail to befriend him.

It was perhaps with a view to the probability that he might soon wish to leave Denmark that Tycho disposed of his portion of the family property of Knudstrup, which, since the death of his father, he had possessed jointly with his brother Steen, and which his sons, as born of a " bond-woman," could not have inherited.   The date of this sale is not known, but it must have been previous to the 10th August 1594, on which day he signed a document by which he reserved to himself the right to continue to call himself " of Knudstrup," without any injury to the rights of his brother or his brother's heirs.[3]

---

[1] *Astr. inst. Mechanica*, fol. G. 6 ; Delambre, *Astr. moderne*, i. p. 253.

[2] Gassendi, p. 131. I have not succeeded in finding Gassendi's authority for this. Curtius is not mentioned in the Meteorological Diary, so he can hardly have been at Hveen.

[3] *Danske Magazin*, 4th Series, ii. p. 325. It is characteristic of the careless

Among the causes which finally induced Tycho to leave
Denmark, the quarrel with his former pupil, Gellius Sascerides, is supposed to have been an important one. We have
mentioned that Gellius spent about two years in Italy. On
the return journey he was at Basle for some time (1592–93),
where he became a Doctor of Medicine, and he reached
Denmark some time in 1593. He soon after began to
visit Uraniborg, and eventually became engaged to Tycho's
eldest daughter, Magdalene, at that time about nineteen
years of age. Gellius would hardly have thought of aspir-
ing to the hand of any other nobleman's daughter, but the
peculiar position of Tycho's children, by many people not
considered to be legitimate, may have given him courage.
Tycho does not appear to have objected to the proposed
marriage, and may have thought that the undoubted learning
of Gellius made up for any supposed deficiency in lineage.[1]
But the pleasant relations between Tycho and Gellius did
not last long, probably because the latter during his long
absence abroad had become unaccustomed to the imperious
manner of Tycho, and the quarrel commenced in earnest in
the following year, when the wedding began to be talked
about. It appears that Tycho did not care to have festivities and expense in connexion with the ceremony, and
further demanded that Gellius, after the wedding, should
remain at Hveen for a while to assist in the work; and not
content with this, he made certain stipulations as to the
manner in which Gellius was to provide his wife with
clothes, &c. On the other hand, Gellius is said to have

---

manner of spelling names which prevailed in those days, that Tycho Brahe's
name in the document is spelt Tygge Brahe, in the signature Thyghe Brahe.
In Latin or German he always wrote Tycho, in Danish generally Tyge.

[1] The following account is taken from Dr. Rördam's paper in the *Danske
Magazin*, 4th Series, ii. p. 16 *et seq.*, which is founded on documents in the
archives in the Copenhagen University which were not accessible to Langebek
(*D. M.*, ii. p. 285 *et seq.*).

expected a dowry with his bride, while Tycho refused this, adding that if he would not take the girl for her own sake, he should not have her at all. In the autumn of 1594 the end of all this disagreement was that Gellius broke off the match. Still he seems about that time to have been frequently at Hveen, and Tycho wrote to his sister that all might yet be well if Gellius did not become vacillating again. But during an interview between Gellius and Tycho they quarrelled again about the matter, in consequence of which Tycho sent two friends to Gellius to demand a clear answer to the question whether he would accept the proposed terms or not. At first Gellius would not give a decisive answer, but during the next few days (in December 1594) he told one of the intercessors, Professor Krag, more than once, that he did not want Tycho's daughter; and on learning this, Tycho and his daughter sent Gellius a formal notice that the engagement was at an end.[1] In a letter to a friend (which was afterwards produced), Magdalene Brahe expressed herself thankful that all was over.

Gellius was greatly blamed by many people, but he tried to shift the blame on others, particularly on Krag, saying that he himself was joking or drunk when he spoke to the latter, and that his words were not intended to be carried further. Tycho, therefore, in the beginning of January 1595, got Krag to give him a written account of all that had happened between him and Gellius, as he particularly wished his sister Sophia to have an unbiassed explanation.

[1] Krag was perhaps hardly a safe person to employ in a delicate mission. He had recently been appointed royal historiographer, and had the following year the meanness to accept all the materials laboriously gathered by Tycho's friend Vedel, whom the Government forced to deliver up all his collections, because he had delayed the writing of his Danish history so long. Krag told Tycho in a pointed manner that he was glad that it had only fallen to his lot to describe the youth of Hveen and not its decay, by which he meant that his history was to stop at the death of King Frederick II. Wegener's *Vedel*, p. 200.

At the same time (11th January) Tycho wrote to the Rector
of the University, and requested a statement from him and
the professors to prevent Gellius from throwing all the blame
on him and his daughter.   This led to an agreement being
drawn up five days later between the parties, which was
signed and sealed by the rector and four professors,[1] and
Tycho now seemed content.   But the affair had of course
been talked about, and Gellius continued his attempts to
place himself in the best possible light.   Tycho in the end
got tired of this, and in February 1596 he again requested
the University to investigate the whole affair, and let all
documents laid before the academic senate by himself or
his adversary be registered by the notary.[2]   About the
same time he drew up a list of all the accusations of Gellius,[3] and invited him to prove them before the professors.
Gellius made several attempts to prevent further proceedings, but failed to do so, and formal conferences before the
academic senate were commenced on the 25th February.
They were continued off and on till the month of July,
when everybody was probably so thoroughly sick of the
wearisome twaddle, which could not lead to anything, that
the matter was allowed to drop.   The details of the proceedings[4] give scarcely any information about the origin of
the quarrel, but it can hardly be doubted that Gellius would
not have dared to trifle with Tycho and his daughter if he
had not seen how unpopular his former master had become;

[1] Alluded to in *Danske Magazin*, ii. p. 292 (Weistritz, ii. p. 250), as a
"contract;" it does not seem to be in existence now.

[2] On the 5th February 1596, Tycho had procured a royal order to the
Chapter of Lund to judge the matter, as Gellius had obtained a medical
appointment in Scania, and therefore in matrimonial matters was under the
jurisdiction of the said Chapter; but it is not known what action the Chapter
took.

[3] Printed in *Danske Magazin*, ii. p. 286 (Weistritz, ii. p. 239).   The charges
of Gellius relate to the demands that he should stay at Hveen, keep his wife
in fine clothes, &c.

[4] *Danske Magazin*, ii. pp. 291–307 (Weistritz, ii. pp. 248–281).

and on the other hand, it is probable that Tycho's domineering manner first brought about the difference between him and Gellius which led to this unpleasant affair.[1]

During the years when all these annoyances happened to the astronomer, his scientific work continued to be carried on, and the years 1594 and 1595 are considerably richer in observations than those immediately preceding. Most of his observations for determining accurate places of fixed stars were made before the end of 1592, and the results were embodied in a catalogue of 777 stars for the end of the year 1600, which is printed in his *Progymnasmata*. After 1590 it was especially the planets which were observed (though they had always been regularly attended to), and in 1593 extensive series of observations of Mars, Jupiter, and Saturn were made. In 1595 observations of fixed stars were resumed in order to bring the number of stars in the catalogue up to 1000, and even in the first two months of 1597, immediately before leaving Hveen, some observations were taken in hot haste to make up the thousand (*pro complendo millenario*), mostly only depending on a single measure of the declination and the distance from one or two known stars, and sometimes with a rough diagram to identify the star. It must therefore be taken *cum grano salis* when Tycho already in January 1595 wrote to Rothmann that he had now finished " about a thousand stars," and when he writes in his *Mechanica* that the great globe was quite finished in 1595, exhibiting a thousand stars.[2] It has been suggested that it was this

---

[1] Gellius married in 1599, became Professor of Medicine in the Copenhagen University in 1603, and died 1612 (Rördam, *l. c.*, p. 31). Magdalene Brahe went with her father to Prague, and apparently never married.

[2] Longomontanus says in his *Astronomia Danica*, p. 201, that the work on the star-catalogue was commenced in 1590, and went on for five years (" Ego . . . huic de fixis cœlitus restituendis negocio et exsecutioni non solum interfui, sed etiam præfui ").

completion of Tycho's star-catalogue which he wished to
commemorate by the striking of a medal (or rather two,
slightly different) bearing the year 1595. This is quite
possible, and he may have wished, in the midst of all his
worry and vexation, to have a memorial of the work carried
on for nearly twenty years at Hveen.[1] A more lasting
memorial of his activity and of the respect with which he was
treated by any one able to value his work was the collection of letters exchanged between him, the late Landgrave
of Hesse, and Rothmann. Rantzov had long ago suggested
the publication of this series of scientific essays, and copies
of some of them had been sent to Hagecius and Peucer,
who had expressed a similar wish. They were printed in
Tycho's own office, and form a quarto volume of 310 pp.,
and 38 pp. of laudatory poems, dedication, and preface.[2]
The title shows that Tycho intended afterwards to publish
letters to and from other astronomers, an intention which
he did not live to carry out, so that only some of these
letters have of late years been published. None of Tycho's

---

[1] One of these medals (in silver) is in the royal numismatic collection in
Copenhagen, described and figured in *Danske Magazin*, ii. p. 161, and Weistritz, ii. p. 14. It is about 1¼ inch in diameter, and shows on one side Tycho
Brahe's portrait, and round it "Effigies Tychonis Brahe O. F. æt. 49"
(O. F. means Ottonis Filii), on the other his coat of arms, and round this his
motto : "Esse potius quam haberi. 1595." The other medal is in a collection at Prague (Friis, *T. Brahe*, p. 363), and is a quarter inch more in diameter ; the only other difference is the inscription round the arms : "Arma
genus fundi pereunt, durabile virtus," (and inside this) "Et doctrina decvs
nobilitatis habent."

[2] "Tychonis Brahe Dani, Epistolarum astronomicarum Libri. Quorum pri-
mus hic illustriss. et laudatiss. Principis Gulielmi Hassiæ Landtgrauij ac ipsius
Mathematici Literas, vnaque Responsa ad singulas complectitur. Vraniburgi.
Cum Cæsaris & Regum quorundam priuilegiis. Anno MDXCVI." Colophon
is the vignette with "Svspiciendo despicio," and underneath : "Vranibvrgi Ex
officinâ Typographicâ Authoris. Anno Domini MDXCVI." The portrait of
Tycho which appears facing the title-page is from 1586, and is engraved by
Geyn of Amsterdam. There is a copy of it in Gassendi's book. The printing
must have commenced before 1590, as Gellius had given Magini a few
printed leaves of the book (Carteggio, p. 233).

other letters can, however, compare in importance with the lengthy essays exchanged between Hveen and Cassel, which give a most instructive picture of the revolution in practical astronomy effected by Tycho. The dedication to Landgrave Maurice alludes to the origin of Tycho's acquaintance with Landgrave Wilhelm, the renewal of it through Rantzov in 1585, praises the Landgrave for not having studied astronomy in books but in the heavens, and quotes from a funeral oration in which the hope had been expressed that the correspondence of the deceased with Tycho Brahe might be published, as it would show the world the merits of the Landgrave's scientific work. In the preface Tycho refers to the length of time necessary to form a complete series of observations by which the restoration of astronomy might be accomplished. Though the solar orbit may be sufficiently investigated in four years, the intricate lunar course requires the study of many years, while it takes twelve years to follow the oppositions of Mars and Jupiter round the zodiac, and even thirty years to see Saturn move round the heavens. He had commenced his own observations at the age of sixteen, though the results of the first ten years' work were less accurate than the later ones. Ptolemy and Copernicus had not observed for such a length of time, and consequently the numerical values of astronomical constants had not been well determined by them. As already remarked, most of the letters are in Latin, only those of the Landgrave and some of Tycho's replies to him being in German, with a liberal sprinkling of Latin words and sentences, which almost render unnecessary the Latin translation which always follows.[1] As also mentioned

[1] Here is a specimen from the Landgrave's first letter (to Rantzov) : " Darneben wollen wir euch auch nicht verhalten, das vff angeben *Paul Vvitichij*, wir vnsere *Instrumenta Mathematica* dermassen verbessert, dass, da wir zuuor kaum 2 Min. scharff, wir jetzo $\frac{1}{2}$ ja $\frac{1}{4}$ einer min. *obseruiren* können. Haben **vns** derhalben vff die Art *Quadrantem Horizontalem* desgleichen ein

above,[1] there is towards the end of the volume a description of the instruments and buildings at Hveen, with woodcuts of the latter. Of the instruments, seven were already figured in Tycho's other books, and it appears that to a few copies of the *Epistolæ* he added an appendix of eleven leaves, with figures of some of the instruments, and on the last leaf a short note promising that a complete account of all of them should soon appear. This appendix was doubtless only printed in a very few copies, as it was soon to be rendered superfluous by the publication of Tycho's special book on his instruments.[2]

While the printing of Tycho's correspondence was being completed important events took place in Denmark. Tycho's last remaining influential friend, Jörgen Rosenkrands, died in April 1596, and the young king, who was now in his twentieth year, was soon afterwards declared of age, and was crowned on the 29th August at Copenhagen.[3] He had appointed Christian Friis of Borreby, Chancellor, and Christopher Valkendorf to the office of High Treasurer, which had been vacant since the death of Tycho's connexion, Peder Oxe, in 1575; but King Christian had both the will and the ability to govern himself, and soon made his authority felt and respected. He was personally of an

---

Sextantem, ad obseruandas distantias Stellarum inter se, lassen zurichten, jedes von gutem Messing vnd *bicubital.* Halten auch drei Gesellen, *Astronomiæ & Obseruationum peritos ad iustificanda loca Stellarum Fixarum."* Letters from learned men, if not written altogether in Latin, were generally written in this mongrel tongue.

[1] See above, p. 211.

[2] This appendix or pamphlet ("Icones instrumentorum qvorvndam Astronomiæ instaurandæ gratia a Tychone Brahe Dano diligentia, impendioqve inestimabili elaboratorvm") is mentioned by Friis, *Tyge Brahe,* pp. 363-364. In 1889 I tried in vain to get a look at it at the Royal Library at Copenhagen, but it was not there. These pictures are alluded to in Tycho's letter to the Chancellor of the 31st December 1596.

[3] Tycho attended the coronation (Meteor. Diary), and a few days after he was visited by Johann Müller, "Mathematicus administratoris Brandenburgensis." See also Gassendi, p. 153.

economical disposition, and at once began to introduce
reductions in various branches of the administration.
Among others who were made to feel the change of government was Tycho Brahe, who lost the Norwegian fief "immediately after the coronation," as he tells us himself.[1] As
this was a serious loss to Tycho, he made an effort to recover the fief, or at least to be allowed to keep it till the 1st
May, the usual time for giving up possession of beneficiary
grants.   On the 31st December 1596 he therefore wrote
a lengthy letter in Latin to the new Chancellor, Friis,[2]
pointing out how deeply interested King Frederick had been
in his work, and how death alone had prevented him from
carrying out his intention of permanently endowing the
observatory at Hveen; how much he had done for the
advancement of astronomy, as might be seen from the
correspondence just published, and of which he would have
sent King Christian a copy if the king had not been
absent in Jutland.   For the present, he only asked to have
the Norwegian estate restored, or at least to let him keep
it till May, as his steward would then have paid him the
rents.   With this letter Tycho sent a copy of his *Epistolæ*
and a copy of the declaration of the Privy Council of 1589,
promising to advise the king to endow Tycho's observatory
in a permanent manner.   In reply, the Chancellor, who was
with the king in Jutland, on the 20th January 1597 wrote
in a short, business-like manner, that he had laid Tycho's
petition before the king, but that his Majesty did not see
his way to pay anything from the Treasury towards the
maintenance of the instruments, and that it was impossible to postpone the surrender of the Norwegian fief,
as the main fief of Bergen (to which that of Nordfjord
belonged) could not spare the income from it.   But if the

---

[1] In his Latin account of how he left Denmark. Barrettus. *Hist. Coel.*, p. 801
[2] Tycho had first applied to Valkendorf, but in vain (*l. c.*).

Chancellor could oblige him in any other way, he should be happy.[1]

It is very difficult to form an idea of the motives which dictated this changed behaviour of the king and the Government to the great astronomer, but there can hardly be any doubt that Tycho had made himself more than one enemy among the nobles, and these found in his own conduct faults enough which they could point out to the king, hinting that this self-willed man, who would hardly condescend to obey the royal authority, had been petted long enough, and that there was no necessity for continuing to spend great sums of money on his instruments, the more so as it could not be a secret that he was by no means devoid of pecuniary resources himself. When they had reminded the king of Tycho's persecution of the tenant near Roeskilde, of his having not only neglected to attend to his duty of keeping the chapel of his prebend in repair, but also turned a deaf ear to repeated injunctions about this matter, it was probably not difficult for his enemies to influence the young king. Who these enemies were is not known with absolute certainty. Tradition mentions among them the king's physician, Peder Sörensen, with whom Tycho had, about twenty years before, exchanged friendly letters, but who is said to have become jealous of Tycho's dabbling in medicine, and particularly of his having distributed remedies against various diseases without payment. But Tycho himself considered Christopher Valkendorf and Christian Friis as having been the principal instigators in the events which led to his expatriation; at least he did so some time afterwards, when he mentioned them as such in several letters.[2] As early as about fifty years after these

---

[1] The two letters are printed in Hofman's *Portraits historiques des hommes illustres de Danemarc* (1746), vi. pp. 14-16, and in *Danske Magazin*, ii. pp. 310-314 (Weistritz, ii. pp. 289-297).

[2] In a letter to Professor Grynæus at Basle, dated 8th October 1597, Tycho

events it was currently believed that the ill-feeling between
Valkendorf and Tycho arose from a quarrel about a dog,[1]
but the story is told in different ways. We have already
alluded to the version of the story according to which the
quarrel occurred on the occasion of the visit of the young
king to Hveen, and it was pointed out that Valkendorf was
not at Hveen at that time. The well-known French writer,
Pierre Daniel Huet, who was at Copenhagen in 1652, tells
the story differently.[2] According to him, an English envoy
had a dog which Tycho wanted for a watch-dog at Uraniborg; but as Valkendorf also coveted it, and the envoy
wished to keep friendly with both of them, he promised to
send them each a dog when he went home. But when the
dogs came, one was much finer and larger than the other,
and the king, who was asked to arbitrate between them,
gave the large one to Valkendorf, which roused Tycho's ire
and caused the enmity between them. But all this probably

says that "duo Dynastæ," either from ignorance of science, or from hatred
and malice towards him, or from both causes, got his endowments taken from
him. In a letter to Magini, dated 3rd January 1600, Tycho speaks in stronger
terms. He wanted Magini to get some Italian writer to compose a panegyric
on him, and had sent Magini some materials for this, but he mentions that he
does not want his country, nor the king, nor the nobility at large to be abused,
as most of these had nothing to do with his exile. "Perstringendi vero solum-
modo pro merito Cancellarius modernus et Aulæ Magister; qui cum patriæ
honorem ex officio promovere debuissent, eum potius ob avaritiam et sorditiem
pari invidia, malignitate et odio coniunctum (cum ipsi liberalibus scieintis vel
nihil vel admodum parum tincti essent) impediverunt et exterminarunt.
Nomina eorum invenies in iis quæ de caussis discessus mei Latine exarata
nunc mitto." He adds that their names are as well worth preserving as that
of Herostratos who burned the temple of Diana at Ephesus! (F. Burckhardt,
Aus Tycho Brahes Briefwechsel, Basel, 1887, pp. 10 and 14. Magini printed
the letter in his *Tabulæ primi Mobilis*, but left out the above passage, which,
therefore, does not occur in the Carteggio, p. 418.)

[1] *Danske Magazin*, ii. p. 322, quotes Th. Bartholin, *De medicina Danorum*
(1666). Gassendi (p. 140) also knows the story.

[2] *Pet. Dan. Huetii, Episcopi Abrincensis, Commentarius de rebus ad eum
pertinentibus.* Amstelodami, 1718, 12mo, p. 90. Huet was on his way to
Queen Christina of Sweden when he visited Copenhagen and Hveen. As he
mentions the Danish savant Ole Worm, he may have had the story from
him.

rests on no other foundation than rumour only; and though Valkendorf as Treasurer may have been instrumental in depriving Tycho of some of his income, he can hardly have been his declared enemy, and a letter which Tycho wrote to him in May 1598 does not look as if there was any hostility or even coldness between them. But it is a necessity for human nature to have a scapegoat, and, with a rare unanimity, astronomical historians have told their readers that Valkendorf was the sole cause of Tycho's exile, and several of them indulge in very pretty expressions of indignation against that monster.[1] Of course they are not aware that Valkendorf's name is in very good repute in Denmark, where he distinguished himself not only as a statesman, but also as a promoter of learning by founding a college for poor students in connexion with the University.[2] It is far more likely that Friis, the new Chancellor, was an active enemy of Tycho's, and we shall see that he reaped a pecuniary advantage from the disgrace of Tycho.[3] As to the young king, there is every excuse for him, for it is really not strange that he should have thought it desirable to diminish the annual burden to the Treasury, which was without precedent, and which undoubtedly might be reduced without seriously interfering with Tycho's scientific work.

The Norwegian estate was not the only endowment which Tycho lost before leaving Hveen. On the 18th March 1597, Valkendorf received the king's order that Tycho's annual pension of 500 daler from the Treasury

---

[1] For instance: "Son nom doit être cité pour être reservé à l'infamie et devoué à l'exécration des savans de tous les âges." Lalande's *Astr.*, i. p. 196 (2nd edit.).

[2] "Valkendorf's Collegium" (founded 1589) is still in existence. Valkendorf died in January 1601.

[3] If Friis was really so great an enemy of Tycho's, it is very curious that he should a few years after act as Mæcenas to Longomontanus, Tycho's favourite pupil. See the preface to his *Astronomia Danica*.

should cease.[1] If Tycho had not already commenced his
preparations for leaving Hveen, he did so at once after
this last blow. Though certainly not a poor man (for he
was able six months later to invest 10,000 daler, or about
£2200, a very considerable sum at that time), he would
have been unable in future to maintain a large staff of
observers, printers, and other assistants; the extensive
buildings would require some outlay to keep them in repair,
and the idea of retrenching could not be pleasant to him.[2]

These considerations, added to the natural feeling of disgust at the want of appreciation he had met with, and
the wish again to enjoy the society of congenial minds,
overcame the regret he must have felt at leaving the happy
home where he had lived for fully twenty years, the buildings he had raised, and which had been the wonder of the
age, and the hitherto obscure little island on which he had
conferred imperishable fame. The observations, which had
been progressing as usual, were discontinued on the 15th
March (on which day the last ones, of the sun, moon, and
Jupiter, were recorded), and the dismantling of the instruments, and the removal of these and other property to his
house at Copenhagen, were rapidly proceeded with. Under
the 21st March we read in the Meteorological Diary : " We
catalogued all the Squire's books; " and we can picture to
ourselves the desolation which soon reigned in the hitherto
crowded library and observatories.

But Tycho was not allowed to leave Hveen without
further annoyance. When the peasants on the island found
that their master was not in favour at court, they drew up

---

[1] See Tycho's account, "De occasione interruptarum observationum et
discessus mei," Barretti, *Hist. Coel.*, p. 801. The date is given in Friis, *Tyge
Brahe*, p. 229.

[2] In addition to Hveen, he still held the prebend of Roskilde and the eleven
farms in Scania; the rent from the latter was barely 200 daler a year (Weistritz, i. p. 170).

a memorial complaining of his oppression and ill-treatment of them. On the 4th April the king, therefore, commanded the Chancellor and Axel Brahe (apparently a brother of Tycho's, who in June 1596 had become a privy councillor) to proceed to Hveen on Saturday the 9th April, in order to examine on the following day into the complaints of the tenants, to inspect the land, and also to see "if he has dared to act against the ritual, as you, Christen Friis, are aware." The report of this expedition is not known, but proceedings were at once taken against the clergyman at Hveen for having acted contrary to the Church ritual. On the 14th April the following commission was issued to a privy councillor, Ditlev Holk: "Know you, that whereas a minister, by name Jens Jensen, has dared during the service in church to act against the ritual, and he for such audacious conduct is to appear before our beloved the honourable and learned Dr. Peder Winstrup, superinten-dent [1] of this diocese of Seeland, on the 22nd April: We order and command that you arrange to be present here in this town at the same time, and afterwards with the said Peder Winstrup in the said case to judge according to what is Christian and right." [2] The judgment of these two commissioners is not known, but in an old diocesan record it is stated that "the minister of Hveen was dismissed in disgrace for not having kept to the ritual and prayer-book in the form of baptism ("I adjure thee"), but acting differently; also for not having punished and admonished Tyge Brahe of Hveen, who for eighteen years had not been to the Sacrament, but lived in an evil manner with a concubine." [3]

In other words, the clergyman had omitted the exorcism

---

[1] After the Reformation the Danish Bishops were for some time styled superintendents, but the old name soon came into use again.

[2] *Danske Magazin*, ii. p. 316 (Weistritz, ii. p. 300).

[3] Ibid., p. 317.

in the baptismal service, a great crime in a Lutheran
country, because it had been omitted by Zwingli and
Calvin, but retained by Luther.[1] The "concubine" would
a few years earlier have been called Tycho's lawful wife, as
we have already shown, and though Tycho may not have
been a regular attendant at the church of Hveen, he was
unquestionably a man of a religious mind, as many passages in his writings show very clearly.[2] Bishop Winstrup
was not very friendly to Tycho (as had appeared during
the proceedings about Gellius before the University), and
the minister of Hveen was probably not a very desirable
person, as he afterwards, while staying with Tycho in Holstein, tried to make mischief between him and the steward
left behind at Hveen.[3] That Tycho should not generally
have conformed to the usage of the Lutheran Church seems
unlikely when we remember his intimate friendship with
Vedel, as well as the fact that there were several future
clergymen, and not less than four future bishops, among his
resident pupils.

[1] It is curious that King Christian IV. already in 1606 desired Bishop
Winstrup, when a little princess was being christened, to leave out the
exorcism. *D. Mag.*, ii. 319.

[2] Riccioli quotes *Progymn.*, pp. 712, 777, to show that Tycho had too much
veneration for Luther, Melanchthon, and Chytræus, "those pests of the human
race" (Kästner, *Gesch. der Math.*, ii. p. 407). Gassendi, on the other hand,
by several extracts shows how full of true religious feeling Tycho always was
when speaking of the Creator of the Universe (p. 190 *et seq.*).

[3] *Danske Mag.*, ii. p. 318 (Weistritz, ii. p. 305), quoting a letter from Tycho
to Holger Rosenkrands. In the above-mentioned letter to Grynæus, Tycho
thus describes the incidents narrated above : "Accesserunt et aliæ non pau-
culæ tribulationes, quibus abitum meum eo citius promoverunt, adeo ut ne
quidem a Parocho meo in mei contumeliam et despectum persequendo ab-
stinuerint, quod is detestandum et impium istum Exorcismum in Pædobaptismo
meo conscio omiserit. Ideoque officio privatum, et per integrum mensem
citra latam sententiam incarceratum, parum abfuit, quin etiam capite plecte-
rent, nisi ego cum meis Amicis apud reliquum Regni senatum tantam sævitiam
avertissem. Quin et Rusticos tam contra me quam eundem Parochum meum
clancularie excitatos tantum aberat, ut secundum leges (prout urgebam) eorum
iniustam perfidiam et rebellionem refrænare voluerint, ut potius horum im-
merita defensione suscepta in malitia illos confirmarint. Ego autem Parochum
tandem ex istis afflictionibus liberatum in Germaniam mecum recepi."

It is not quite certain whether Tycho was still at Hveen during the month of April, while his treatment of the tenants and the conduct of the clergyman were being investigated. By the end of March the removal of his instruments, printing-press, and furniture had been completed, and only four of the largest instruments were left behind for a while, as too troublesome to move.[1] Shortly after Easter, Tycho Brahe and his family left their home at Hveen for ever, and took up their residence temporarily at Copenhagen.[2]

[1] These were: Armillæ maximæ (with the equatorial arc belonging thereto), and Quadrans chalybeus magnus, both at Stjerneborg; the great Mural quadrant and Semicirculus magnus azimuthalis, the latter of which was in the southern observatory at Uraniborg. See Tycho's account, *De occasione interrupt. obs.*, Barrettus, p. 801.

[2] The diary and the account in the observing ledger (Barrettus, p. 801) differ as to the date of Tycho's departure from Hveen. In the latter he says that he left the island with his family "statim a Paschatis Festo die 29 Aprilis" (most distinctly written in the original MS.). But Easter was the 27th *March*. The diary is silent from the 22nd March to the 10th April inclusive, "propter alias occupationes observasse aut notasse non potuimus," and under April 11 it has: "Primum ingressi sumus novum Musæum Hafniense." On the 17th April: "Profectus est Tycho Roschyldiam." The diary stops abruptly on April 22nd at the middle of a page, and was never taken up again. Probably it was on the 29th *March* that Tycho left Hveen, and this is confirmed by his German account, in which he says that he was at Copenhagen "in die dritte Monat," *i.e.*, more than two months.

CHAPTER X.

TYCHO'S LIFE FROM HIS LEAVING HVEEN UNTIL HIS ARRIVAL AT PRAGUE (1597-1599).

WHEN Tycho arrived at Copenhagen in April 1597, he probably did not intend to make a long stay there, but merely to watch events for a short time. He can hardly have intended to settle in his house at Copenhagen and continue his work there, as he had the Isle of Hveen for life, and might as well have stayed there if he had any wish to remain in Denmark, unless, indeed, the troubles at Hveen had risen to such a height that the island had become odious to him. He had brought his instruments, chemical apparatus, and printing-press with him, but he does not appear to have commenced astronomical observations at the tower on the rampart close to his house. Probably he had not time to get any of the larger instruments mounted, as he tells us in the account of his leaving Denmark, as well as in several of his letters, that the Treasurer, acting in the name of the king, who was absent in Germany, forbade him to take observations in the tower on the rampart. He does not say on what pretext this was done, but possibly the Government did not wish him to settle permanently on any part of the fortification.[1] He is also said to have been for-

[1] In the account "De interruptione," &c. (Barrettus, p. 801), as well as in a letter to Vedel in 1599 (Weistritz, i. p. 171), Tycho says that the order not to observe on the rampart was given by Aulæ Magister (*i.e.*, Valkendorf), though he had been one of the four protectors who had granted him the use of the tower in 1589. See also a letter to Vincenzio Pinelli of Padua (*Aus T. Brahe's Briefwechsel*, p. 12).

bidden by the mayor, Carsten Rytter, to make chemical
experiments in his own house, and Gassendi adds that he
and his clergyman were subjected to personal annoyance,
and that he was not able to obtain legal reparation; but
this doubtless refers to the troubles at Hveen, and not to
anything which happened at Copenhagen.[1]   But an event
which at first sight looks even more strange took place
soon after.   On the 2nd June, Thomas Fincke, Professor of
Mathematics (afterwards of Medicine), and Iver Stub, Professor of Hebrew, were ordered to proceed to Hveen, as the
king had learned that the peasants had damaged the instruments; they were to examine into this matter and report
on it.[2]   Their report is not known, and this expedition is
not mentioned in any of Tycho's accounts of his expatriation, except in his poem *Elegia ad Daniam* (which will be
mentioned farther on), and a garbled account of it may
have reached him after his departure from Denmark.   According to Gassendi, the two professors declared that the
instruments were not only useless, but even noxious curiosities,[3] which probably only referred to the chemical apparatus.   Fincke had in 1583, at Basle, published a Geometria
Rotundi, in the preface to which he had addressed some highly

[1] Tycho says (Barrettus, p. 802): "Taceo nunc, quæ circa reprobos istos
Insulares et Parochum in odium mei evenerunt" (compare footnote 3 on page
237).   In a letter to Paschalius Mulæus (Claus Mule, one of his pupils), of
unknown date, but found among the MSS. of Longomontanus, Tycho says
(after describing how he had lost his endowments and had been forbidden by
the mayor to carry on his *exercitia*): "I shall also pass over what happened
to my clergyman from hatred to me, also the insolence shown to me by those
who were instigated to it, also that I was forbidden to take legal proceedings
against them" (Bang's *Samlinger*, ii. p. 493; Weistritz, i. p. 155).   Gassendi,
p. 140, uses almost the same words, and has them probably from the same
source.

[2] Friis, *T. Brahe*, p. 234, quoting from the original document in the archives
at Copenhagen.

[3] Gassendi (p. 140) evidently knows very little beyond the allusion to the
trip in the *Elegy;* he only knows the name of one of the emissaries, and mis-
spells it Feuchius.   He does not mention that any damage had been done to
the instruments.

complimentary sentences to Tycho, and the book is the earliest in which the words *secant* and *tangent* are proposed, while several new fundamental formulæ of trigonometry occur in it for the first time, so that the author must have been a man of considerable ability.[1] The mission of the two professors was no doubt caused by some disturbances at Hveen, which, perhaps, had more to do with Tycho's departure than we are aware of, and it is much to be regretted that we do not possess any account of these transactions except Tycho's own. Gassendi thinks that the report of the professors was the cause of Tycho's chemical experiments being forbidden ; but this cannot have been the case, as the expedition of the two learned professors must have taken place after the 2nd June, and Tycho must have left Copenhagen either on that date or immediately after it, as he arrived at Rostock during the first half of June.

After having spent two or three months at Copenhagen, Tycho must have felt that there was nothing to be obtained by delaying his departure from Denmark any longer, and early in June 1597 he sailed for Rostock with his family, some students and attendants, about twenty persons in all, taking his instruments, printing-press, &c., with him. His principal assistant of late years, Longomontanus, who wished to study at German universities, had obtained his discharge with a kind testimonial from Tycho, dated at Copenhagen on the 1st June.[2] Among those who accom-

---

[1] See particularly pp. 77–78, and p. 292, rule 15. About this book, compare R. Wolf, *Handbuch der Astronomie*, pp. 173, 179, and Catalogue of *Crawford Astr. Library*, p. 188. Kästner (i. p. 629) does not seem to have perceived the valuable parts of the book. Fincke (1561–1656) was first physician to the Duke of Holstein-Gottorp, then Professor of Mathematics, and from 1603 of Medicine at the University of Copenhagen. He had studied at Strassburg and Padua, and corresponded for some years with Magini. According to Lalande and Poggendorff, he wrote previous to 1603 several tracts on astronomical subjects, but after 1603 he devoted himself only to medicine.

[2] This testimonial is printed by Gassendi, pp. 140–141. Tycho calls himself "Dominus hæreditarius de Knudstrup et arcis Uraniburgi in insula Daniæ Venusia Fundator et Præses."

panied Tycho was a young Westphalian gentleman, Franz
Gansneb Tengnagel von Camp, who had been with him at
Hveen since 1595, and who afterwards became his son-
in-law.

At Rostock Tycho had still friends from former days,
though his correspondent Brucæus had died four years
previously. But Chytræus was still alive, and on the 16th
June he wrote a friendly letter, regretting that the state of
his health prevented him from paying his respects to Tycho.[1]
But the exiled astronomer found that though he was at
once welcomed to Germany, he had not improved his position in Denmark, for immediately after his departure, on
the 10th June, he was deprived of the prebend of Roskilde,
which was conferred on the Chancellor, Friis, although the
latter already enjoyed the best prebend in the chapter, and
though the rules were that nobody could hold more than
one prebend in any cathedral, that they were tenable for lifetime, and that the heirs of a prebendary should enjoy *annum
gratiæ* after his death.[2] But here it must in fairness to
the Government be recollected that Tycho had for years
showed the most complete disregard of his obligations as
a Prebendary, and that he had apparently left the country
for ever in order to obtain employment abroad wherever
he could get it. There was, therefore, some excuse for
depriving him of this lucrative sinecure ; but it certainly,
on the other hand, seems to point to Friis as an enemy of
Tycho's, since he made this an occasion for feathering his
own nest.

When Tycho Brahe had been about a month at Rostock,
he took a step which he probably ought to have taken long

---

[1] Letter in *Danske Magazin*, ii. p. 325 (Weistritz, ii. p. 318).

[2] Ibid., p. 325. That Friis already had another prebend is stated by Tycho
himself, ibid., p. 348 (Weistritz, ii. p. 346). Tycho says (*Hist. Coel.*, p. 802,
that he was "vix e patria egressus" when this happened. He must, therefore, have left Copenhagen between the 1st and 10th June.

before, and addressed himself directly to King Christian IV. As it is of great interest, we shall give a translation of the letter, keeping as closely as possible to the words of the Danish original.[1]

" Most puissant, noble King, my most gracious Lord! with my willing and bounden duty most humbly declared. I beg most humbly to inform your Majesty, that whereas I had no opportunity of appearing before your Majesty before my departure, neither knew whether it might be agreeable to your Majesty or not, I am now obliged shortly to let your Majesty know in writing what I should otherwise humbly have stated verbally.

" Whereas from my youth I have had a great inclination thoroughly to study and understand the laudable astronomical art, and to put it on a proper foundation, and for that purpose formerly hoped to remain in Germany in order conveniently to do so, then your Majesty's father of laudable memory, when H. M. learned this, graciously desired and induced me to undertake and carry out the same at Hveen. Which I have done for more than twenty-one years with the greatest diligence, and at great expense, believing to have thereby shown that I liked best to do it to the honour of my own Lord and King and of my country. And your Majesty's father graciously intended and promised that whatever I started in the said art should by a foundation be sufficiently endowed and perpetuated on several good conditions which were graciously promised me, which your Majesty's Lady mother, my most gracious Queen, doubtless still remembers, and formerly has stated to the Privy Council of Denmark. For that I have received the public act of the Privy Council on parchment, confirming and further assuring me of this. Therefore I have

---

[1] The original is printed in *Danske Magazin*, ii. pp. 327-330, translated in Weistritz, i. p. 122 *et seq.*

since incurred great trouble and expense, even more than formerly, hoping that your Majesty when coming to the Government would be graciously pleased to let me and mine profit thereby. But it has turned out differently from what I had believed, about which I shall now only state the following. Your Majesty is doubtless aware that I have been deprived of what I should have had for the maintenance of the said art, and that I have been notified that your Majesty does not intend further to support it, in addition to much else which has happened me (as I think) without my fault or error. And whereas I, by the grace of God, shall have to carry to an end what I once with so much earnest and for so long have worked at, which is also known to many foreign nations and greatly desired, and I have not myself means for this, as I have been so reduced that I, notwithstanding the fiefs I held, have been obliged to part with my hereditary estate ; therefore I trust that your Majesty will look to my necessities, and not be displeased with this my departure, as I for these and other reasons am greatly in need of seeking other ways and means, that what has been well begun may be properly finished, and that I may maintain my good name and reputation in foreign countries. But I have not departed with the intention of totally leaving my native land, but only to look for help and assistance from other princes and potentates, if possible, so that I may not too much be a burden to your Majesty and the kingdom. If I should have a chance of continuing my work in Denmark, I would not refuse to do so, for I should still as formerly much prefer to do as much as I can to the honour and praise of your Majesty and my own native land in preference to any other potentates, if it could be done on fair conditions, and without injury to myself. And if not, though it be ordained that I am to remain abroad, I shall always be subject to your Majesty with all respect and humility and humble capacity. Submitting also to the gracious consideration of your Majesty, that it is by no means from any fickleness that I now leave my native land and relations and friends, particularly at my age, being more than fifty years old and burdened with a not inconsiderable household, which I, at great inconvenience, am obliged to take abroad. And that which is still left at Hveen proves that it was not formerly my purpose and intention to depart from thence. Hoping, therefore, humbly, that when your Majesty considers these circumstances, your Majesty will be and continue my gracious Lord and King, and with all royal favour and grace incline toward me and mine. I shall always be found humbly true and dutiful to your Majesty to the best of my ability, wherever the Almighty sends me. The same good God who rules all worldly government grant your Majesty during your reign happiness, blessing, good counsel and design. Datum Rostock the 10th July 1597."

The same day Brahe wrote a letter (in Latin) to a young friend, Holger Rosenkrands (afterwards known as a writer on religious subjects), in which he thanked Rosenkrands for a letter he had just received, which showed that Ovid's words, "quam procul ex oculis, tam procul ibit amor," could not be applied to him. He had desired a painter to send a portrait of himself to Rosenkrands. He would like to know what was going on in Denmark, and what people said about his departure. He was still staying at Rostock, waiting for the return of the Danish embassy,[1] in order to speak to his brother Steen, and he had been advised by some people versed in state affairs not to apply to any foreign Government before he was assured as to the

---

[1] Probably this was an embassy to *Cöln an der Spree* (Berlin) in connection with the approaching marriage of the king with Anna Catharina of Brandenburg.

intentions of the Danish king ; but if he found that his
Majesty was unfavourable to him and his studies, he expected confidently to find advice elsewhere.

It would almost seem that Tycho already regretted having
left Denmark, as he now made every effort to influence
King Christian in his favour, though he had neglected to
approach the king personally while he was still in the
country. On the 29th July he wrote a letter in German
to Duke Ulrich of Mecklenburg-Güstrow, the maternal grandfather of the king, reminding him of the visit which the
Duke had once paid to Uraniborg, and stating that he
had been obliged to leave Denmark for reasons which he
did not wish to put in writing. For the present he had
taken up his abode at Rostock, which he hoped was not
displeasing to the Duke, who doubtless would regret that
work, which was progressing well and which was valued by
learned men all over Europe, should be so suddenly interrupted and almost come to nought. He therefore begged
the Duke to advise him how this work might be continued,
if not in Denmark, then somewhere in the Roman Empire,
and promised in future publications gratefully to acknowledge any assistance the Duke would give him.   At the
same time Tycho wrote to the Duke's chancellor, Jacob
Bording, whose father had been physician to King Christian III. of Denmark, and asked the chancellor to speak for
him to the Duke. Bording answered at once, assuring Tycho
of the good-will of the Duke, who would in a few days
write to him as well as to the king. On the 4th August
Duke Ulrich wrote to Tycho, expressing his sympathy, and
asking whether Tycho would wish him to send off a letter
to King Christian, of which he enclosed a copy. He could
not express an opinion as to how the astronomical work
might be carried on, but it would require the patronage of

---

some great potentate. In his letter to the king the Duke merely asked his grandson not to allow Tycho's work to be interrupted, as it did great credit to the late king and the country, and was renowned among all nations.[1]

While Tycho Brahe was still at Rostock awaiting the result of his own and the Duke's letters to the king, he occupied himself in investing the ready money which he had brought with him from Denmark. As he repeatedly states that he had been obliged to part with his hereditary estate on account of the great outlay on buildings and instruments, which all his endowments did not cover, it would almost seem certain that his aunt and foster-mother, Inger Oxe, who died in 1591, must have left him a very considerable sum of money.[2] He found a very convenient way of investing his money, as the Dukes Ulrich and Sigismund August, as guardians of the young Dukes Adolph Friedrich I. and Johann Albrecht II. of Mecklenburg, happened to require money, and were willing to borrow from Tycho. In the summer of 1597 they opened negotiations with him for the loan of 10,000 " harte Reichsthaler " (*i.e.* of full value, not clipped). As a prudent man, Tycho wanted proper security, and demanded a bond, by which ten well-known men should declare themselves and their heirs bound to him in the sum of 10,000 thaler; but as it was not customary in Mecklenburg for sureties to bind their heirs, he had to give up that point. As it took time to procure the consent and the signatures of the sureties, Brahe agreed to pay the money on receiving a temporary receipt from the two ducal guardians, and a mortgage on the county of Doberan; but when this was settled and two officials came

---

[1] These letters are printed in the *Danske Magazin*, ii. pp. 330–336 (Weistritz, ii. p. 323 *et seq.*).

[2] Several letters between Tycho and his kinsman Eske Bille (from the years 1599–1600) seem to show that Tycho had some dispute with several other heirs of his aunt. See *Breve og Aktstykker*, pp. 49 and 99.

for the money, he would not pay it, as the receipt did not
contain a certified copy of the bond to be given by the ten
men, and did not specify the interest to be paid. At last
everything was settled and the bond was delivered, dated
the 24th August 1597, to "Tycho Brahe auf Knustorf
im Reiche Dänemark erbgesessen," after which the money
was paid.[1]

In the meantime the plague had appeared at Rostock,
but Tycho still lingered there, awaiting the reply to his
letter to the king. If, before he took the decisive step
of removing his family, his great treasure of observations,
and nearly all his instruments out of the Danish dominions,
Tycho had addressed himself to the king, who was of an
open, generous nature, it is not unlikely that he might
have been treated very differently; but to an impartial
observer it is not strange that the king should be offended
with a subject whose previous behaviour had been far from
faultless, who had left the country in a huff in order to
carry his talents to the most profitable market, and who
now declared himself willing to forget the past and come
back if it was made worth his while. Of the interference
of his grandfather the king took no notice whatever,[2] but
to Tycho's own letter he sent on the 8th October the following answer, which we also translate literally.[3]

"Christian the Fourth, by the grace of God of Denmark
and Norway, the Vends and the Goths, King, &c. Our
favour as hitherto. Know you that your letter which you
have addressed to us sub dato Rostock the 10th day of July
last, has been humbly delivered to us this week, in which

---

[1] G. C. F. Lisch, *Tycho Brahe und seine Verhältnisse zu Meklenburg*, pp.
9-10 (*Jahrbücher des Vereins für meklenburgische Geschichte*, xxxiv.).

[2] See Tycho's letter to Vedel of September 1599 (Weistritz, i. p. 172).

[3] The Danish original in *Danske Magazin*, ii. p. 336 *et seq.*, translated by
Weistritz, i. p. 126 *et seq.*

among other things are counted up, first, that you had no
opportunity to speak to us before you left this kingdom,
neither knew whether it were convenient to us or not;
therefore you have humbly wished to let us know your case
in writing, and [you add] that we are doubtless aware that
you have lost whatever allowance you hitherto have had for
the maintenance of the astronomical art, also that we will
not continue to support the said art, and other things which
unexpectedly have occurred and have happened to you without any fault or error of yours, as you think. Furthermore,
that you have not yourself the means to perfect the said
art at your own expense, and even though you had your
former benefices, you have yet been so reduced as to have
had to part with your estate. And whereas you for the
said reasons are obliged to seek in other places from foreign
potentates and lords help, assistance, and counsel to promote
the astronomical art, then you beg that we will not with
displeasure look upon your journey, particularly as you will
not altogether leave your native land. Furthermore, you
state that if it may be granted to you in this kingdom to
continue your work, then you would not refuse it, but grant
that honour to us and your native land, if it could be done
on fair conditions and without injury to you, as your
lengthy letter furthermore details it. Now we would
graciously not withhold from you, first, as regards that you
have not had an opportunity of speaking to us before
you left the kingdom, and that you were not aware whether
it would be agreeable to us or not: You must well remember that you were staying for some weeks in our city
of Copenhagen before you left the kingdom, and not only
did not ask authority from us to leave the country, as you
should have done, but never even spoke to us except on the
one occasion when the peasants of Hveen and you were in
court before us, and you were commanded and ordered to

appear before us at the castle.   And although you do
not blush to make your excuse for this in a manner as
if you were our equal, we desire in this letter to let you
know that we are aware of that, and that we expect from
this day to be respected by you in a different manner, if
you are to find in us a gracious lord and king.   As regards
your not doubting that we are aware that you have lost
some fiefs you had held, and your thinking that it happened
through no fault or error of yours ; you remember well
what complaints our poor subjects and peasants at Hveen
have had against you, how you have acted about the church
there, of which you for some years took the income and
tithes and did not appoint any churchwarden, but let it
stand ruinous ; also took the land from the parsonage ; and
partly pulled down the houses, and the parson who should
live there and use the land to keep himself and his wife,
him you have given some pennies per week and fed him
with your labourers, so that there have been during some
years many parsons, who yet did not receive a call from the
congregation in accordance with the ordinance, nor were
lawfully dispossessed.   In what way the words of baptism
for a length of time have been omitted, against the established usage of these kingdoms, with your cognisance, is
too well known to everybody.   Which things, as well as
others, which have occurred on that poor and small land,
and were known to us for a good while before it became
publicly known, have caused us to grant our tenants and
the crown's in fief to others who would keep them under
the law, right, and established custom.[1]   With regard to
your not being wealthy enough to promote the astronomical
art by your own means, but sold your hereditary estate

---

[1] This refers to the fief of Nordfjord and the estate of the Roskilde prebend.
The Island of Hveen could not be taken from him as he had got it for life,
and we shall see that Tycho continued to keep a steward there, and received
rent and produce from the island.

while you yet held your fiefs, so that you have left the
kingdom to ask for help from foreign potentates, and not
intending to leave your native land altogether, which
journey you humbly ask us not to take umbrage at : there
is great doubt whether you have spent the moneys you
received for the property you sold on astronomical instruments, as it is said here that you have them to lend in
thousands of daler to lords and princes, for the good of
your children and not for the honour of the kingdom or
the promotion of science.   Also it is very displeasing to us
to learn that you seek for help from other princes, as if we
or the kingdom were so poor that we could not afford it
unless you went out with woman and children to beg from
others.   But whereas it is now done, we have to leave
it so and not to trouble ourselves whether you leave the
country or stay in it.   Lastly, as you humbly state that if it
might be permitted you to finish your work in this kingdom
you would not refuse if it could be done without injury to
you ; now we shall graciously answer you that if you will
serve as a mathematicus and do what he ought to do, then
you should first humbly offer your service and ask about it
as a servant ought to do, and not state your opinion in such
equivocal words (that you will not refuse it).   When that
is done, we shall afterwards know how to declare our will.
And whereas your letter is somewhat peculiarly styled, and
not without great audacity and want of sense, as if we were
to account to you why and for what reason we made any
change about the crown estates ; and we besides remember
how you have published in your epistles various nonsense
about our dear father, to the injury both of his love and
of yourself ; now we by this our letter forbid you to issue
in print the letter you wrote to us, if you will not be
charged and punished by us as is proper.   Commending

you to God.   Written at our Castle of Copenhagen the 8th October Anno 1597.   Under our seal,

CHRISTIAN."

(*Address*)—"To our beloved, the honourable and noble Tyge Brahe of Knudstrup, our man and servant."

The harsh and angry tone of this letter shows how completely the king's mind had become estranged from Tycho; and no matter how badly Tycho may have treated his inferiors, the fact remains that he was in his turn treated with severity and a want of appreciation of his great scientific merit which is inexcusable.   It could not be expected that the king or his advisers should have been able to appreciate the true value of Tycho's scientific labours, but they could not help being aware that he enjoyed a world-wide reputation, such as no Dane had ever acquired before; and if he was a bad landlord, they might have endowed him in some other way.   But this is neither the first nor the last time that a Government has given science the cold shoulder, since even in later and much more enlightened times statesmen of all nations not unfrequently have distinguished themselves by a sovereign contempt of science. But all the more let us admire the truly enlightened mind of Tycho's great benefactor and friend, King Frederick the Second, whom he had unfortunately lost too early. King Christian seems to have felt personally offended with Tycho Brahe for having first retreated to a distance and then attempted to make terms with him.   But it is not impossible that Tycho may have thought of Vedel, who in 1595 had not only been deprived of his office of historiographer for delaying too long to write the Danish history, but had even been forced to deliver up all the materials which he had been collecting for years.   Possibly Tycho wished to bring his great treasure of observations out of the reach of envious people, who might suggest that it had been gathered at the public expense, and therefore was public property; but by doing so he destroyed the bridge behind him, and could now only look abroad for a place to continue his labours.

As Tycho had no reason to remain any longer at Rostock, where the plague besides made the stay unpleasant if not dangerous, he now accepted the invitation of Heinrich Rantzov to reside for a while in one of his castles. Of these, Wandsbeck, which had been rebuilt not long before, seemed to Tycho the most convenient, as it was situated close to Hamburg (only two or three miles north-east of it), and the intercourse with foreign countries, therefore, was easy. As Rantzov, who was a very wealthy man, had spent great sums on accumulating books and treasures of art in his various castles in Holstein and Slesvig, Tycho found at Wandsbeck (or Wandesburg, as the new castle was called) not only a comfortable dwelling, but also one in which the owner's refined tastes had created a home which might to some extent bear comparison with the one he had left for ever. Tycho removed with his family and belongings to Wandsbeck about the middle of October 1597, and met a former acquaintance there in the person of Georg Ludwig Froben (Frobenius) from Würzburg, who six or seven years before had visited Uraniborg after studying at Tübingen and Wittenberg. He was at that time probably employed by Rantzov at Wandsbeck in literary work, and he settled in the year 1600 as a printer at Hamburg, where he remained till his death in 1645.[1]

Tycho could now think of resuming the observations which had been interrupted seven months before. On the 20th October he wrote a short statement of the causes of this interruption and of his departure, which we have

---

[1] Jöcher's *Gelehrten Lexicon*, vol. ii.

already quoted,[1] and a long poem "Ad Daniam Elegia," in which he taxes his native land with having rewarded him with ingratitude. It begins thus :[2]—

> " Dania, quid merui, quo te, mea Patria, læsi
> Usque adeo ut rebus sis minus æqua meis ?
> Scilicet illud erat, tibi quo nocuisse reprendar,
> Quo majus per me nomen in orbe geras ?
> Dic age, quis pro te tot tantaque fecerat ante,
> Ut veheret famam cuncta per astra tuam ? "

The writer next inquires who is to make use of the precious things which he has left behind. "Somebody has been sent to Hveen who was believed to know Urania's secrets ; he came, and when he beheld the great sights (though but a few are left), he stared with wonder. What could an ignorant man do, who had never seen such things ? He inquires their name and use, but lest he should seem to have been sent thither in vain, he sneers at what he does not comprehend, probably instructed by my enemy, who already before has injured me." The poem further alludes to all he has done for science, and how little his Herculean labours have been valued ; how he has cured the sick without payment, and suggests that this perhaps has roused the envy of his enemies. He regrets that his ungrateful country shall lose the honour which he conferred on it, but he looks to the future without fear, as the whole world will be his country and he will be appreciated everywhere. He exonerates the king from all blame, but there are a few others whom he never injured, but who yet have done him all the harm they could. Finally, he thanks Rantzov for having so hospitably received him.

The statement about the interruption of the observations

---

[1] "De occasione interruptarum observationum et discessus mei." *Historia Cœlestis*, pp. 801-802.

[2] Ibid., p. 802, also in *Resenii Inscript. Hafnienses* (1668), p. 347, and in Casseburg, *Tychonis Brahe relatio de statu suo, &c.* Jena, 1730, less correctly given by Gassendi, p. 143.

and the elegy were copied into the volume in which the observations of the years 1596 and 1597 were written, and copies of the poem were sent to various correspondents. Though it was probably not intended for the eye of King Christian, it fell into his hands by accident. On a copy of the poem which Tycho in the following year sent to Joseph Scaliger he added a note to the following effect:—Rantzov got a copy of the poem as soon as it was written, and had it stitched in a calendar of his,[1] and when the king in the course of the winter paid a visit to Rantzov at one of his other castles in Holstein, he happened to find the book lying open on the library table. The king took it up, and when he saw the poem with Tycho's signature underneath, he read the whole of it thoughtfully and slowly, though he on other occasions would not have been affected by such things.[2] Having read it, he silently put down the book and never spoke to Rantzov about it, nor did he in conversation allude to Tycho Brahe. When Rantzov was told that the king had seen the poem, he was much vexed, but Tycho on hearing it only hoped that the king had understood all the allusions, and expressed himself ready to send the king a copy.[3]

Though Tycho Brahe had been unsuccessful in his application to the king and in his attempt to use the influence of Duke Ulrich of Mecklenburg, he still tried to bring all the influence he could to bear on him. In December 1597 he went on a visit to Rantzov at Bramstedt in Holstein, where he met Margrave Joachim Frederic, who shortly afterwards became Elector of Brandenburg, and his consort, who were on their way home after attending the wedding

---

[1] *Ranzovianum Calendarium*, printed at Hamburg in 1590, described by Kästner, ii. p. 413.

[2] "Qui alias talibus rebus non afficitur."

[3] "Quod et adhuc facere paratus sum." This copy of the poem (2½ pp. folio) is now in the University Library at Leyden. See also *Danske Magazin*, i. 340 (Weistritz, ii. p. 334).

of their daughter and King Christian at Haderslev in Slesvig,
on the 27th November. On the 22nd December Tycho
handed the Margrave a letter in which he expressed his
regret to find that the king was displeased with him for
leaving Denmark, though any one might know that he would
not without cause have left his home with wife and children,
and at the age of fifty. But as it perhaps had been so
ordained by God, he was content, and had no wish to be
reinstated, and even if that should be done, he would be
very unwilling to live any longer at Hveen, and always to
stay there.[1] But he would ask the Margrave to write to
the king that he would, though abroad, continue to do all he
could for the honour of his country, and it might perhaps
elsewhere be done as well, if not better, and much more
conveniently and quietly than in Denmark. If the king
would carry out his father's intention, and would permanently endow Uraniborg, Tycho would see that the work
should be carried on well, if not by himself, at least by one
of his [family], and he would let the four great instruments
remain there, and supply others as well. In that case he
hoped the king would endow the observatory with canonries
in accordance with the promise of the Government during
the interregnum. But if the king did not desire to keep
up the observatory, he hoped he might remove the four
instruments, and that he might receive some compensation
for all the trouble and expense he had gone to.[2]

With this letter has been preserved another memorandum
of Tycho's reasons for going abroad, which he doubtless

---

[1] " Dan ich darum keinen Verlangen trage, nunmehr vor mein Person in
Dennemarck zu sein und gerestituiret zu werden, und wan das schon geshehen
solte, so ist es mir doch sehr ungelegen auf der Insel Huen lenger zu wohnen,
und stets zu bleiben, wovon ich an einem anderen Ort meine Ursachen ver-
zeichnet habe." I believe there is not any document extant in which these
reasons for not living at Hveen are stated.

[2] *Danske Magazin,* ii. pp. 342–344 (Weistritz. ii. p. 336 *et seq.*).

gave the Margrave with the letter.[1] In this memorandum
it is stated that Tycho had, at the wish of King Frederick,
settled at Hveen, where he had erected a number of costly
buildings and constructed more than fifty fish-ponds, which
were a great boon to the island, as often there was formerly
a scarcity of fresh water. All this, as well as his instruments, had cost over 75,000 daler, though the king and
Council had only paid 10,500 daler towards it.[2] When
the Privy Council, shortly after the king's death, had pledged
itself to recommend the young king, when he attained his
majority, to perpetuate the observatory, Tycho had in the
following eight years even expended more than before.
But after the coronation he lost first his Norwegian fief,
which had brought him in about 1000 daler annually, and
soon after that he also lost his pension of 500 daler. His
removal to Copenhagen is then mentioned, and how he was
during the king's absence forbidden to continue his work
there. Then, when he left for Germany, the Chancellor
got his prebend, which was worth about 700 daler and ten
Danish læster corn.[3] King Frederick had under his hand
and seal promised him the first vacant prebend in the Cathedral of Lund, but this had been ignored afterwards. He
had met with these and other troubles, which he did not
wish to put in writing, and he could only conclude that
there was no good-will in Denmark towards him or his
science, though he was willing to excuse the king, and to
believe that all arose from the envy and hatred of his enemies.
He would therefore leave all to God, and pray for His help
and blessing to continue his work.

[1] *Danske Magazin*, ii. pp. 344–349. " Die Vrsachen warumb Tycho Brahe sich
aus Dennemarck in Teutschlandt begeben, kürtzlich zu vermelden, sein diese."
[2] He must mean exclusive of his annual income from the various endowments.
[3] About 300 hectolitres. In ready money Tycho, therefore, had 2400 daler
(£533) a year, including the rent from the eleven farms at Kullen. See above,
p. 235, footnote.

On the 25th January 1598, the Elector of Brandenburg (who had just succeeded to this dignity on the death of his father) wrote to King Christian enclosing Tycho's letter, which he asked his son-in-law to consider favourably. He also wrote to his daughter, and asked her to put in a good word for Tycho. These letters were sent under cover to Friis and Valkendorf, with a short note asking them to do whatever they could in this matter. On the 4th February the Electress wrote to the king asking him to give a gracious answer to Tycho Brahe's petition, and to her daughter the queen she wrote in nearly the same terms, asking her to use her influence with the king to that effect.[1] What answers were sent to these letters is not known, but at any rate they did not lead to anything.

In the meantime Tycho had resumed the observations at Wandsbeck, the first being made on the 21st October 1597. During the first few months he only employed a radius, as in the early days of his youth, before he had got a number of good instruments together, and he was even obliged to observe the important opposition of Mars in this manner, as he had not yet got the heavier instruments transported to Wandsbeck and erected in suitable places. By the beginning of February 1598 this was done,[2] and he was again able to use quadrants for determining the time by altitude observations, instead of (as during the previous months) by watching when the pole-star and another star were in the same vertical. He also laid aside the radius for the more accurate sextant, and set up an equatorial armilla for observing the sun. On the 25th February 1598 a solar eclipse took place, which was total in the middle of Germany, while in Holstein about nine-twelfths of the solar diameter was eclipsed. Tycho observed this eclipse, and

---

[1] *Danske Magazin*, ii. pp. 349-351 (Weistritz, ii. pp. 348 *et seq.*).
[2] *Barretti Hist. Cœl.*, p. 822.

received observations from his former pupils, Longomontanus, who at that time was staying at Rostock, and Christen Hansen of Ribe, who observed it in Jutland, and who had formerly observed the comet of 1593 at Zerbst. It appears that Tycho got some kind of information about this eclipse from somebody at Hveen, perhaps from David Petri (Pedersen), whom he had left in charge of the buildings and other property on the island, as Tycho afterwards wrote both to Magini and Kepler that the eclipse had been observed at Hveen from beginning to end (while only the beginning was seen at Wandsbeck owing to clouds), and that the time of beginning and end agreed well with his own tables.[1] With the exception of this eclipse of the sun and two of the moon, and a few meridian altitudes of the sun, the planets only were observed at Wandsbeck. Tycho felt that the thousand star-places were enough to have to show to the world, and he felt that observations of the planets were of greater value to complete the material accumulated at Hveen. He was assisted at Wandsbeck by Johannes Müller, mathematician to the Elector of Brandenburg, who had visited him at Hveen in 1596, and whom he was requested by the Electress to train not only in chemistry but also in the

[1] In the letter to Magini (28th November 1598, Carteggio, p. 222, also p. 238), Tycho says that the middle of the eclipse at Uraniborg was observed at 11h. 5m. A.M., magn. of eclipse between 9 and 10 digits. In the letter to Kepler he wrote (Dec. 9, 1599, Opera, i. 225) that the observer at Hveen found by the large armillæ the beginning, end, and middle, in accordance with Tycho's tables. In his Optics Kepler made use of this observation, and gave the contacts as having occurred at 10.3 and 12.32 (Opera, ii. 367), but in the Tab. Rudolph., p. 110, he says that Origanus had observed 10½ and 12.32, and that the figures given in the Optics must have been copied from Origanus, putting 10.3 for 10½ (compare Opera, ii. 441). But if so, this is no fault of Tycho's, as he did not give any observed contacts. There is nothing about this observation in the Historia Cœlestis, nor could I find it in the original volume for 1596–97. Tycho does not mention the name of the observer at Hveen, only in the letter to Kepler he says the observation was made "a quodam istic relicto studioso."

preparation of medicines.[1]   The distinguished astronomer
David Fabricius of Ostfriesland also visited Tycho at Wandsbeck, but probably only for a short time.

In addition to the observations, Tycho devoted his time
at Wandsbeck to the preparation of the illustrated description of his instruments, which he had for years intended
to publish, and which it seemed particularly desirable to
issue now, in order to sustain his reputation and impress
learned and influential men with the magnitude of his
scientific work and its great superiority over that of previous
observers.   Woodcuts of a number of the instruments had
already been prepared at Uraniborg, and some of them had
been inserted in his books on the new star and the comet
of 1577.   Some engravings were now made of other instruments not yet figured, and the text was soon put together
by enlarging the account formerly prepared for the Landgrave.   As Tycho had brought his printing-press with him,
he was able to have the book printed under his own eyes
at Wandsbeck by Philip von Ohr, a printer from Hamburg.
Early in 1598 the *Astronomiæ instauratæ Mechanica* was
ready, a handsome thin folio volume, slightly larger than
the reprint of 1602, and now extremely scarce, so that the
number of copies printed can hardly have been considerable.[2]

---

[1] Letter from the Electress to Tycho of 14th February 1598.   *Danske
Magazin*, ii. p. 352 (Weistritz, ii. p. 353).

[2] *Tychonis Brahe Astronomiæ Instauratæ Mechanica*, in the centre the
vignette "Suspiciendo despicio," underneath, "Wandesburgi, Anno ⅭⅠƆ ⅠƆ ⅡƆ.
Cum Cæsaris et Regum quorundam Privilegiis."   Colophon: Vignette
Despiciendo suspicio, and under that: "Impressum Wandesburgi | in Arce
Ranzoviana prope Hamburgum sita, | propria Authoris typographia | opera
Philippi de Ohr Chalcographi | Hamburgensis | Ineunte Anno MDIIO."   This
original edition now only exists in a few great libraries.   In the Royal Library
of Copenhagen are two copies with all the pictures beautifully illuminated and
gilt, the one presented to Grand Duke Ferdinand de Medici, the other to the
Bohemian nobleman "Peter Vok Ursinus, Dominus a Rosenberg;" in the
Strahöfer Stiftsbibliothek at Prague is one presented to Baron Hasenburg
(*Astr. Nachr.*, iii. p. 256); in the British Museum is a copy presented to

The book was dedicated to the Emperor Rudolph II., whom Tycho was now specially anxious to interest in his labours. The dedication, which is dated the 31st December 1597, refers shortly to the instruments of the ancients and the limited accuracy attainable with them, and gives a summary of the contents of the book. Then follow (after a poem by Holger Rosenkrands) figures and descriptions of the seventeen principal instruments used at Uraniborg and Stjerneborg; of the sextant used in 1572–73 (two figures), of the great quadrant at Augsburg, and of a mounting once used for the largest azimuthal quadrant, and superseded by the one figured as No. 7. We shall not here dwell on these descriptions of Tycho's instruments, as they will be considered in some detail in the last chapter, and some of them have already been alluded to in previous chapters. It was natural that Tycho should at that time, with an uncertain future before him, point with some satisfaction to the convenient construction even of the larger instruments, which enabled him to take them asunder for the sake of transportation to different parts of the world. For an astronomer must be cosmopolitan (" Oportebit enim Astronomum esse κοσμοπολιτήν"), as among statesmen there are rarely found any who admire his studies, but frequently those who despise them. But the student of this divine art should not care about the opinions of ignorant people, but only think of his studies, and if interfered with by politicians or others, let him move himself and his belongings to some

Hagecius, &c. On the front cover of these presentation copies is Tycho's portrait stamped in gold, with the inscription round it :

" Hic patet exterior Tychonis forma Brahei,]
Pulchrius eniteat, qvæ latet interior."

The back shows his coat of arms (a golden pale on azure ground), with the distich round it :

" Arma, genus, fundi pereunt, Durabile virtus
Et doctrina decus nobilitatis habent."

other place, preferring his heavenly and sublime endeavours
even to his native soil, and remembering that—

> "Undique terra infra, cœlum patet undique supra
> Et patria est forti quælibet ora viro." [1]

After the illustrated description of instruments follows a
short account of six smaller portable instruments and an
engraving and description of the great globe. Tycho next
gives a sketch of his life from his youth onwards, his travels,
and how he became settled at Hveen, and passes in review
the principal results of his observations; [2] the improved
elements of the solar orbit; the discovery of a new inequality in the moon's motion; the variability of the inclination of the lunar orbit and of the motion of the nodes; the
observed accurate positions of a thousand fixed stars; the
explosion of the time-honoured error about the irregularity
in the precession of the equinoxes (*trepidatio*); the accumulation of a vast mass of carefully planned observations of
the planets in order to have new tables of their motions
constructed; and lastly, the observations of comets proving
them to be much farther away from the earth than the
moon. This was indeed a proud record of the twenty
years' work at Hveen, and was sufficient to show the world
that Tycho Brahe was worthy to rank with Hipparchus,
Ptolemy, and Copernicus.

After this review of his labours, Tycho prints a letter
from the late Imperial Vice-Chancellor Curtius and several
from Magini,[3] and a short abstract of a letter from Padua
(of December 1592), "from a certain Doctor of Medicine
then staying there" (he did not like to add, "of the name of
Gellius"). From this it appeared that the Government of

---

[1] *Astr. inst. Mech*, fol. A. 6, and fol. D. *verso*.

[2] He divides his observations into "pueriles et dubitæ" (at Leipzig),
"juveniles et mediocriter se habentes" (up to 1574), and "viriles, ratæ et
certissimæ" (from 1576).

[3] See above pp. 213 and 223.

Venice intended to send an observer to Egypt, and Tycho takes occasion to address a suggestion to the Venetians that they should cause the latitude of Alexandria to be redetermined, to see whether there had been any change in this quantity since the time of Ptolemy, as maintained by some, and he offered to assist them in this undertaking with instruments and advice. The book is then wound up with a description, with views and plans, of Uraniborg and Stjerneborg (to which he adds some remarks about the necessity of a good site for an observatory), a map of Hveen, and a short account of his transversal divisions and improved sights.[1]

In the original edition of this book there was no engraved portrait of Tycho, but in several of the copies which he presented to distinguished or influential persons a portrait in water-colours is pasted on the back of the title-page. This portrait is much larger than any published portraits, and represents him bareheaded, very bald (with a small tuft of hair over the middle of the forehead), and a very woe-begone countenance. It does not offer much resemblance to the well-known engraving by Geyn of Amsterdam of 1586, which appears in Tycho's *Epistolæ* and in the edition of the *Progymnasmata* of 1610, which represents him standing in a kind of arch on which the arms of the families of Brahe and Bille, and of the families connected with them, are suspended. This engraving has been reproduced in Gassendi's book.[2] Another

---

[1] The figures in our Chapter V. are reduced copies of Tycho's figures. The principal contents of the *Mechanica* are given in the introduction to Flamsteed's *Hist. Cœl. Brit.*, vol. iii., and the figures of the instruments are copied in the *Mémoires de l'Académie* for 1763.

[2] In the first issue of the *Progymnasmata* (1602) there is quite a different portrait, not resembling any other, but standing in the same arch. In Hofman's *Portraits historiques* there is another engraving by Haas of Copenhagen, apparently a copy (reversed) of Geyn's, which is reproduced in Weistritz's book.

portrait of Tycho Brahe, but of unknown date, was an oil-painting in the historical portrait gallery at Frederiksborg Castle, which was destroyed in the great fire of that castle in 1859.[1] In the letter (quoted above) which Tycho wrote to Rosenkrands from Rostock, he mentioned that he had ordered a painter to paint his portrait, and would send it to Rosenkrands when it was ready. This picture is probably the same which in the following century was preserved in the library of King Frederick III., and in the corner of which was an emblematic design with the following inscription:—

> "Stans tegor in solido, Ventus fremat, ignis & unda.
> Vandesbechi
> Anno MIƆXCVII, quo post diutinum in patria
> Exilium demum pristinæ libertati restitutus fui
> Tycho Brahe Ot."[2]

This portrait (or a copy of it) was found in England in 1876, and now belongs to the Royal Observatory, Edinburgh.[3] A full figure portrait occurs in *Baretti Historia Cœlestis*, representing Tycho leaning on a large sextant; the face resembles the engraving by Geyn, and the picture is apparently copied from a water-colour drawing on parchment in a copy of Tycho's *Progymnasmata* in the Strahof Monastery at Prague.[4]

Tycho was not content with issuing the description of his instruments, but as the first volume of his book (*Pro-*

---

[1] There is a copy of this portrait in Friis' book, *Tyge Brahe* (1871).

[2] The first line ("I am protected, standing on solid ground, let wind, fire, and waves rage") is evidently intended to express Tycho's trust in the future, notwithstanding the threatening aspect of the time. Ot. means Ottonides. The inscription is given in *Resenii Inscriptiones Hafnienses*, p. 335, Weistritz ii. p. 334, and identifies the picture.

[3] It was first noticed by Dr. S. Crompton (*Proc. Manchester Lit. and Ph. Soc.*, vol. vi., 1876), and was in 1881 purchased by the Earl of Crawford, who in 1888 presented it with his great astronomical library and all his instruments to the Royal Observatory, Edinburgh. See frontispiece.

[4] *Vierteljahrsschrift der astron. Gesellschaft*, xvi. p. 273 (1881), Safarik.

gymnasmata), in which the catalogue of 777 stars occurred, was still unfinished, he thought it desirable to distribute a limited number of manuscript copies of his catalogue of stars. It was probably for this purpose only that he had before leaving Hveen got a number of stars hastily observed in order to exhibit the places of a thousand stars, and not be inferior to Ptolemy with his 1028 stars. This catalogue of longitudes and latitudes of 1000 stars for the year 1600 was now neatly copied on paper or parchment by his assistants, and to it were added tables of refraction and precession, of the right ascension, and declination of a hundred stars for the epoch 1600 and 1700, and a catalogue of longitude, latitude, right ascension, and declination of thirty-six stars according to Alphonso, Copernicus, and himself, for the sake of comparison.[1] The lengthy introduction to this manuscript work was in the form of a dedication to the Emperor Rudolph II., dated the 2nd January 1598.[2] In this Tycho reviews the successive star-catalogues of Hipparchus and his successors down to and including "incomparabilis vir Nicolaus Copernicus," and he remarks that in reality nobody after Hipparchus has observed any great number of stars, but that Ptolemy, Albattani, Alphonso, and Copernicus had merely added precession to the longitudes, which circumstance in connexion with the limited accuracy of the catalogue of Hipparchus, and the numerous great errors which had crept into it, made it desirable to have a new star-catalogue prepared, in which the positions of the stars were given with the greatest accuracy now attainable. This Tycho had done, and offered it as a New Year's gift to the Emperor. The catalogue and the printed book, *Mechanica*, were sent to the Emperor by the hands of Tycho's eldest son, who also was the bearer of a

---

[1] The three first-mentioned tables are printed in the *Progymnasmata*.

[2] This introduction is printed by Gassendi, pp. 247–256.

letter, dated 2nd January 1598, in which Tycho stated that he had been obliged to leave his country and had come to Germany, where he hoped it might be granted him to complete his labours under the auspices of the Emperor.[1] About the same time Tycho sent magnificently bound copies of the star-catalogue to the Archduke Matthias, to the Vice-Chancellor Corraducius, to Wolfgang Theodore, Archbishop of Salzburg,[2] the Bishop of Lübeck, and to other influential men in Austria and Germany, to the King of Denmark, Prince Maurice of Orange,[3] Joseph Scaliger,[4] Magini, Kepler, two years later to the Elector of Saxony, &c. As already remarked, the *Progymnasmata*, which was not published until after Tycho's death, only contained 777 stars, but Kepler in 1627 published the thousand star-places in his *Tabulæ Rudolphinæ ;* while it is most significant that Longomontanus, Tycho's principal assistant, in his *Astronomia Danica*, only inserted the 777 stars, doubtless because he knew well how worthless the additional star-places were. The handsome manuscript volumes entitled " *Tychonis Brahe Stellarum octavi orbis inerrantium accurata restitutio*, Wandesburgi, Anno CIƆIƆIIC," were chiefly intended as advertisements, and it would be perfect waste of time to collate the various copies with a view to correcting Kepler's edition.[5]

When Tycho sent a copy of this catalogue to King Christian,[6] he probably also sent a letter to the king, of

[1] Printed in *Breve og Aktstykker*, p. 31 (from two draughts in the University Library at Basle).

[2] This is the copy which afterwards came into the possession of Gassendi, who gives (pp. 257–259) a list of remarkable discrepancies between star-places in it and in the *Tab. Rudolph.*

[3] *Astron. Jahrbuch fur* 1786, p. 216.

[4] This copy is now in the University Library at Leyden ("Descriptio stellarum octavi orbis inerrantium"). There is a copy of the catalogue in the Bodleian Library, presented to a Venetian nobleman.

[5] As suggested by Baily in his reprint of the catalogue, *Mem. R. Astron. Soc.*, xiii. ; compare his *Account of the Rev. J. Flamsteed*, p. 368.

[6] Now in the Royal Library, Copenhagen. In a letter from Henrik Ramel

which a draught is now preserved in the University Library
at Basle, dated the 7th February 1598.[1] In this Tycho,
after offering his congratulations on the king's marriage,
remarks that the troubles which he had met with in the preceding year were perhaps ordained by fate, since it was the
third *annus climactericus* (*i.e.*, the twenty-first year), since the
foundation of Uraniborg. He, however, thanked the king
for not having impeded his journey when he found it
necessary for his studies to go abroad, though he regretted
that his letter from Rostock had not been found satisfac-
tory ; but to show his feeling for his country and king, he
now forwarded two books which had been recently completed.

While Tycho in this manner paid his respects to the king,
notwithstanding the want of consideration with which the
latter had treated him, he did not hesitate to write to
Valkendorf to try to obtain some arrears of rent still due to
him.   In this letter, dated the 28th May 1598, Tycho first
thanks the Treasurer for all the kindness he has shown him,
and for the help he has given the steward at Hveen, who
had informed Tycho that he had in several cases concerning
the tenants there been supported by the authority of the
Treasurer.   " If it were known how contrary and disobedient the peasants on that little land are, and what I have
suffered from them all the time I lived there, and yet had
patience with them, and been more kind to them than they
deserve, then perhaps people would think differently about
them than they have done."   Tycho next asks the Treasurer
to instruct the Governor of Bergen to order half a year's
rent of the Nordfjord estate to be paid to him or his agent,

to Sophia Brahe (of 20th September 1599) the former writes that he would
have sent her the books, but had to ask the king first, and his Majesty had
said that though he did not understand or care much about them, still he
would keep them as they were presented to him by Tycho Brahe (*Breve og
Aktstykker*, p. 39). These books were possibly the *Mechanica* and the *Catalogue
of Stars.*   [1] *Breve og Aktstykker*, p. 34.

as it is still owing to him; and in conclusion he apologises
for giving so much trouble, but he expects everything good
from Valkendorf, and is sure that the latter will help him in
everything just, and right, and feasible.[1] The whole tone
of this letter seems to show with certainty that Valkendorf
cannot have been a declared enemy of Tycho's, as the latter
was of too haughty a disposition to condescend to write so
pleasantly to an avowed and open enemy; but on the other
hand, this does not prove that Valkendorf did not assist in
depriving Tycho of his great endowments.

Some time before this last appeal was dispatched to
Denmark, Tycho had on the 24th March 1598 written to
Longomontanus. He had heard from the Jesuit Monavius
of Breslau that Longomontanus had arrived there and had
had a look at Wittich's books, and Tycho therefore wished
to know whether there were any manuscripts among them,
and whether they were to be sold. He also inquired
whether Longomontanus had seen the recent slanderous
publication of Reymers Bär, which was too far beyond the
limits of decency to deserve a refutation; still it might be
well for Longomontanus to put in writing all he had heard
from his colleagues at Hveen about that person and his
doings, as he himself might have forgotten some of the
circumstances through all the troubles he had met with.
Finally, he desired Longomontanus to come to him at
Wandsbeck as quickly as possible, as he had something
very important to discuss with him, and if he had not
sufficient money, he was to borrow some or pawn something,
and Tycho would settle about it afterwards, and he would
not detain him long, as he did not himself intend to
remain long at Wandsbeck. He had Johannes Müller from
Brandenburg with him in charge of his observatory, but
he hoped Longomontanus would not disappoint him, and

---

[1] This letter is printed in the *Danske Magazin*, 3rd Series, iii. pp. 79-80.

he might bring with him a copy of Everhard's *Ephemerides*, which he had seen mentioned in a Frankfurt book-list, but which could not be had at Hamburg.[1] Gassendi suggests that Tycho may have wanted the help of Longomontanus to complete the chapter of the *Progymnasmata* on the lunar theory, where some sheets were still unfinished, while the recent eclipses had shown that this theory was still capable of further improvement.

While Tycho Brahe was living at Wandsbeck, his host not only tried to make his stay there agreeable,[2] but also did his best to assist him in finding a permanent abode, and the pecuniary support necessary to enable him to resume his labours on the same scale as formerly. Rantzov wrote to the Elector of Cologne, and asked him to use his influence with the Emperor in favour of Tycho, and to endeavour to interest the Austrian Privy Councillor, Barwitz, in the cause of the exiled astronomer. At the same time Tycho wrote himself to his friend Hagecius, and explained how he was situated, in order that the physician to the Emperor might speak to his master, and also enlist the sympathy of the Vice-Chancellor Corraduc. In order not to neglect any chance, Tycho also sent one of his disciples, Franz Teng-nagel, a native of Westphalia, to Prince Maurice of Orange to present copies of the *Mechanica* and the star-catalogue to the Prince, together with a letter from the author. The Prince answered that he would endeavour to persuade the States General to invite Tycho to settle in the Netherlands, and a similar answer was sent by the Advocate of Holland

---

[1] *Martini Everarti Ephemerides novæ et exactæ* 1590–1610 *ex novis tabulis Belgicis.* Lugduni Batav., 1597.

[2] In the Museum of Northern Antiquities at Copenhagen there is a watch which is said to have been presented to Tycho Brahe by Rantzov. It is oval in shape, has two dials, one for hours and one for minutes, and Tycho's name, arms, and the motto, " Qvo fata me trahunt, A.D. 1597," are engraved on the inner case. In the same museum is a wooden easy-chair which is supposed to have belonged to T. Brahe.

(or Grand Pensionary, as he was afterwards called), Olden
Barneveld, to whom Tycho, as a prudent politician, had also
written and sent his books. Joseph Scaliger, who five years
before had been called to Leyden as a professor, also wrote
that he would do his best, but he feared that the slow pro-
cedure of the States General would deprive the country of
so great an honour and himself of the pleasure of being
associated with a great man. In the meantime the Emperor had desired Corraduc to answer Tycho that he would
willingly receive him and see that he should want nothing
for the furtherance of his studies. In the course of the
summer Tycho not only learned this from Corraduc, but
also received a letter from Hagecius urging him to come to
Bohemia as soon as possible, while the Elector of Cologne
replied to Rantzov that he had every hope of Tycho's being
well received by the Emperor, and added that if Tycho,
against all expectation, should not find his work liberally
enough supported by the Emperor, then he would himself
promote it to the best of his ability. Tycho therefore, on
the 23rd August, wrote to Scaliger, sending him his books
(even the unfinished one), and thanked him for his kindness, and assured him that he would not have been disinclined to go to Holland, but that he had now been invited
by the Emperor and would soon set out for Prague. But
if this journey should not lead to the expected result, and
the States would make him a liberal offer, then he would
willingly come to them with his astronomical apparatus.[1]

Tycho was still at Wandsbeck on the 14th September
1598, on which day he wrote to Duke Ulrich of Mecklenburg to thank him for a letter of recommendation to the Emperor, and to ask him to accept a copy of the star-catalogue,

---

[1] Gassendi, pp. 156–157. In return for these books, Scaliger some months
later sent Tycho a copy of his *Conjecturæ et notæ in Varronem*, which Tycho
gave or lent to Taubmann, Professor of Poetry at Wittenberg. Kästner,
Gesch. d. Math., ii. p. 409.

with the same favour with which he had received his book on instruments.[1] Not long afterwards Tycho left Wandsbeck with his sons, his students, and a few small instruments, leaving for a while longer his wife and daughters and the greater part of his luggage in the kind charge of his host, who, however, died on the 1st January following. He travelled himself as far as Dresden, where he learned that there was pestilence and dysentery at Prague, and that the Emperor had retired with his court to Pilsen ; and when he wrote to Corraduc to announce his arrival, the Vice-Chancellor, at the Emperor's command, requested him to remain at Dresden until the epidemic was over. From Dresden Tycho wrote on the 28th November to Magini, with whom he had held no communication for about seven years, and told him that he had not finished his book yet, as the theories of the planets were not yet complete. He also gave a short account of the cause of his leaving Denmark, and added in a postscript that Tengnagel, who was the bearer of the letter, would verbally communicate something secret. This turned out to be that Tycho would like some Italian to write a eulogy of him, for which Magini two years later recommended Bernardino Baldi, who was going to write the lives of great mathematicians.[2]

Tycho did not remain long at Dresden, but preferred to spend the winter at Wittenberg, where he had still friends from his two former visits. In the first week of December 1598 [3] he went to Wittenberg with his sons and assistants, entered his own name and those of his two sons on the roll of students in the University,[4] and was lodged in the house

---

[1] Letter (in the archives at Schwerin) printed in Friis, *Tyge Brahe*, p. 319.

[2] *Carteggio inedito di Magini*, pp. 217 and 230. Baldi's *Delle Vite de' Matematici* was never published (Kästner, ii. 140).

[3] He observed the meridian altitude of the sun on the 9th December at Wittenberg.

[4] *Mulleri Cimbria literata*, vol. ii. p. 105.

which formerly had belonged to Melanchthon, and now
belonged to his son-in-law, Peucer, and where the physician
Jessenius (Johannes Jessinsky) lived at that time. In the
meanwhile Longomontanus had proceeded to Wandsbeck,
but on his arrival he only found Tycho's wife and daughters
there. He remained with them until Tycho's servant Andreas
arrived with letters requesting them to set out for Wittenberg, upon which Longomontanus accompanied the ladies as
far as Magdeburg, and then returned to Denmark, where he
observed the lunar eclipse on the 31st January following in
his native village. On the 31st December 1598 Tycho wrote
to him in reply to a letter he had just received, in which
Longomontanus had informed him that a printer at Hamburg, who had been intrusted with the printing of the sheets
relating to the lunar theory, had performed his task very
badly, so that it would be necessary to do it over again.
Tycho therefore wrote that he would get it done at Wittenberg.[1] He thanked Longomontanus for his attention to the
ladies, and offered to supply him with means for studying
at some German University until he had himself become
quite settled at the Emperor's court. He also expressed his
pleasure at hearing that Longomontanus intended to write
a refutation of the so-called defence of the Scotch opponent,
and he wished that it might be finished soon, so that it
might be printed at Wittenberg as an appendix to the
volume on the comet of 1577.[2] On the 11th January
1599 Tycho again wrote to Longomontanus asking him

[1] He afterwards abandoned this idea, because the eclipse of January 31,
1599, did not agree with his theory, although he had expected that it should
agree as well as that of January 1582, as they both took place near the apogee
and at the same time of year. (Letter to Longomontanus of 21st March 1699,
Gassendi, p. 159). This shows that he had at that time an idea of the existence of the annual equation. (See next chapter.)

[2] We have already mentioned (p. 209) that this refutation was never published. It appears from a letter to Scultetus, written in January 1600, that
Tycho was still thinking of adding an appendix to the book on comets.

soon to come to Wittenberg at his expense, and offering to get him the professorship at Prague now held by Reymers Bär, who would doubtless soon make himself invisible; or if Longomontanus would prefer a post at Wittenberg, Tycho would see that a professor there, who was not disinclined to go to Prague, was appointed to Reymers' post, and Longomontanus might then get the post vacated at Wittenberg. None of these proposals were, however, accepted, and Longomontanus did not join his old master until the latter had been at Prague for some time.[1]

It was not difficult for Tycho to foresee that Reymers would not care to await his arrival at Prague. When the former swineherd saw the expressions which Tycho and Rothmann had used about him in their letters, and which were made public by the printing of these, he naturally became furious, and in 1597 he published at Prague, where he had in the meantime become Professor of Mathematics, a book *De astronomicis hypothesibus*, in which he gave his fury full play.[2] The title-page shows the motto (in Greek), "I will meet them as a bear bereaved of her whelps" (Hosea xiii.), and indeed the language of the author is bearlike enough. First he tells how he discovered the Tychonic system on the 1st October 1585, and explained it to the Landgrave on the 1st May 1586, after which a brass model of it was made by Bürgi, and he suggests that Tycho may have heard of it through Rothmann (or, as he calls him throughout the book, Rotzmann, *i.e.*, Snivelman). After-

---

[1] Gassendi, pp. 158, 159.

[2] "Nicolai Raimari Vrsi Dithmarsi S. Cæs. Maj. Mathematici de astronomicis hypothesibus seu systemate mundano tractatus astronomicus et cosmographicus scitu cum iucundus tum vtilissimus. Item astronomicarum hypothesium a se inuentarum, oblatarum et editarum contra quosdam eas sibi temerario vel potius nefario ausu arrogantes, vendicatio et defensio. . . . Pragæ Bohemorum apud auctorem. Absque omni priuilegio. Anno 1597." 78 pp., 4to. Kästner, iii. p. 469. Delambre, *Astr. mod.*, i. p. 294. I have not seen this book myself.

wards he maintains that Tycho had merely imitated the system of Apollonius of Perga, and that Helisæus Röslin had recently with equal coolness claimed the same as his own.[1] He attacks Tycho and Rothmann with the coarsest abuse, and is very anxious to disprove that he was ever in Tycho's employment, as Rothmann had believed, and tells how he came to Hveen with Erik Lange. It appears that Tycho cut him short during a dispute with the remark that "those German fellows were all half-cracked,"[2] and that he generally went by the appellation of "Erik's Dreng" (*i.e.*, Erik's boy), and he adds proudly, "Jam non sum Jerix Dreng sed Imp. Rudolphi II. Mathematicus." To Tycho's accusation that Reymers had stolen the idea of the new system during his stay at Uraniborg, he answers that in that case it would have been stolen from him again, since Tycho, before his departure, got somebody to search his papers at night, when nothing was found but some plans of the buildings. The only way he could ever have spoken ill of Tycho must have been by joking about his nose, of which the upper part had been cut off, and he indulges in some scurrilous remarks about the facilities which Tycho possessed for taking observations through his nose without sights or instruments. But other parts of the book, like the "Fundamentum astronomicum,"[3] showed that Reymers was a very skilful mathematician, who deserves every credit for having by his own exertions, and apparently without enjoying the advantages of regular teaching, raised himself from the position of a swineherd to

---

[1] Helisæus Rœslinus in 1597 published a book, *De opere Dei Creatoris*, in which he stated that he had independently found the same system as Tycho Brahe, and in a later publication he stated in detail how he did this after reading Ursus' book of 1588. (See Frisch in vol. i. p. 228 of his edition of *Kepleri Opera*.)

[2] Reymers writes this in broken Danish : "Den Tyske Karle er allsammell all gall" (should be : "de Tydske Karle ere allesammen halv gale ").

[3] See above, p. 183.

that of a professor at Prague. It is easy to understand that his venomous attack must have been doubly annoying to Tycho at the particular moment when it was published and when he was anxiously seeking a new home. Tycho therefore began to collect evidence to show that his enemy had really behaved in a suspicious manner while at Hveen, and a document has been preserved, drawn up and signed before a notary at Cassel by Michael Walter, secretary to Reymers' former master, Lange. In this the writer states that Reymers, when Lange at his urgent request had consented to take him to Hveen, continued to poke and pry among Tycho's instruments and books whenever nobody was near, and to make drawings of everything; that one of Tycho's pupils warned his master about this, and mentioned it to a certain Andreas [1] who then went to sleep at night in the room with Reymers, and while the latter slept took a handful of papers out of one of his breeches-pockets, but was afraid to search the other for fear of waking him; that Reymers on discovering his loss behaved like a maniac, upon which he received back those of his papers which did not concern Tycho Brahe. The secretary also states that Reymers, after Lange's return to Bygholm Castle in Jutland, continued to behave more and more like a madman, and told everybody that Lange was going to hang him, until his master got tired of all this and dismissed him.[2]

Though it could not possibly be proved that Reymers had copied the idea of his planetary system from Tycho Brahe, it must be conceded that the latter had good reasons for suspecting him, even before he published his system in 1588, and we must remember that the scientific men of the age were always afraid of being robbed of their discoveries, and often took great pains to secure priority by

---

[1] Perhaps No. 5 on the list of pupils (Note B.).
[2] This document is printed in *Kepleri Opera*, i. p. 230.

hiding them in anagrams.   But on the other hand, Reymers certainly was an original thinker, and he may quite independently have come to the same idea which Tycho had already conceived.   But the whole question is not of much consequence, and we have merely dwelt so long on it because it attracted a good deal of attention at the time, and because the steps which Tycho afterwards took against Reymers throw considerable light on his own character.

Having spent the winter 1598–99 at Wittenberg, where his family had joined him, Tycho was further delayed by the illness of his eldest daughter; but shortly after Easter he at last set out for Prague, letting his family, however, stop for a while half way, at Dresden, until he had himself seen the state of things at Prague.

CHAPTER XI.

TYCHO BRAHE IN BOHEMIA—HIS DEATH.[1]

THE German Emperor, Rudolph the Second, whose service Tycho Brahe was now about to enter, was a man deeply interested in science and art, personally of a most amiable disposition, but most singularly unfit for the exalted and difficult position he had to fill. Totally devoid of energy and taking no interest in political matters, he let public affairs drift in whatever direction they liked, ignorant or careless of the fact that his apathy was hastening the catastrophe which a few years after his death plunged Central Europe into the war which turned Germany into a desert and almost annihilated the Imperial power. The times were certainly most serious, and the difficulty of settling the religious question almost overwhelming, but a monarch of spirit and determination might have done much to counteract the intrigues of the Spanish and Jesuitical party, who blindly pursued their narrow-minded policy, and finally brought on the Thirty Years' War. But, regardless of the duties imposed on him by his station, the Emperor reluctantly devoted a moment to business of any kind, while he willingly gave his time and the limited pecuniary means of his impoverished dominions to collecting art treasures and promoting science—the real science represented by Tycho and Kepler, as well as the imaginary ones

[1] In addition to Gassendi and Tycho's letters to Vedel and Longomontanus, the sources for this period are : Frisch's *Vita Kepleri*, in vol. viii. of *Joh. Kepleri Opera Omnia*, and Joseph v. Hasner, *Tycho Brahe und J. Kepler* in *Prag. Eine Studie*, Prag, 1872.

taught by the disciples of Cornelius Agrippa and Cardanus. Prague, where he usually resided, was not a very favourable place for the growth of science and art, as Bohemia had never settled down since the Hussite disturbances. The Germans and the Czechs were sharply separated by race and language; Catholics were opposed to Lutherans, Moravians, and Utraquists, the last-mentioned differing from the Catholics only by partaking both of bread and wine in the Eucharist. But notwithstanding this state of things and the miserable condition of the University, Rudolph succeeded in bringing together a number of men of culture in Prague, and for a short time he made the city one of the centres of civilisation—a distinction which was unfortunately destined to be but very short-lived. Long before his death, the Emperor's mind had been so persistently influenced by the intrigues of the Spanish party, that he had no feeling but distrust and suspicion for his surroundings, and scarcely felt relieved from the burden of government in the circle of his scientific friends. But while Tycho Brahe lived, Rudolph was still comparatively free from political anxiety, and ready to do his utmost to befriend the distinguished foreigner who had sought shelter under his roof.

When Tycho arrived at Prague in June 1599, the Emperor sent the Secretary Barvitz to conduct him to the house of the late Vice-Chancellor Curtz, where the widow was still residing. He had only a few instruments with him, as most of those he had brought away from Hveen were still at Magdeburg. He was shortly afterwards received in audience by the Emperor, who welcomed him to Prague, begged him to let his family come from Dresden, and conversed with him for a long time in Latin. Tycho presented the Emperor with three volumes of his works, and was afterwards told by Barwitz that the Emperor often read in

them till very late at night. As Tycho left it to the
Emperor to fix the amount of his salary, it was settled that
this was to be 3000 florins a year, in addition to some
" uncertain income which might amount to some thousands."
Tycho tells all this in a long letter to his old friend Vedel,
which he wrote on the 18th September following, in which he
adds that some councillors were against these arrangements,
pointing out that there was nobody at court, not even
among counts and barons of long service, who enjoyed so
large an income ; but as the Emperor insisted on it, and
neither the Secretary of State, Rumph, nor the Chamberlain,
Trautson, spoke against it, it was settled, and 2000 florins
were at once paid to Tycho. The Emperor even ordered
that the salary should date from the time when Tycho had
been invited to Prague, as he had not accepted service elsewhere since then. The Emperor also of his own accord
promised him an hereditary estate whenever one should fall
to the Crown, in order that he and his family might feel
secure.[1] It was afterwards ordered that 2000 florins a
year were to be paid to Tycho from the Treasury, and 1000
from the estates of Benatky or Brandeis, both dating from
the 1st May 1599.[2]

In the meantime Tycho had unpacked the few instruments he had brought with him, which were examined
with great interest by Corraduc, Hagecius, and other men
of learning, as well as by the Emperor, who desired him to
send for the remainder as soon as possible. Wishing to
display the same taste and elegance in his arrangements as
formerly at Uraniborg, Tycho had a pedestal made on which
instruments might be placed, and the four sides of this

---

[1] Bang's *Samlinger*, ii. p. 511 ; Weistritz, i. p. 175.
[2] One florin (*schock meissn.*) = 5 mark 81 pf. ; 1000 florins therefore about
£300 ; but the value of money in Bohemia appears at that time to have been
about four times as great as now.

pedestal were adorned with pictures of King Alphonso, with
Ptolemy and Al Battani sitting below him; Charles the
Fifth with Copernicus and Apianus; Rudolph the Second,
and below him Tycho, seated at a table looking towards the
Emperor; and lastly, Frederick the Second with Uraniborg.
Under the last picture was an epigram by the Imperial
poet-laureate.[1]

In his as yet unsettled state Tycho was not able to commence observations with the vigour of former days, the only
observation of any interest made at this time being one of
the end of a small solar eclipse at sunrise on the 22nd July
(new style, which Tycho used from henceforth). One of his
pupils, Johannes from Hamburg, observed this eclipse with
the little gilt quadrant, from the tower of a neighbouring
college.[2]

But Tycho did not wish to settle within the city of Prague.
In his letters he states that he did not like that the widow
of Curtius should leave her house for his sake, and he feared
to be too much disturbed by visitors. Tradition speaks of
his being annoyed by the constant ringing of bells at night
in the neighbouring Capuchin monastery,[3] but this may more
probably refer to his stay in the city during the last year
of his life, or it may never have happened; at any rate, it
is not mentioned by Tycho himself. But he was accustomed
to a country residence, with plenty of fresh air, and he probably longed to get away from the city, which was not very
clean, if we may believe Fynes Moryson, who had visited it
only seven years before Tycho's arrival, and who gives the

---

[1] Gassendi, p. 161, where another poem composed on the same occasion is
also given.

[2] Ibid., p. 162; *Barrettus*, p. 844. The quadrant is described above, p. 102.

[3] Mädler, *Pop. Astronomie*, 1st edit., 1841, p. 561 (not in the latest editions), and Heiberg, *Urania, Aarbog* for 1846, p. 131. According to another
tradition, the monks did not like the neighbourhood of the heretic, and got up
an apparition of a ghost to persuade the Emperor to turn him out.

following description of it :—" On the west side of Molda is
the Emperour's castle, seated on a most high mountaine,[1] in
the fall whereof is the suburbe called *Kleinseit* or little side.
From this suburbe to go into the city, a long stone bridge
is to be passed over Molda, which runnes from the south to
the north and diuides the suburbe from the city, to which
as you goe, on the left side is a little city of Jewes, compassed with wals, and before your eies towards the east is
the city called new Prague, both which cities are compassed
about with a third, called old Prague.   So as Prague consists of three cities, all compassed with wals, yet is nothing
less than strong, and except the stinch of the streetes driue
backe the Turks or they meet them in open field, there is
small hope in the fortifications thereof.   The streets are
filthy, there be diuers large market places, the building of
some houses is of free stone, but the most part are of
timber and clay, and are built with little beauty or art, the
walles being all of whole trees as they come out of the wood,
the which with the barke are laid so rudely as they may on
both sides be seen."

When the Emperor learned that Tycho Brahe wished to
reside outside Prague, he gave him his choice between the
three castles of Lyssa, Brandeis, and Benatky, " zur Exer-
cirung seines Studii."   Having seen them all, and having
learned that Brandeis (which was situated rather low) was
the Emperor's favourite hunting-lodge, Tycho selected
Benatky on the River Iser (a tributary to the Elbe), about
twenty-two miles north-east of Prague.   The Castle of
Benatky, which the Emperor had recently purchased from
Count Dohnin, had been erected in 1522 in the place of
an older castle which had been destroyed during the Hussite
wars.   It has since Tycho's time been considerably enlarged,
so that the building inhabited by him now only forms the

[1] Hradschin, where the house of Curtius was situated, west of the castle.

western wing. The present church tower is also a later
addition. The castle is situated close to the town of Nové
Benatky (in German, Neu Benatek) on the right bank of
the Iser, on a hill raised about two hundred feet over the
river. The castle commands a fine view of the vineyards
and orchards on the hilly northern (right) bank and tilled
fields and pasture-lands on the southern, which latter are
not seldom flooded by the Iser, so that the inhabitants on
such occasions are surrounded by a lake.[1] This may account
for the name of Venetiæ Bohemorum by which Benatky has
frequently been called, though Tycho believes that the
general beauty of the surroundings has also contributed to
the use of this name.[2] On the way to Benatky, Tycho sent
from Brandeis a letter to Longomontanus at Rostock, dated
the 20th August, in which he mentioned that the road was
level, and that the journey took about six hours; an official
from Brandeis was that day or the next to conduct him and
his belongings to Benatky, where he expected to remain
until he got the estate which the Emperor intended to
confer on him in fief.[3] As soon as Tycho arrived at
Benatky he set about altering the building and constructing an observatory and a laboratory. As usual, he expressed his pleasure at having at last found a resting-
place in various Latin poems, two of which were inscribed
over the entrances to the observatory and the laboratory.[4]
The principal instruments were to be placed in separate rooms,
as at Hveen, all connected with each other, and with the
laboratory and residence, and a separate entrance was to be

---

[1] Description of Benatky by David. See Zach's *Monatliche Correspondenz*,
vi. p. 475 (1802). On the appended plate the wing in the centre and the
church-spire were added after Tycho's time.

[2] Letter to Pinelli, *Aus Tycho Brahe's Briefwechsel*, p. 12. Tycho always
calls the place Benach.

[3] Gassendi, p. 163; Bang's *Samlinger*, ii. p. 501 (Weistritz, i. p. 164).

[4] Gassendi, p. 164.

CASTLE OF BENATKY.

made for the Emperor, who had reserved an adjoining house for his own use whenever he visited Benatky.[1] Whether Rudolph ever came to Benatky while Tycho was there is not known, but it is not likely, as he was again in the autumn of 1599 driven from Prague to Pilsen by the plague, and did not return till July 1600. Tycho also left home for the same reason towards the end of 1599, and lived for six or seven weeks at an Imperial residence at the village of Girsitz, a few miles south of Benatky, where some observations were made in December.[2] It was during this new visitation of the plague that the Emperor desired Tycho to give him the prescription for his "elixir" against epidemic diseases, as already mentioned.[3]

In the meantime the family had arrived from Dresden, and as everything now appeared to promise Tycho that he had found a haven for the remainder of his days, he sent about the end of September his eldest son, together with a certain Claus Mule, to Denmark, to remove the four large instruments which were still at Hveen, and took this opportunity of sending a number of letters to his family and friends. Among these letters was one to Valkendorf, asking him to facilitate the transport of the instruments,[4] one to his own brother, Axel, to the same purport, another to Longomontanus, and a very long one to his old friend Vedel. In this he gave a very full account of his doings since he left Hveen, which he asked Vedel to incorporate in his Danish history, so that it might be handed down to posterity, whether printed or not.[5] Tycho's daughter Magdalene

[1] Letter to Sophia Brahe in *Breve og Aktstykker*, pp. 85–86.

[2] *Barrettus*, pp. 850 and 856 ; *Breve og Aktstykker*, pp. 98 and 108.

[3] See above, p. 130.

[4] The letter is not extant, but Tycho alludes to it in the letter to Longomontanus (Gassendi, p. 167 ; Weistritz, i. p. 186). The Emperor had directed Barwitz to write to the Danish Privy Councillor, Henrik Ramel, on the same matter.

[5] This letter was published at Jena in 1730 (23 pp. 4to) by G. B. Casseburg,

sent a letter to Claus Mule's mother, in which she also
described the travels of the family.[1] To his kinsman Eske
Bille, who seems to have done his best for Tycho in the
way of executing commissions and looking after his affairs
at Hveen and elsewhere in Denmark, Tycho also wrote on
this occasion. Bille had some months before sent him 700
daler, which however did not reach Prague till a short time
before Christmas, and he was to receive some money which
Tycho's cousin, Axel Gyldenstjerne (Governor of Norway),
owed him; and on the other hand, he was to pay 5000
daler which Tycho owed to the widow of Heinrich Rantzov,
which he did in the course of the year 1600.[2] Tycho's son
got the instruments at Hveen dismounted and sent by sea
to Lübeck, after which he returned to Bohemia, where he
arrived in January 1600 with a supply of salt fish from
Hveen, which island Tycho continued to hold in fee till
his death. He was also the bearer of a great many letters
from relatives and friends—among others, of one from
Tycho Brahe's aged mother.

At that time the instruments were still at Lübeck, probably because Tycho's agent there was unable to get them
sent on to Hamburg, where they did not arrive till the
following April. On the 8th September 1599 the Emperor had written to the Burgomaster and Senate of Hamburg, desiring them to forward the instruments by ship on
the Elbe, and Tycho himself had written to them on the
29th September, requesting them to get the instruments
under way before the Elbe froze over, but these letters were

Tychonis Brahe Relatio de statu suo post discessum ex patria, and more
accurately in the Dänische Bibliothek, iii. 1740, p. 180 et seq. Translated in
Weistritz, i. p. 169 et seq.

[1] Danske Magazin, ii. p. 359 ; Weistritz, ii. p. 365. Claus Mule was a son
of the Burgomaster of Odense in Denmark. A letter from Tycho to him (apparently written while Mule was abroad, perhaps at Rostock, as it alludes to
Professor Caselius) is quoted above p. 240, footnote.

[2] Breve og Aktstykker, pp. 50, 57, 93, 101, 117, 148.

not read in the Senate till the 21st April following, when
the agent at Lübeck had at last forwarded the instruments to Hamburg.[1] A similar delay occurred with the bulk
of the instruments, books, &c., which Tycho had himself
brought from Denmark as far as Magdeburg. About the
transport of these to Prague by the Elbe the Emperor had
also written in September 1599 to the civic authorities at
Magdeburg, and he wrote a reminder to them some time
after; but the Town Council coolly replied that they were
unable to do anything, and, among other excuses, they
mentioned the great damage done to the town when the
celebrated Elector Maurice of Saxony, as commander of the
Catholic forces, had besieged it some fifty years before.
Having, to the disgust of his Austrian councillors, swallowed
this affront, which showed how little the Imperial authority
was respected in the North of Germany, Rudolph addressed
himself to the Chapter of Magdeburg, and Tycho forward-
ing this letter by a servant of his in April 1600, also wrote
to the Chapter begging them to help him in the matter.[2]
It appears from a letter which Tycho wrote in September
1600 to Duke Otto of Brunswick (who wanted his horoscope prepared) that the instruments and books had then
only got as far as Leitmerits, in Bohemia, and in November
1600 Tycho wrote to Landgrave Maurice that he had at
last got all his twenty-eight instruments at Prague.[3] But
he had then long ago left Benatky.

While the instruments were on their way to Bohemia,
Tycho was endeavouring to push forward the alteration of

---

[1] The two letters (in the city archives of Hamburg), printed in Friis, *Tyge
Brahe*, pp. 320 and 324. Letter from Tycho to Vincent Müller, Burgomaster
of Hamburg, of April 24, 1600, in *Breve og Aktstykker*, p. 125.

[2] Tycho's letter to the Chapter, *Aus Tycho Brahe's Briefwechsel*, p. 21;
compare *Breve og Aktstykker*, p. 114.

[3] *Breve og Aktstykker*, pp. 141 and 143. Tycho wrote to Eske Bille on
November 16, that on looking over his things, he noticed that some articles
were missing which might still be at Copenhagen or at Lübeck. *Ibid.*, p. 149.

and addition to the Castle of Benatky; but he had a
good many obstacles to contend with, which must often
have made him think with bitter regret of the easy times
he had had in Denmark, where an order on the Exchequer
was at once exchanged for cash without any trouble. At
Benatky everything had to be done through Kaspar von
Mühlstein, manager of the crown estates of Brandeis
and Benatky; and as the estates were in a sad condition,
and the Bohemian Exchequer was always empty, the
manager was in a bad plight, as Tycho wanted money,
and thought the Emperor's orders should produce the
money immediately. On the 2nd December 1599, Mühlstein wrote to the President of the Bohemian Treasury,
announcing that the building operations were in progress,
but that Tycho continued to make new plans, so that the
cost would very far exceed the estimate. As Mühlstein
could not consent to this without orders from the Treasury,
Tycho had threatened him with the Emperor's displeasure,
and said that he would leave Bohemia again, and let the
world know the reason why. Mühlstein had now received
a letter from Barwitz, in which the latter informed him that
his Majesty had taken Tycho into special favour, and ordered
to let him, in addition to the buildings commenced, erect a
wooden dwelling-room and a furnace, for which eighty florins
were granted. Mühlstein wrote back that he could not do
this without an order from the Treasury. Tycho had also
shown him a communication from Barwitz to the effect that
the Emperor granted him a thousand florins from the estate of
Benatky, and Tycho now demanded the money. Mühlstein
answered that he had neither got instructions from the
Emperor nor from the Treasury, and even if he had, he
did not know where to get so much money from, and it
would be much better to spend it on improving ponds,
stocking the land, &c. He had also to mention that he

had every week to supply Tycho with wood from the Imperial forest, and with charcoal for distilling water. Mühlstein therefore requested the President to consider all this; he would soon send a specification of the outlay already incurred.

The matter was referred by the Treasury to the Emperor, who from Pilsen on the 10th December issued a decree, countersigned by Barwitz, in which he informed the Treasury that he had taken the mathematician Tycho Brahe into his service, and granted him the Castle of Benatky for his use until further orders, and directed that he was to be paid one thousand florins annually from the 1st May 1599 from Benatky or Brandeis, and the cost of building some small rooms (but not more than already granted, as was known to the manager at Benatky).[1] This decree having pacified the conscience of Mühlstein, the building operations were proceeded with, and Tycho and he seem to have got on better afterwards; at least Tycho went to Prague in the following spring to attend Mühlstein's wedding.

While the new observing rooms were being prepared,[2] Tycho kept up his correspondence with scientific men, and endeavoured to enlist assistants for the new observatory. In the above-mentioned letter to Longomontanus, Tycho wrote (after requesting him to help in packing the instruments) that he hoped his old pupil would come back to him; he was expecting Johann Müller from Brandenburg, and he had got the Emperor to write to the Elector to permit Müller to go to Prague, as they had agreed at

[1] Hasner, p. 7 et seq.

[2] Where these were situated is not known, and there are no remains of Tycho's buildings or inscriptions, &c., as the Castle of Benatky has changed owners many times since then. In March 1801 Professor Aloys David determined the latitude of Benatky and found 50° 17' 24" (Tycho gives 50° 18' 15") and longitude 50m. 0s. east of Paris. *Monatl. Correspondenz*, vi. (1802), p. 477. Tradition attributes a still existing sundial at Benatky to Tycho, but there is no proof of its having been constructed by him.

Wittenberg. He also hoped that David Fabricius, from
Ostfriesland, whom Longomontanus had met at Wandsbeck,
would come to act both as domestic chaplain and as observer ;
and he was getting two students from Wittenberg, who had
offered themselves through Jostelius, as he hoped to start
again an astronomical school for the benefit of posterity and
for the glory of God and the credit of the Emperor. Possibly Christopher Rothmann would also come, as he had
recently written to Rollenhagen of Magdeburg (a well-
known writer on astrology and many other things), so that
he was not dead, as Tycho had for some time believed ; but
that bear-like and Dithmarsian brute (*ursina ista et Dithmar-
sica bestia*) had told a double lie when he had spread the
rumours that Tycho had fled from Denmark for some great
act of villainy, and that Rothmann had died from debauchery.
The same Reymers had secretly absconded lately from Prague,
but he would yet meet the punishment he deserved.[1] The
sheets which were still wanting in the first volume of the
Progymnasmata, and which the Hamburg printer had done
so badly, were soon to be printed, and when Longomontanus
came, all might be settled, so that the book might be issued
together with the second volume (on the comet of 1577),
while the third volume on the other comets might follow.[2]

Several of the collaborators whom Tycho in this letter
hoped to secure did not put in an appearance. Longomontanus arrived in January 1600 with Tycho's son,[3] but Rothmann never came ; Fabricius only came in June 1601 for a
couple of weeks,[4] and Müller did not arrive till after March

---

[1] In the letter to Vedel, Tycho also mentions this, and adds that Reymers
had left his wife behind, who (of course) enjoyed an evil reputation.

[2] In January 1600 Tycho inquired from Scultetus whether the printing
could not be done at Görlitz (*Aus Tycho Brahe's Briefwechsel*, p. 16).

[3] *Kepleri Opera*, viii. p. 715.

[4] Apelt, *Die Reformation der Sternkunde*, p. 271. Fabricius went with a
message from the Count of Ostfriesland to his envoy at Prague. A letter
which he wrote to Kepler from Prague is printed, *Opera*, i. p. 305.

1600,[1] and left again in the spring of 1601, after which he disappears from the history of science altogether. But in the meantime negotiations had been entered into with a far greater man than any of these, which terminated in the removal of Kepler from Gratz to Prague, an event which produced the happiest results.

Johann Kepler was born on the 27th December 1571, at Weilderstadt, in Würtemberg, and studied from 1589 at the University of Tübingen under the talented mathematician Michael Mästlin, through whom he became acquainted with the Copernican system, and convinced himself of its being the only true representation of the planetary system. He completed his studies in the faculty of Arts, and took the Master's degree in 1591, after which he entered the theological faculty, and spent the next two years in studying the intensely narrow-minded dogmas which then prevailed in the Lutheran Church, and which were so distasteful to him that he was soon known among theologians as one unfit for a clerical career. When, therefore, in 1594 the post of " provincial mathematician " of Styria was offered to him, he was urged by his friends to accept it; and though he hesitated somewhat, as he had not particularly devoted himself to the study of mathematics, he yielded in the end, as it might not be easy for him to find suitable employment in Würtemberg, while the lively intercourse between the numerous Protestants in Styria and their co-religionists at Tübingen helped to bridge over the distance of Gratz from his home. In Gratz the young professor lectured less on mathematics than on classics and rhetoric, while from 1594 he prepared annual calendars, with the usual meteorological predictions and hints on the political events of the coming year. In 1596 his first great work appeared, *Prodromus Dissertationum Cosmographicarum continens Mysterium Cos-*

[1] *Breve og Akstykker*, p. 110.

mographicum, in which he set forth a relation between the five regular polyhedra and the distances which then were assumed between the planets and the sun in the Copernican system. The genius of the writer was conspicuously displayed in this book and at once attracted attention. Kepler had already in November 1595 addressed a letter to Reymers, in which he explained the ideas contained in his forthcoming work, but the "Cæsarean mathematician" took no notice of the letter of the unknown young man until June 1597, when he had probably heard the book well spoken of, and wrote to Kepler to ask for a copy.[1] In the mad book which he published in the same year, Reymers inserted Kepler's letter of 1595, at which Tycho did not feel particularly pleased, though he had sense enough to acknowledge that Kepler had merely been civil to a man whom he only knew through his scientific writings. In a letter which Tycho wrote from Wandsbeck on the 1st April 1598, to thank him for a copy of the *Prodromus* (which Kepler had recently sent with a respectful letter), he expressed himself to that effect. At the same time he gave due praise to the ingenious speculations of Kepler, though he had some doubts as to the numerical data employed, and of course he could not help regretting that the Copernican system was the foundation on which Kepler had built. He expressed, however, the hope that Kepler would yet adopt something similar to the Tychonic system, which made Kepler (who throughout furnished the letter with marginal notes) remark: "*Quilibet se amat.*"[2] Shortly afterwards Tycho

---

[1] Ursus had just published a work on chronology, *Chronotheatrum sive Theatrum temporis annorum* 4000, of which he sent Kepler a copy with the letter (the full title is given by Hanisch, *Epist. ad I. Keplerum,* p. 90; it must be an extremely scarce book). Kepler was so little aware of the enmity between Tycho and Ursus that he even asked Ursus to forward a copy of the *Prodromus* to Tycho (*Opera,* i. p. 233).

[2] *Epist. Kepleri,* p. 102; *Opera,* i. p. 43 and p. 219; Kepler's marginal notes, p. 189.

also wrote to Mästlin (to whom he had ten years previously sent his book on the comet of 1577 without hearing from Mästlin since then), and repeated some of the doubts he had already expressed to Kepler.[1] The latter was, however, not discouraged by these doubts, and wrote to Mästlin that he could in no way accept the Tychonic system, and that Tycho had abundance of riches which he did not use properly, as was generally the way with rich people, and it would be well to extort his riches from him by getting him to publish all his observations.[2] To Tycho himself Kepler addressed a letter in which he, with manly and unaffected eloquence, protested against the crafty use which Reymers had made of his complimentary letter, which he had written simply because he had read Reymers' *Fundamentum astronomicum* with much profit, had been advised by some Styrian noblemen to make a friend of this man on account of his influential position (though they called him a new Diogenes), and had felt a desire of communing with a mathematician, since there were none in his own neighbourhood.[3] The whole letter evidently made a good impression on Tycho, as Kepler's open and noble mind is reflected in every line, and Tycho wrote in reply·that he had not required so elaborate an apology.

The literary intercourse which had thus been opened between Tycho Brahe and Kepler was soon to become a personal one. The very numerous Protestants in Styria had hitherto been perfectly unmolested by their Catholic rulers, but during a pilgrimage to Loretto which Archduke Ferdinand (afterwards the Emperor Ferdinand II.) undertook in 1598, he vowed to root out the heretics from his dominions, and on the 28th September all preachers and the teachers at the Gymnasium of Gratz were ordered to leave

[1] *Opera*, i. p. 45 *et seq.*   [2] Ibid., p. 48 *et seq.*
[3] Ibid., p. 220 *et seq.*; *Epist. ed. Hanschius*, p. 106.

the town before sunset. Kepler had to leave his family (he had eighteen months before married a young widow, who was the mother of a girl seven years old) and depart for Hungary. He was, however, recalled within a month, as some of the Jesuits were much interested in his scientific work, and hoped that he might be persuaded to change his faith. He soon saw that he could not hope to be left in peace very long, and he made vain attempts to obtain some employment at Tübingen. Mästlin was, however, unable to help his former pupil, and Kepler saw no other opening elsewhere. Meanwhile Tycho had been invited to Prague, and Kepler, who had already been anxious to meet him, was now more than ever desirous of doing so, and thought of undertaking a journey to Wittenberg for the purpose of conferring with Tycho. In August 1599 he learned from Herwart von Hohenburg, Chancellor to the Duke of Bavaria, who was a correspondent of Kepler's, and had frequently consulted the rising astronomer on matters connected with chronology,[1] that Tycho had arrived at Prague and was to have a salary of 3000 florins. Herwart ended the letter by saying, " I wish you had such a chance, and who knows what fate may have in store for you." Kepler now consulted a number of friends and some men of influence at Prague, among whom was Baron Hoffman, a privy councillor who was well acquainted with Tycho, but who would at first give only an evasive answer. The most sensible advice was given by Papius, a physician, who had been obliged to leave Gratz as a Protestant, and was then practising his art at Tübingen. He suggested that Kepler should make all possible inquiries at Prague about the con-

---

[1] Tycho had also for some time corresponded with Herwart, to whom he, on the 31st August, wrote a letter explaining his lunar theory, and particularly the calculation of eclipses. About this letter and Herwart's answer, see Gassendi, p. 165.

ditions on which he might become associated with Tycho,
and that he should let his literary productions be shown
at Prague in order to pave the way for him there.[1]   The
latter part of the advice was certainly superfluous, and Tycho
himself was more than willing to accept Kepler's services.
In a long letter which Tycho wrote from Benatky on the
9th December 1599, he expressed his hope of soon meeting
Kepler, though he did not wish that the latter should be
driven to him by misfortune, but by his own free will and
his love of science, and he assured Kepler that he would
find in him a friend who would always stand by him with
help and counsel.[2]

Tycho's letter did not find Kepler at Gratz.   He had
at last made up his mind to examine the state of things
at Prague with his own eyes, and, encouraged by Baron
Hoffmann, he started with this nobleman from Gratz on the
6th January 1600, and arrived at Prague about a fortnight later.   On the 26th Tycho wrote to Hoffmann that
he had with great pleasure heard of their arrival, and
thanked Hoffmann for being the means of introducing Kepler
to him.   Tengnagel (who had just returned from his home
in Westphalia) and Tycho's eldest son were the bearers of
this letter, as well as of another for Kepler, in which Tycho
apologised for not welcoming him in person, but he rarely
went to Prague except when called by the Emperor ; the
oppositions of Mars and Jupiter were now to be observed,
and the other three planets and a lunar eclipse likewise,
so that he did not like to interrupt his work, but he would
receive Kepler, not as a guest, but as a dear friend and
colleague.[3]   On the 3rd February Kepler arrived at Benatky
with a civil answer from Hoffmann, warmly recommending

[1] *Kepleri Opera*, viii. p. 709.
[2] *Epist. ed. Hanschius*, p. 108 *et seq.* ; *Opera*, i. p. 223 and p. 47.
[3] *Kepleri Opera*, viii. p. 716 ; *Aus Tycho Brahe's Briefwechsel*, p. 18.

him to Tycho.[1] Within a few days some preliminary arrangements were made with regard to the distribution of work between the various assistants. Tycho's younger son, Jörgen, was to have charge of the laboratory ; Longomontanus had the theory of Mars in hand ; Kepler at first had to put up with the promise of the, next planet which was taken up, but afterwards Mars was intrusted to him, as he was particularly eager to attack this most difficult planet, while Longomontanus undertook the lunar theory.[2] But though Tycho was most cordial to Kepler, he did not enter very much into learned discourses, so that Kepler had often to coax him into answering some question while they were at table. He had a feeling that he was not looked upon as a man of recognised scientific standing, but merely as an ordinary assistant to the world-famed Tycho Brahe, and yet he felt that he ought to have full access to the great treasure of observations which Tycho possessed. In a document which Kepler drew up for the information of his friends,[3] he remarks that Tycho had hitherto, by the magnitude of his undertakings, been prevented from discussing his observations, and now that old age was approaching and soon would enervate him, he would hardly be able to undertake that great work himself. If the journey from Gratz was not to have been made in vain, either Tycho should allow him to copy the observations, which he doubtless would refuse, since they were his treasure to which he had devoted all his life, or he should admit Kepler to a share in the working out of the results from them. And the Emperor should do as King Alphonso had done, and associate others with Tycho. He himself ran the risk of losing his post at Gratz, for if Tycho took offence at some-

---

[1] *Opera*, pp. 716, 717 ; *Briefwechsel*, p. 19.
[2] Kepler, *De Stella Martis*, chap. vii. (*Opera*, iii. p. 210).
[3] *Opera*, viii. p. 718 *et seq.*

thing in Bohemia and went away, or if something happened to him, what would then become of himself (Kepler), and perhaps the observations would be lost or become inaccessible.

Influenced by these considerations as well as by the possible difficulty of getting the consent of his wife's relations to her removal from Gratz, where she had a small property near the town, to Benatky, where she would have to live among foreigners, Kepler drew up several different proposals for a formal agreement with Tycho Brahe, in which he most carefully tried to secure his future position, both as regards the lodging of his family at Prague, or at least in an upper storey at Benatky, with separate kitchen, supply of fuel and victuals, &c., as also with regard to his scientific work.[1] He added that he would not be content with general promises, which was a rather superfluous remark, since the minuteness with which he had specified his demands made this very evident. On the 5th of April the matter was discussed verbally between Tycho and Kepler in the presence of Jessenius of Wittenberg, and in answer to Kepler's written demands, Tycho partly read himself, partly let Jessenius read, a written answer which followed Kepler's demands point for point.[2] Tycho took the whole matter far more quietly than might have been expected from a man of his hot temper and imperious ways, but though he offered to bear part of Kepler's travelling expenses, and to do his utmost to get him settled at Prague (if he absolutely wanted to live there), or in a separate house in or near Benatky, he was unable to guarantee anything about salary or the keeping open of Kepler's Styrian post, until he could communicate with the Emperor and with Corraduc and Barwitz. Though Tycho begged Kepler to wait until his servant Daniel came back from Pilsen with replies to letters which

[1] *Opera*, viii. pp. 721-724.   [2] Ibid., p. 725.

Tycho had written to Corraduc, Kepler refused to listen to
reason, and left Benatky the following day with Jessenius to
return to Baron Hoffmann at Prague.[1]

There had evidently for several weeks been some misunderstanding between the two astronomers, as Tycho
already, on the 6th March, had written to Hoffmann that
as soon as he could find time from other occupations, they
would both drive to Prague to discuss with Hoffmann the
question as to Kepler's position.   It cannot, however, have
been Kepler's uncertain prospects alone which brought about
the crisis on the 5th April, for it appears that Kepler on the
following day wrote a very violent letter to Tycho, of which
the latter took no notice beyond sending it to Jessenius.[2]
It seems, therefore, probable that Kepler, as we hinted
above, felt himself treated too much as an inferior and a mere
beginner, while he, conscious of his genius, expected to be
regarded as an independent investigator.   Tycho, however,
always expressed himself most kindly of Kepler in his letters,
and it probably never occurred to him that he ought not to
place Kepler on the same footing as his assistants.   He now,
on the 6th April, wrote a short letter to Hoffmann, in which
he referred him to Jessenius for information as to the difference between Kepler and himself, and expressed the hope
that Hoffmann, with his prudent advice, would endeavour to

---

[1] Among the Kepler MSS. in the Hofbibliothek at Vienna is a declaration
written and signed by Kepler on the 5th April 1600, in which he, having been
hospitably received by Tycho Brahe, "auch diese gantze zeit vber aller müg-
licheit nach also tractirt worden, das ich mich hingegen iederzeit, zue aller
Vnderthaniger Danckbarkheit schuldig erkhennen"—pledges himself to keep
secret all observations or inventions which Tycho Brahe had communicated or
might communicate to him. (Friis, *Tyge Brahe*, p. 327). This MS. is not
mentioned by Frisch. If the date is correctly given, this document may have
been an attempt on Kepler's part to conciliate Tycho, in order that the latter
might make some concession as to the scientific work.

[2] Kepler's letter is only known from Tycho's letter to Jessenius. *Opera*,
viii. p. 728. In this letter Tycho asks Jessenius to find out whether Kepler
had now taken up with Reymers Bär, who had returned to Prague.

settle the matter.   He was not disappointed, for the remon-
strances of Hoffmann, who was anxious to·see Tycho and
Kepler co-operate in the service of science, succeeded in
softening Kepler, to which Jessenius, as a friend of both
parties, also contributed.   About three weeks after his departure from Benatky, Kepler therefore wrote a repentant
letter to Tycho, in which he acknowledged that he had met
with nothing but kindness from Tycho; and begged to be
forgiven for his conduct, which was the result of a youthful
and choleric temper and his shaken health.[1]   The two
astronomers met at Prague, were reconciled, and went back
to Benatky together, where Kepler now stayed four weeks,
until at the beginning of June he left Bohemia for a while to
settle his affairs at Gratz.   At parting, Tycho gave him a
most flattering testimonial, in which he spoke in the highest
terms of the manner in which Kepler had devoted himself
to scientific work at Benatky.[2]

Kepler had hoped to be able to retain his appointment at
Gratz and get leave for a year or two to work with Tycho.
To settle permanently with him he was not inclined, but he
soon had very little choice in the matter.   Early in August
an Ecclesiastical Commission arrived at Gratz, and every
official had to appear before it and to state whether he would
become a Roman Catholic or not.   Those who refused were
ordered to dispose of their goods and to leave the Austrian
provinces within forty-five days.   Among these was Kepler,
who again applied to Mästlin and Herwart for advice.
But at Tübingen there was no opening, and Herwart
strongly advised him to go to Prague.   There seemed to
be no help for it now, and no matter what doubts Kepler
might have as to the feasibility of living in the same house
with Tycho and his family, or of preparing planetary tables
in concert with a man from whom he differed on the most

[1] *Opera*, viii. p. 729.   [2] Ibid., p. 730.

fundamental questions, he had no choice but to go to Prague.

In the meantime the Emperor had returned from Pilsen to Prague in July 1600, and about the same time, or shortly afterwards, Barwitz advised Tycho to leave Benatky and move to Prague, as the Emperor would like to have his astronomer near him. Probably Tycho was not sorry to leave Benatky, where he and Mühlstein still kept up a running fight about money matters and building operations. He therefore left Benatky and took up his quarters temporarily in the hotel *Beim goldenen Greif*, on the Hradschin,[1] while his instruments were placed in Ferdinand I.'s villa, not far from the castle.[2] A few days after his arrival, Tycho was received in audience by the Emperor, who conversed with him for an hour and a half. The Emperor inquired about Tycho's work, upon which Tycho remarked that he necessarily required more help, and suggested that Kepler might be attached to the observatory for a year or two. The Emperor nodded his consent to this, and desired Tycho to mention this proposal in a memorial about his requirements, which was to be sent to Barwitz. Tycho afterwards spoke to Corraduc, and asked that the Styrian authorities might be requested officially to give Kepler leave of absence for two years, and let him retain his salary, to which the Emperor would add a hundred florins on account of the expense of living at Prague. Tycho wrote to Kepler

[1] Hasner, p. 10. The house "Zum goldenen Greif" is still in existence (Neuweltgasse, No. 76), but is no longer an inn or a hotel. This quarter of Prague was then the most aristocratic one, being close to the castle. It was thoroughly devastated by the Prussians in 1757 by bombardment, and has since been the poorest part of the city.

[2] Now called the Imperial Belvedere. On January 24, 1601, Tycho wrote to Magini that he had now all his twenty-eight instruments "non longe ab Arce, in Cæsaris quadam magnifice extructa domo." Carteggio, p. 241. The observations at Benatky had been stopped at the end of June 1600, and they were not resumed till the 2nd December, "in domo Cæsaris horto vicina ubi instrumenta mea adhuc disponebantur." Barrettus, p. 860.

VILLA OF FERDINAND I.

on the 29th August, and gave him an account of all this,[1]
and as Tycho's removal from Benatky to Prague promised
to do away with some of the difficulties, Kepler, though
still hesitatingly, set out for Prague early in September.
Troubled not only by his anxiety for the future, but also by
an intermittent fever which clung to him for nearly a year,
he left most of his luggage half-way at Linz, in case he
should yet want to go to Würtemberg. Ill and miserable,
he arrived with his family at Prague in October, and was
hospitably received in Hoffmann's house. His first communication with Tycho was by a letter (on the 17th October [2]), in which he wrote that his hopes of retaining his
Styrian salary were now at an end, and their former agreement consequently also ; but as Tycho had laid the matter
before the Emperor, he had thought it best to come. He
had, however, very little money left, and could only wait
four weeks, and if his position at Prague could not be made
secure within that time, he would have to look out for himself elsewhere.

Tycho was much pleased to see Kepler return to Prague,
the more so as he had lost his most experienced assistant,
Longomontanus, who had wished to return to Denmark, and
had received his discharge on the 4th August, when Tycho
at parting gave him a very kind letter of recommendation.[3]
It took a long time to get the question about Kepler's
salary settled by the Government, but he and his family
soon removed from Hoffmann's to Tycho's house, and he
began work. This was probably not until the Emperor had
purchased Curtius' house from the widow for 10,000 thaler,

---

[1] *Opera*, viii. p. 732.   [2] Ibid., pp. 734–737.
[3] Printed by Gassendi, p. 174. In 1603 Longomontanus became head-
master of Viborg school, in Jutland (where he had been educated himself);
in 1605 Professor at the University of Copenhagen ; in 1607 Professor mathematum superiorum. He died in 1647, before the University Observatory on
the *Round Tower* (which existed till 1861) was finished.

and Tycho had taken possession of it, which he did on the
25th February 1601.[1] Kepler still could learn nothing
about his salary, and continued, though in vain, to look
out for an appointment in Germany, while Tycho now and
then helped him with money. His health also gave him
cause for anxiety, as he could not get rid of the intermittent
fever, and early in 1601 he was troubled with a bad cough,
which even made him fear that he was consumptive. In
April he was obliged to go to Gratz to arrange some affairs
connected with his wife's property, whence he did not return
until August, having failed to accomplish his object, but
having recovered his health. A curious letter has been
preserved[2] which Kepler's wife wrote to him on the 31st
May, in which she tells him that Johann Müller had left
again; that Tengnagel had not yet given her any money,
but that he and Tycho were friends again, and that his
wedding (with Tycho's second daughter, Elisabeth) was to
take place a week after Whitsuntide.[3] This cannot have
been the first complaint Kepler received from his wife about
her getting no money, for he had already on the 30th May
written an indignant letter to Tycho, blaming him for not
having given her the twenty thaler which had been
promised. Tycho did not trouble himself to answer this,

---

[1] Gassendi, p. 176. The site of Curtius' house on the Loretto Place is
now occupied by the Černin Palace. Canon David determined the geographical position, lat. +50° 5′ 28″, 32° 3′ 37″ east of Ferro. In 1804–5 an
old tower was pulled down, which probably had been part of Tycho's obser
vatory.

[2] *Opera*, viii. pp. 739–741.

[3] "Der hanss Miller ist den 29 Mai mit seiner frau darvon vnd haim, der
diho Prei (Tycho Brahe) hat jm abgeförtigt vnd hat jm göben was er jm hat
zuegesagt, aber vom khaiser ist jm khain heler nit worten. Der Diho hat jm
sein hantl verdörbt beim khaiser er het sonst woll ein guette Verehrung
bekhumen so hat ehr des Diho müessen engelten. Der franz (Tengnagel) vnd
der Diho sint witer einss sie rihten ietz zu der hochzeit zue, der franz hat
mier noh khein gelt göben." . . . Müller left on the 26th, according to a letter
from Eriksen to Kepler (*Epistolæ, ed. Hanschius*, p. 176).

but let one of his pupils, Johannes Eriksen,[1] write to
Kepler that he had unasked, through his daughter, promised
her ten thaler soon after Kepler's departure, which she
also got on asking for them, and when she a fortnight
later again requested ten more, Tycho sent to her by
Eriksen six thaler and promised her more, though he had
not much cash at the time. All this he had done without
grumbling, and both he and his family had been kind and
obliging to Kepler's wife and her daughter. Tycho therefore
desired Eriksen to beg of Kepler to have more confidence in
him, and to conduct himself in future with more prudence
and moderation towards his benefactor, who had been very
patient with him, and wished him and his well.[2] This
letter had probably the desired effect, and Kepler, who at
heart was most generous and noble, but whose weak point
it was always to complain to everybody about money
matters, no doubt acknowledged having been too hasty.
When he returned to Prague in August, Tycho brought
him to the Emperor, who congratulated him on his recovery,
and promised him the office of Imperial mathematician on
condition that he should work jointly with Tycho on the
new planetary tables, which Tycho begged the Emperor's
permission to call the Rudolphean or Rudolphine tables.[3]

It was mentioned above that Tengnagel was engaged to
be married to Tycho Brahe's second daughter, Elisabeth.
On the 5th April 1601, Tycho wrote a letter (in Danish) to
his friend Holger Rosenkrands, inviting him to the wedding,
which was to take place between Easter and Whitsuntide,
and the following day he wrote another letter (in Latin), in

---

[1] This name occurs here for the first time. Perhaps he had come to Tycho
on the 15th August 1596, as we read in the diary : " Rediit Tycho Hafnia, cum
eo duo studiosi, alter Germanus commendatus a Landtgravio, alter Danus,
Joannes nomine."

[2] *Opera*, viii. p. 741.   [3] Gassendi, p. 177.

which he mentioned that he expected his sister Sophia.[1]
Neither she nor Rosenkrands came, however, to the wedding,
which was celebrated on the 17th June, after which the
married couple set out for Westphalia, the home of the
bridegroom, accompanied by Eriksen.[2]

Tengnagel does not seem to have occupied himself much
with astronomy, and probably did not take an active part in
the scientific work in Tycho's house. At Prague, Tycho
had not as many assistants as at Hveen. In addition to
Longomontanus, Müller, and Eriksen, he was assisted for
some time by Melchior Joestelius, Professor of Mathematics
at Wittenberg;[3] by Ambrosius Rhodius, who left Prague
shortly before Tycho's death, and likewise became Professor
at Wittenberg; by a certain Matthias Seiffart, who afterwards for some years assisted Kepler in computing and
observing; and from June 1601 by a young Dane, Poul
Jensen Colding.[4] It appears also that Simon Marius
(Mayer), who afterwards obtained some notoriety by laying
claim to various discoveries and inventions long after they
had been published by others, spent some time at Prague
with Tycho and Kepler in the summer of 1601.[5] The
Imperial physician, Hagecius, with whom Tycho had corresponded for so many years, died on the 1st September 1600,

---

[1] The Danish letter is printed in *Danske Magazin*, ii. p. 360 (translated in
Weistritz, ii. p. 366) ; the Latin one is only alluded to ibid. and iii. p. 23.

[2] *Epist. ed. Hanschius*, p. 179, and *Opera*, viii. p. 741.

[3] Joestelius must have returned to Wittenberg before June 1600, when he
observed the solar eclipse there (Kepler, i. p. 56).

[4] Son of a wealthy citizen at Kolding, in Jutland ; born 1581, died 1640 as
a clergyman in Seeland. Came to Tycho in June 1600, and was with him till
his death ; wrote an elegy on him, which is printed by Gassendi, p. 241.
About him see *Norsk Historisk Tidskrift*, ii. p. 338 (1872).

[5] On the 27th May 1601 Eriksen wrote to Kepler that Marggravii Anspach-
ensis Mathematicus, Simon Marius, was expected in a day or two, and would,
the writer hoped, relieve him of some of the observing (*Epist. ed. Hanschius*,
p. 176), but I cannot find any evidence that Mayer really came. He had
already in 1596 published a small pamphlet of the ordinary type on the comet
of that year.

after a prolonged illness, but Tycho found other scientific friends at Prague, among whom were Martin Bachazek, Rector of the University, Peter Wok Ursinus of Rosenberg, Baron Johan von Hasenburg (who was an ardent alchemist), and the Jewish chronologist, David Ganz.

As Tycho at Benatky or at Prague had never more than a few assistants at a time, and most of his instruments did not reach him till October or November 1600, the observations made in Bohemia cannot compare in fulness and extent with those made during an equal period of time at Hveen, to which disparity the interruptions caused by Tycho's various removals also contributed. In December 1600, and the first week of January 1601, observations were made in the Emperor Ferdinand's villa, and on the 3rd March the work was resumed in Curtius' house, where Tycho had just become settled. Kepler hardly took an active part in the observations, but he began preparing for the great work to which he afterwards devoted his life. When he arrived at Benatky in February 1600, Mars had just been in opposition to the sun, and a table of the oppositions observed since 1580 had been prepared, and a theory worked out which represented the motion in longitude very well, the remaining errors being only about 2'.[1] On this a table of the mean motion of Mars and the mean motions of the apogee and node for 400 years had been founded (as was done for the sun and moon in the *Progymnasmata*). But the latitudes and annual parallax at opposition (or the difference between the heliocentric and geocentric longitude) gave trouble, and Longomontanus was just then occupied with this matter.[2]

---

[1] Kepler, *De Stella Martis*, cap. viii. ; *Opera*, iii. p. 210. Apelt (*Reformation der Sternkunde*, p. 276), quoting Gassendi, believes his "duo minuta" to be a misprint for "duodecim;" but Kepler distinctly says "intra duorum scrupulorum propinquitatem."

[2] The greatest drawback of the Tychonic system was the difficulty of distinguishing between the real and apparent orbit of a planet; the greatest

Kepler therefore began to consider whether the theory might not after all be wrong, though it represented the longitudes so well; but during the short time he was at Benatky he was unable to make any progress in this problem, which it eventually took him four years to solve.

During Kepler's residence with Tycho at Prague between October 1600 and April 1601, he seems to have been mainly occupied with a piece of work which cannot have been congenial to him—a refutation of the book of Reymers Bär. Shortly after his arrival in Bohemia, Tycho began to take legal proceedings for libel against this person, who had fled to Silesia, from whence he, however, secretly came back some time afterwards.[1] In the summer of 1600 Tycho learned that there was *periculum in mora*, as Reymers was very ill; but even this did not soften Tycho's heart, and he persisted in having the poor wretch punished, and persuaded the Emperor to appoint a commission of four members, two barons and two Doctors of Law, to try the libeller. But just as the trial was about to commence, Reymers died, on the 15th August 1600. The Emperor directed the Archbishop of Prague to have every obtainable copy of the book confiscated and burned, while Tycho, who was rather unduly proud of his system of the world, wished to publish a book which was to contain all the documents on the subject of the alleged plagiarism, as well as a scientific refutation of Reymers' book. The preparation of the latter had to be undertaken by Kepler, who, while battling with intermittent fever in 1601, wrote his *Apologia Tychonis contra Ursum*, in which he showed that neither Apollonius of Perga nor

observed latitude at opposition was naturally assumed to be the inclination of the orbit, and this turned out to have a different value at different times, as if the Ptolemean oscillations of the orbit really existed.

[1] Kepler spoke to him at Prague in January 1600, without revealing his own name (*Opera*, i. p. 237).

any one else before Tycho had proposed the Tychonic system. Tycho's death made this memoir superfluous, and Kepler laid it aside, so that it has only recently been published in the complete edition of his works.[1] The same was the case with an unfinished reply to the attack of Craig on Tycho's book on the comet of 1577.[2]

In 1601 Kepler also occupied himself with the theories of Mercury, Venus, and Mars, and noticed that it was not possible to represent the apparent motion of the planets by assuming for the orbit of the sun (or the earth) a simple excentric circle with uniform motion, as had always hitherto been done, but that it would be necessary to have recourse to an equant as in the planetary theories of Ptolemy. When Kepler asked Tycho if he would not mention this in the *Progymnasmata*, he declined to do so, as it would take time to investigate the equal motion, and he wanted the book published at once. The subject was therefore merely alluded to by Kepler in the Appendix with which he wound up the book after Tycho's death.[3]

While Kepler was thus reconnoitring the ground for his future work on the planets, Longomontanus had before his departure finished the lunar theory and tables, the incomplete state of which had so long delayed the publication of the *Progymnasmata*. It is much to be regretted that the

---

[1] *Opera*, i. pp. 236-276. It is curious that Gassendi should have fallen into the same error with regard to Apollonius (*Vita Copernici*, p. 297). Kepler had already, in March 1600, at Benatky, written a short refutation, which is printed i. p. 281 *et seq.*

[2] *Opera*, i. pp. 279-281.

[3] In a letter to Longomontanus, Kepler wrote in 1605 :—"Ab Octobri 1600 in Augustum 1601 quartana me tenuit. Interim scripsi contra Ursum jubente Tychone, & alia ipsius studia pro ipsius arbitrio & meis viribus adjuvi. Speculatus sum, indignante Tychone, in Venere, Mercurio, Luna, in illis utiliter, in Luna plane frustra : Speculatus sum et in Marte, correxi inæqualitatem primam, . . . A Septembri, inquam, coepi laboriosissime inquirere excentricitatem solis, in quo labore Tycho mortuus est." *Epist. ed. Hanschius*, p. 171. About the solar excentricity, see below, Chapter xii.

account of Tycho's lunar theory is very short, and gives no
account of the successive steps which led Tycho to his great
discoveries in this branch of astronomy.   When he wrote
the report on his labours at Hveen for his *Mechanica,* he was
already in possession of the discovery of the third inequality
in longitude (variation), and of the periodical change of the
inclination and of the motion of the node.   During his stay
at Wittenberg (if not before) he had from observations of
eclipses perceived the necessity of introducing an equation
in longitude with a period of a year, but the theory had
already required so many circles and epicycles that Longomontanus thought it simplest to allow for this equation by
using a different equation of time for the moon ; and when
Tycho did not appear very delighted with this makeshift,
the pupil answered his master somewhat rudely, that he
might try to find another method himself, which would agree
better with the observations.[1]

It must have been a great satisfaction to Tycho to see his
researches on the moon reach at least a temporary conclusion, as his mind had of late years been so full of anxiety for
the future that he could doubly enjoy a ray of sunshine.
The Emperor appears always to have been most friendly to
him, and Tycho wrote in March 1600 to his sister Sophia
that Rudolph had not only been very kind to him while
the plague was raging during the previous winter, and had
offered to send him and his family to Vienna while it lasted,
but that the Emperor also took the greatest interest in his

[1] "Tu ergo ipse aliud inveni, quod cum tuis consentiat observatis " (see a
letter from Kepler to Odontius on Tycho's lunar theory, *Opera,* viii. p. 627).
That Longomontanus had enjoyed the advice of Kepler on many points in the
lunar theory appears from the letters exchanged between them in 1604 and
1605, in one of which Longomontanus counts up the various steps in the
work, while Kepler after each item put a mark, and wrote in the margin,
"Vide etiam atqve etiam, hæc me svadente et præeunte exemplo." *Epist. ed.
Hanschius,* p. 165. We shall consider the lunar theory in more detail in
Chapter xii.

work, had read the unfinished *Progymnasmata*, and had consented to let it be dedicated to him.[1]   But however pleasant
his relations with the Emperor were, Tycho had often practical experience of the scarcity of money in Bohemia, and he
could not be blind to the shaky condition of the Imperial
Government, caused by the religious and political flames
which, though as yet only smouldering, were certain ere long
to burst out in their fury, and for the quenching of which
the weakness of the Emperor did not promise well. Perhaps
he may sometimes have wondered in his own mind whether
it might not have been wiser to have remained in peaceful
Denmark, even without an endowment for his observatory,
instead of coming to the stormy Bohemia, where he had no
guarantee for the continuance of his salary but the life of his
patron, just as in the old days at Hveen.   His health would
also seem to have become shaken, if we may judge from
Kepler's remark that the feebleness of old age was approaching, since he would hardly have said so of a healthy man
only fifty-three years of age.

But the die was cast, and Tycho Brahe could only try to
make himself as much at home in Bohemia as possible. On
the 9th February 1601 the Emperor wrote to the Bohemian
Estates that Tycho Brahe and his sons desired to be naturalised,
and to have their names entered on the roll of the nobility.
It is not known whether this matter was considered by the
Estates, but the name of Brahe does not occur in their proceedings, so that Tycho must have died before he could get
his wish fulfilled.[2]   In several of his letters Tycho alludes to
his intention of buying landed property in Bohemia, and in
order to do so he took steps to get back the money which he

[1] *Breve og Aktstykker*, p. 85.
[2] F. Dvorsky, *Nové zpravy o Tychonu Brahovi a jeho rodine* (New particulars
about Tycho Brahe and his family), *Časopis Musea Královstvi Českeho*, vol.
lvii., 1883, pp. 60–77.

had lent in 1597 to the Dukes of Mecklenburg. In accordance with the terms of the bond, he had, about Michaelmas 1600, through his kinsman Eske Bille, given notice to the agent of the Dukes to repay the capital sum of 10,000 thaler at Easter 1601 ; and for fear of his letter to Bille having been lost, he took the further precaution of giving notice himself to the "Landrentmeister," Andreas Mayer, to pay the money at Michaelmas 1601, if Bille had not already carried out his instructions.[1] He stated expressly that he required the money for buying property in Bohemia, and wrote to this effect to Duke Ulrich in April 1601, as he had been informed that Mayer had stated that he could not get the money together so soon. At Easter the money was not forthcoming, as Mayer "at the last moment" was disappointed about some money which should have been paid to him, but Tycho's money was promised for St. John's Day. Bille now sent his own servant to Doberan, where Duke Ulrich was then staying, to receive the money, but he was again disappointed, and Tycho had on the 18th July to send off another reminder, to which the Duke answered that the money had been ready, but that Bille's servant had not called again. Before the 9th August Bille had himself arrived at Rostock to receive the money, which the Duke on that day ordered to be paid to him "before his departure on Tuesday morning." This must have been done, as the cancelled bond is still in the archives at Schwerin.[2]

There appears to be nothing known as to whether Tycho

---

[1] *Breve og Aktstykker*, p. 145. It appears from this letter that an annual interest of 6 per cent. was paid on the capital to Tycho's friend, Professor Backmeister, in Rostock (in whose house the quarrel with Parsbjerg had begun in 1566 which led to the loss of part of Tycho's nose).

[2] Lisch, *T. Brahe u. seine Verh. zu Meklenburg*, pp. 11, 18. In the letter of 10th April 1601, Tycho mentions that all his instruments are now conveniently placed in Curtius' house.

actually purchased land in Bohemia after receiving back his
money from Mecklenburg, but it is not likely that he did so,
as his life terminated very suddenly soon after.   On the 13th
October 1601 (new style) he was invited to supper at the
house of the Baron of Rosenberg,[1] and went there in company
with the Imperial Councillor, Ernfried Minkawitz.   During
supper he was seized with illness, which was aggravated by his
remaining at table.   On returning home, he suffered greatly
for five days, when he became somewhat relieved, although
sleeplessness and fever continued to harass him.   He was
frequently delirious, and at other times refused to keep the
prescribed diet, but demanded to be given to eat anything
he fancied.   Five more days elapsed in this manner.   During
the night before the 24th October he was frequently heard
to exclaim that he hoped he should not appear to have lived
in vain (" ne frustra vixisse videar ").   When the morning
came, the delirium had left him, but his strength was exhausted and he felt the approach of death.   His eldest son
was absent, and his second daughter and her husband also;[2]
but he now charged his younger son and the pupils to continue their studies, and he begged Kepler to finish the
Rudolphine tables as soon as possible, adding the hope that
he would demonstrate their theory according to the Tychonic
system and not by that of Copernicus.[3]   Among those pre-

---

[1] Kepler dedicated his little book *De Fundamentis Astrologiæ* to "Petro
Wok Ursino, Domus Rosembergicae Gubernatori."

[2] The younger Tycho Brahe had in January 1601 started for Italy in company with Robert Sherley, ambassador from the Shah of Persia to various
European courts.   Carteggio di Magini, p. 237.

[3] "Ego in sequentibus demonstrationibus omnes tres auctorum formas con-
jungam.   Nam et Tycho, me hoc quandoque suadente id se ultro vel me
tacente facturum fuisse respondit (fecissetque si supervixisset) et moriens a
me, quem in Copernici sententia esse sciebat, petiit, uti in sua hypothesi omnia
demonstrarem."   Kepler, *De Stella Martis*, cap. vi. ; *Opera*, iii. p. 193.   Gassendi (p. 179) gives it in these words :—"Quæso te, mi Joannes, ut quando
quod tu Soli pellicienti, ego ipsis Planetis ultro affectantibus et quasi adulan-
tibus tribuo, velis eadem omnia in mea demonstrare hypothesi, quæ in Coper-
nicana declarare tibi est cordi."

sent at his bedside was a namesake of Tycho's, Erik Brahe, Count Visingsborg, a Swede by birth, but in the service of the King of Poland, whom Tycho thanked for all the kindness he had shown to him during his illness, asking him to carry his last remembrances to his relations in Denmark. Soon afterwards he peacefully drew his last breath, amidst the tearful prayers of his family and pupils. He had only reached the age of fifty-four years and ten months; a short span of time (as Gassendi remarks) if we look to the age which he might have attained, but a lengthy one if we consider the magnitude of his works, which will live in the recollection of mankind as long as the love of astronomy remains among us.[1]

On the 4th November the body of the renowned astronomer was with great pomp brought to its last resting-place in the Teynkirche (Týnský kostel), in which a semi-Protestant (utraquistic) service was still tolerated. The funeral procession was headed by persons carrying candles embellished with the arms of the Brahe family; next was carried a banner of black damask with the arms and name of the deceased embroidered in gold; then came his favourite horse, succeeded by another banner and a second horse, after which came persons bearing a helmet with feathers in the colours of his family, a pair of gilt spurs, and a shield with the Brahe coat of arms. Then followed the coffin, covered with a velvet cloth, and carried by twelve Imperial gentlemen-at-arms. Next after the coffin came the younger son or the deceased, walking between Count Erik Brahe and the Imperial Councillor, Ernfried Minkawitz, and followed by councillors and nobles, and Tycho's pupils. Then came the

---

[1] Kepler wrote in the observing ledger a short account of Tycho's last illness, which was printed by Snellius in his *Observationes Hassiacæ*, Lugduni Bat., 1618, pp. 83–84, and in this volume, Note D. Kepler also wrote an elegy over Tycho Brahe, which is printed by Gassendi, p. 235 *et seq.*, and in Kepler's *Opera*, viii. p. 138 *et seq.*

TOMB OF TYCHO BRAHE.

widow, walking between two aged gentlemen of high rank, followed by her three daughters similarly attended. A long funeral oration was delivered by Dr. Jessinsky of Wittenberg, with whom Tycho Brahe had lived for some months two years before. He praised him for having led a life befitting a Christian and a learned man, living happily with his wife and family, keeping his sons to their studies, and his daughters to spinning and sewing, for being civil to strangers, charitable to the poor. He was open and honest in his dealings, was never hypocritical, but always spoke his mind, by which he sometimes made enemies. The speaker also dwelt on his scientific merits, and the favour shown to him by many princes; and it is characteristic of the time that a reference to his deformed nose and the plagiarism and calumnies of Reymers, as well as a detailed account of his last illness, were included in the oration.[1] The tomb is at the first pillar on the left side in the nave, next the chancel, where Tycho's children some years later erected over it a handsome monument of red marble, which still marks the spot. It consists of a tablet standing upright against the pillar, with a full figure in relief of Tycho Brahe clad in armour, with the left hand on his sword-hilt and the right on a globe, underneath which is a shield with the arms of Brahe, Bille, and the families of his two grandmothers, Ulfstand and Rud. The helmet stands at his feet. Round the tablet runs an inscription recording the

---

[1] " De vita et morte D. Tychonis Brahe Oratio Funebris D. Johannis Jessenii, Pragæ, 1601, 4to, Hamburgi, 1610, and reprinted by Gassendi, pp. 224–235. In May 1602 the King of Denmark wrote to the Elector of Saxony to complain of the unfair way in which Jessenius had alluded to the broken nose ("facies decora et aperta, quam ante annos triginta Rostochii quidam noctu ausu prorsus sicario læsit, vestigio ad mortem usque semper conspicuo "). The duel, he added, had been a fair fight, and the two adversaries had always afterwards been good friends, and though Parsbjerg's name had not been mentioned, the story was so well known, that the remarks of the orator were most insulting, and ought publicly to be retracted. Danske Magazin, 4th Series, ii. p. 325.

name and date of the death of the deceased.[1]   Above is a
smaller tablet with his motto, *Esse potius quam haberi*,[2] and
a lengthy Latin inscription recording his life and merits,
and mentioning that his wife is also buried here; while at
the foot stands an inscription from Stjerneborg, *" Non fasces
nec opes, sola artis sceptra perennant."* [3]

It is scarcely worth mentioning that a silly rumour very
soon began to spread that Tycho Brahe had died from
poison, administered by some envious courtier at Prague,
or, as others thought, by his old enemy Reymers Bär.   As
the latter died fourteen months before his supposed victim,
it would indeed have been a remarkably slow poison.[4]

The most important inheritance which Tycho left to
Kepler and to posterity was the vast mass of observations,
of which Kepler justly said that they deserved to be kept
among the royal treasures, as the reform of astronomy could
not be accomplished without them.   He even added that
there was no hope of any one ever making more accurate
observations, for it was a most tedious and lengthy business!
This would have been perfectly true if the telescope had not
afterwards been invented.   It is not here the place to set forth
how Kepler, when Tengnagel had given up pretending that
he was going to work out the theory of the planets, took up the
work, and how his mighty genius mastered it and gave to the
world the great laws of Kepler, at one breath blowing away
the epicycles and other musty appendages which disfigured

---

[1] Anno Domini MDCI die 24 Octobris obiit illustris et generosus Dñus
Tycho Brahe, Dñus in Knudstrup et Præses Uraniburgi, & Sacræ Cæsareæ
Maiestatis Consiliarius, cujus ossa hic requiescunt.

[2] Sometimes he wrote it "Non haberi sed esse."

[3] The inscription is printed in *Danske Magazin*, ii. p. 357; Weistritz, ii.
p. 362, where the tombstone is also figured.

[4] Andreas Foss, Bishop of Bergen, who had visited Tycho in 1596, wrote
to Longomontanus in February 1602 to inquire if the rumour had any foundation (Bang's *Samlinger*, ii. p. 529; Weistritz, i. p. 195).   The astrologer Rollenhagen wrote at the same time to Kepler that Tycho evidently died "per
Ursianum quoddam venenum" (*Epist. Kepleri, ed. Hanschius*, p. 193).

the Copernican system. But Kepler was not only a great genius, he was also a pure and noble character, and he never forgot in his writings to do honour to the man without whose labours he never could have found out the secrets of the planetary motions. On the title-page of his *Astronomia nova de motibus stellæ Martis,* he states that it is founded on Tycho's observations, and on that of the *Tabulæ Rudolphinæ* he mentions Tycho as a phœnix among astronomers.[1] And it was no exaggeration. Archimedes of old had said, "Give me a place to stand on, and I shall move the world." Tycho Brahe had given Kepler the place to stand on, and Kepler did move the world! And so it was with Kepler's labours in other fields, as we may see in that wonderfully interesting book, *Ad Vitellionem Paralipomena, sive Astronomiæ Pars Optica,* where Tycho's name is quoted so constantly as having supplied the materials. Kepler and Tycho had squabbled often enough while the latter was alive, but after his death this was forgotten, and Kepler's mind had only room for gratitude for having become heir to the great treasures left by Tycho. But on the other hand, it must be conceded that it was fortunate for Tycho's glory that his observations fell into the hands of Kepler. Longomontanus would doubtless have hoarded them

---

[1] "Tabulæ Rudolphinæ, quibus astronomicae scientiae, temporum longinquitate collapsæ, restauratio continetur, a Phoenice illo Astronomorum TYCHONE ex illustri et generosa Braheorum in Regno Daniae familia oriundo equite, primum animo concepta et destinata anno Christi MDLXIV., exinde observationibus siderum accuratissimis, post annum praecipue MDLXXII., quo sidus in Cassiopeiae constellatione novum effulsit, serio affectata, variisque operibus, cum mechanicis, tum librariis, impenso patrimonio amplissimo, accedentibus etiam subsidiis Friderici II. Daniæ Regis, regali magnificentia dignis, tracta per annos XXV. potissimum in insula freti Sundici Huenna et arce Uraniburgo, in hos usus a fundamentis exstructa, tandem traducta in Germaniam inque aulam et nomen Rudolphi Imp. anno MDIIC. Tabulas ipsas, jam et nuncupatas et affectas, sed morte auctoris sui anno MDCI. desertas, . . . perfecit, absolvit adque causarum et calculi perennis formulam traduxit Johannes Kepplerus." (Ulm, 1627.)

carefully as a great treasure, but he would most certainly not have discovered the laws of planetary motion, and Tycho's exile thus turned out to be of vast advantage to science.[1]

[1] Delambre has made this remark in a somewhat exaggerated form, *Hist. de l'Astr. moderne*, ii. p. xiv. : " Si Tycho fut resté dans son île, jamais Kepler ne se fut rendu à ses invitations ; nous n'aurions certainement pas la *Théorie de Mars*, et nous ignorerions peut-être encore le véritable système du Monde."

CHAPTER XII.

CONCLUSION.

TYCHO BRAHE'S SCIENTIFIC ACHIEVEMENTS.

AMONG the most important instruments in use at Alexandria were the so-called spheres or armillæ. These are said to have been used in China at an early date,[1] but the invention has doubtless been made independently by the Greek astronomers. They were probably known at the time of Timocharis and Aristillus (about 300 B.C.), and it is certain that Hipparchus employed them. In the complicated form used by him and his successors (called by Tycho " armillæ zodiacales ") the instrument consisted of six circles, of which the largest represented the meridian, and was carefully placed in position on a solid stand. On the inner rim it was furnished with two pivots, representing the north and south poles, on which turned a slightly smaller circle, the solstitial colure, to which was fixed immovably and at right angles another of the same size, representing the ecliptic. The colure was furnished with pivots representing the poles of the ecliptic, and on these turned two circles, one larger than the colure, another smaller than it, while the latter enclosed a sixth circle, which could slide inside it, and was furnished with two sights diametrically opposite to each other. With this instrument the difference of longitude and latitude could be measured, the circles being divided to one-sixth degree, while half that quantity could be estimated.

---

[1] About the armillæ (equatorial) of the Chinese, see *Observations mathematiques, astronomiques, &c. tirées des anciens livres Chinois, red. par Souciet,* ii. p. 5, iii. p. 105, and my paper on the instruments at Peking in *Copernicus,* vol. i. (1881), p. 134 *et seq.*

The Arabs constructed similar instruments, and already
Mashallah, who lived about the year 775 (before the time of
Al Mamun), wrote about astrolabes and armillæ, and these
were used by Ibn Yunis, Abul Wefa, and others.[1]   Alhazen

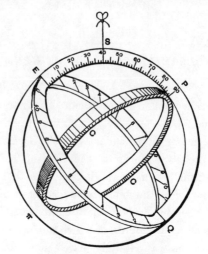

GEMMA'S ASTRONOMICAL RING.

also made use of armillæ for his investigations on refraction,
and it has even been assumed that he is the inventor of the far
simpler equatorial armillæ, which are generally ascribed to
Tycho Brahe, who also considers himself as their inventor.[2]
But in any case, it is certain that equatorial armillæ were not
known in Europe; that Walther, the principal observer before Tycho, only knew zodiacal armillæ; and that the principle
of equatorial ones was first described in 1534 by Gemma
Frisius, who, however, only designed an instrument of very

---

[1] Sédillot, *Prolégomènes des tables astron. d'Oloug-Beg*, Paris, 1847, p. xvi.
Mémoire sur les instruments astron. des Arabes, Paris, 1841, p. 198.   Abul
Wefa used only five circles, the smaller latitude circle being crossed diamet-
rically by a pointer or by a tube carrying the sights.

[2] Sédillot, *Prolégomènes*, p. cxxxiv. ; *Mémoire sur les instr.*, p. 198 ; but the
"armille équatoriale" mentioned in the latter place is evidently nothing but
Ptolemy's instrument for observing the solstices, *i.e.*, a graduated circle in the
plane of the meridian.

small dimensions, intended to be held in the hand.[1] Tycho
remarks that the instruments of the ancients were of solid
metal, and as they had to be very large to allow spaces of 10
to be marked on them, they must have been very cumber-
some;[2] and it is deserving of particular notice that he has an
open eye to the importance of perfect symmetry in the instrument. He points out that the poles of the ecliptic at different
times occupy different positions with regard to the meridian,
and that the instrument therefore must be subject to severe
strains, which would seriously affect the accuracy of the
observations, even if the circles are of moderate dimensions
and not too heavy. For the same reasons he rejected the
clumsy "Torquetum" of Regiomontanus, which had never
been much used.[3]

Although Tycho possessed a zodiacal instrument which had
the advantage of consisting only of four circles, he chiefly
made use of equatorial armillæ, which instruments represent
a great step forward, on account of their comparative simplicity
and perfect symmetry. He constructed three instruments of
this kind, which are all figured in his *Mechanica*. The first
one, which was mounted in the small northern observatory of
Uraniborg, consisted of three circles of steel, of which the meridian and the equator were firmly joined together, and both
the equator and the movable declination circle were furnished
with sights (made of brass), which could be moved along the
circles, and to which the observer applied his eye, while a
small cylinder in the centre of, and perpendicular to, the polar
axis served as objective sight. The second instrument was
placed in the small southern observatory, and only differed

---

[1] *Tractatus de annulo astronomico.* The author possesses a scarce little
book in which the various uses of simple circles, quadrants, and systems of
circles (including Gemma's rings) are described. *Annuli astronomici, instrumenti cum certissimi tum commodissimi usus, ex variis authoribus.* Lutetiæ,
1557, small 8vo., 159 ff.

[2] *Progymn.*, i. p. 140; *Epist.*, p. 9.

[3] *Mechanica*, fol. C. 4; *Progymn.*, i. p. 141.

from the first one by the equator being movable and attached
to a revolving (but not graduated) declination circle, while
a smaller and graduated declination circle carried sights.
The undivided circle might very well have been left out and
the graduated one fixed to the equator. The outer circle
(meridian) was nearly five feet in diameter.[1]

The third and most important instrument of this kind was
mounted in the largest crypt of Stjerneborg and was far
more extensively used by Tycho, who considered it one of
his most accurate instruments. It consisted merely of a
declination circle 9½ feet in diameter, and a semicircle, which
represented the part of the equator below the horizon, and
rested on eight stone piers. The former has two pointers
turning round a small cylinder in the centre of the polar
axis, and perpendicular to the plane of the circle, and each
furnished with an eye-piece sight, while a third sight slides
along the equator. The polar axis (of iron, but hollow) could
be adjusted in inclination and azimuth by screws, which acted
on a square plate in a hole in which the lower pointed end
of the axis fitted. By reversing the circle double observations
of declination might be taken, using first one sight and then
the other, and Tycho remarks that this instrument had the
further advantage over the two others that stars near the
equator were as easily observed as those more distant from it,
as the equatorial arc was at some distance behind the declination circle.

Circles and semicircles had naturally been in use from a
very early date. We need only refer to the astrolabium of
Ptolemy, which consisted of a graduated circle inside which
another circle could slide, carrying two small cylinders diametrically opposite to each other, while the instrument was
kept vertical by a plumb-line. This astrolabium was imitated
by many successive astronomers ; among others, by Abul Wefa,

---

[1] Also figured in *Progymn.*, i. p. 251.

who has described a meridian circle for observing the sun,[1]

ARMILLÆ A EQUATORIÆ MAXIMÆ.

while smaller circles became extensively used, particularly

[1] Sédillot, *Mémoire*, p. 196; Wolf's *Geschichte der Astronomie*, p. 132.

by navigators. Nonius suggested attaching the pointer or alidade to some point on the circumference instead of to the centre, as the divisions on this plan might be made twice as large as usual.[1] I only mention this proposal because Tycho (who does not allude to Nonius) constructed a large semi-circle revolving round a vertical axis, with a long ruler turning on a pivot at one end of the horizontal diameter.

In addition to complete circles, quadrants were also used long before Tycho's time, though not extensively. Ptolemy describes a meridian quadrant attached to a cube of stone or wood, with a small cylinder in the centre of the arc, of which the shadow indicated the altitude of the sun on the gradua-tion.[2] Among the numerous instruments which Nasir al-din Tusi erected in the splendid observatory at Meragha, in the north-west of Persia (about A.D. 1260), was a Ptolemean mural quadrant, made of hard wood, and with a radius of about twelve feet. The limb was of copper, on which three arcs were drawn; the middle one was divided into degrees, while of the others, one showed every fifth degree, the other the minutes (all of them?). Nasir al-din mentions the in-strument of Ptolemy, which evidently had served him as a model.[3]

Tycho Brahe is therefore not (as supposed by some writers) the inventor of the mural quadrant, an instrument which up to the end of the eighteenth century was the most important one in astronomical observatories. Of course he cannot have known anything of the Arabian quadrants, but the descrip-

---

[1] Delambre, *Astr. du moyen age*, p. 399.

[2] As remarked by Delambre (*Astr. ancienne*, ii. p. 75), it appears doubtful whether Ptolemy ever actually used an instrument of this kind, as he only quotes one observation made with it, the difference between the sun's altitude at the two solstices, for which he gives exactly the same value as had been found by Eratosthenes; and as his latitude was 15′ wrong, his quadrant (if he used it) must have been very small.

[3] *Monatliche Correspondenz*, xxiii. (1811), p. 346. A perfectly similar description from an Arabian MS. by Muvayad al-Oredhi of Damascus is given by Sédillot, *Mémoire*, p. 194.

tion of Ptolemy must have been familiar to him. The advantage of meridian observations for many purposes was also well known before his time, particularly for finding the declination of the sun, which gave its place in the zodiac by the tables. Hagecius had even observed the altitude and time of transit of the new star over the meridian,[1] but nobody had erected an instrument permanently in the meridian. The great superiority as to stability which a mural quadrant possessed over the armillæ did not escape Tycho; and as he was the first thoroughly to perceive the influence of refraction in altering the apparent positions of stars, the wish naturally arose to observe the stars at their greatest altitude on the meridian, where that influence was smallest.[2]

From the meridian quadrant to quadrants which could be placed in any azimuth the transition was simple enough, and we find accordingly among the instruments at Meragah an "instrument des quarts de circles mobiles." This consisted of an azimuth circle on which were two quadrants turning on a common vertical axis, by which two observers could find the altitudes and difference of azimuth of two objects.[3] In Europe an azimuth instrument seems to have been first used by Landgrave Wilhelm IV., who observed altitudes and azimuths of the new star of 1572, apparently by setting the instrument to a certain whole or half degree of azimuth, and measuring the altitude when the star reached that azimuth.[4] Quadrants capable of being turned round a vertical axis had

---

[1] *Progym.*, p. 521. Tycho did not approve of this method, as it involved the use of clocks.

[2] For a description of the Tychonic quadrant, see above, p. 101.

[3] *Mon. Corresp.*, xxiii. p. 355. The instrument is doubtless the same as described by Sédillot, *Mémoire*, p. 200. An azimuth circle of copper, 10 cubits in diameter, was in the year 513 after Hedschra erected at Cairo for observations of the sun. Caussin, *Le livre de la grande Table Hakémite* (*Notices des manuscrits*, tom. vii.), Paris, an. xii. p. 21.

[4] *Progymn.*, p. 491. At Kremsmünster observatory there is a small azimuth circle with a vertical semicircle of ivory, dating from 1570. Wolf's *Geschichte der Astronomie*, p. 112.

been known long before,[1] but as it was so much easier to
graduate a straight line than an arc, the triquetrum continued to be the favourite instrument for measuring altitudes
down to the end of the sixteenth century. Tycho did not
think much of this instrument, which he calls "instrumentum parallacticum sive regularum," and he did not make
much use of the two he had constructed, and one of which
was of large dimensions, and furnished with an azimuth
circle 16 feet in diameter.[2] He preferred the "quadrans
azimuthalis," and constructed four instruments of this kind,
which were extensively used, though chiefly for merely
observing altitudes, while the azimuths were rarely taken,
especially during his later years. The largest quadrant
(*quadrans magnus chalibeus*) was enclosed in a square (also
of steel), of which the side was equal to the radius of the
quadrant. Two of the sides were graduated, and the alidade
pointed to these graduations as well as to those on the arc,
so that the instrument was a combination of a quadrant and
the "quadratum geometricum" of Purbach (which the Arabians had also known), which increased the solidity of the
instrument.

An important use to which the quadrants were put at
Uraniborg was the determination of time. At Alexandria
the beginning, middle, or end of the hour was generally the
only indication of time which accompanied the observations
of planets, which was perhaps sufficient, owing to the limited
accuracy of the observations. The time was found by water-
or sand-clocks, which were verified by observing the culmination of some of the forty-four stars which Hipparchus had
selected so well that the time could be determined with an

---

[1] The "instrument aux deux piliers" at Méragah was a modification of
Ptolemy's quadrant (Sédillot, *l. c.*, fig. 113), but it could also be arranged so as
to be movable in azimuth (see also *Monatl. Corresp.*, xxiii. p. 359).

[2] *Progymn.*, pp. 142 and 636 ; *Epistolæ*, p. 75.]

error not much exceeding a minute.[1] An important step forward as regards the accurate determination of time was made by the Arabs in the ninth century. Ibn Yunis mentions a solar eclipse observed at Bagdad on the 30th November 829 by Ahmed Ibn Abdallah, called Habash, who at the beginning of the eclipse found the altitude of the sun to be 7°, while at the end the altitude was 24°. This seems to have been the earliest though crude attempt to use observations of altitude to indicate time, but the advantage of the method was evident, and at the lunar eclipse on the 12th August 854 the altitude of Aldebaran was measured equal to 45° 30′. Ibn Yunis adds that he from this made out the hour-angle to be 44° by means of a planisphere. Ibn Yunis communicates a number of other instances from the tenth century,[2] but the instruments used were very small, and only divided into degrees; and though Al Battani gave formulæ for the computation of the hour-angle, the Arabians generally contented themselves with the approximate graphical determination by the so-called astrolabe or planisphere.

In Europe the use of observations of altitude for determining time was introduced in 1457 by Purbach, who, at the beginning and end of the lunar eclipse on the 3rd September, measured the altitude of "penultima ex Plejadibus."[3] Bernhard Walther was the first to introduce in observatories the use of clocks driven by weights. Thus we find among his observations one of the rising of Mercury. At the time of rising he attached the weight to a clock of which the hour-wheel had fifty-six teeth, and as one hour and thirty-five teeth passed before the sun rose, he concluded that the interval had been one hour thirty-seven minutes. Walther

[1] Schjellerup, *Sur le chronomètre céleste d'Hipparque* (*Copernicus*, vol. i., 1881, p. 25).

[2] Caussin, *l. c.*, p. 100 *et seq.* About some errors of copying, by which some of the observations were affected, see Knobel's paper on Ulugh Beigh's Catalogue, in the Monthly Notices of the Roy. Astr. Soc., xxxix. p. 339.

[3] *Scripta cl. mathematici J. Regiomontani*, Norimb., 1544, fol. 36.

adds that this clock was a very good one, and indicated correctly the interval between two successive noons; but all the same he must have seen how unreliable it was, for though he used the clock during the lunar eclipse in 1487, he at the same time measured some altitudes.[1]

In Tycho Brahe's observatory the clocks never played an important part. Though he possessed three or four clocks, he does not anywhere describe them in detail, while he in several places remarks that he did not depend on them, as their rate varied considerably even during short intervals, which he attributed to atmospherical changes (although he kept them in heated rooms in winter), as well as to imperfections in the wheels. At the side of the mural quadrant he had placed two clocks, indicating both minutes and seconds, in order that one might control the other, and in the southern observatory was a large clock (*horologium majus*) with all the wheels of brass. Whether Bürgi, during Tycho's residence at Prague, supplied him with a pendulum clock, as stated by a later writer,[2] must remain very doubtful, but that Tycho did not possess such a clock at Uraniborg seems certain, as he would not have neglected to describe so important an addition to his stock of instruments. As he found the clocks so uncertain, Tycho also tried time-keepers similar to the clepsydræ of the ancients, which measured time by a quantity of mercury flowing out through a small hole in the bottom of a vessel, in which the mercury was kept at a constant height, in order that the outflow might not vary with the varying weight of the mercury. By ascertaining the quantity of mercury which flowed out in

[1] Ibid., fol. 50 *et seq.* The mere statement, what degree of the zodiac was on the meridian (*medium cœli*) when an observation was made, was, however, still very often the only indication of time given, even by Walther. See, for instance, Tycho's first observation at Hveen, above, p. 86 footnote.

[2] Joachim Becher, *De nova temporis demetiendi ratione theoria* (1680), buoted by R. Wolf, *Geschichte d. Astr.*, p. 370.

twenty-four hours, it was easy to make out the interval which passed between the culmination of the sun and a star by starting the time-keeper when the former passed the meridian, and letting it run until the latter passed, and then weighing the amount which had flowed out. Instead of mercury, Tycho also tried lead monoxide powder, and adds to his account of these experiments some remarks about Mercury and Saturn (lead), and their astrological relations, which naturally suggested themselves to his mind.[1] But he does not seem to have used these clepsydræ except by way of experiment, and his methods of observing made him in most cases independent both of them and of the clocks. In addition to the altitudes (about which he justly remarks that they must not be taken too near the meridian, where they vary very slowly, nor near the horizon, where they are much affected by refraction), he observed hour-angles of the sun or standard stars with the armillæ to control the indications of his clocks, and his observations of the moon, comets, eclipses, &c., where accurate time determinations are indispensable, were thereby doubly valuable. Occasionally azimuths were also observed for the same purpose, the zero of the azimuth circle having been found by observing the east and west elongation of the Pole Star.[2]

For observations of altitude Tycho also used a sextant of $5\frac{1}{2}$ feet radius, turning on a vertical axis, with one end-radius kept horizontal by means of a plumb-line attached to the centre of the radius. We have already mentioned that

[1] "Est in Mercurio, quicquid quærunt Sapientes . . . Sicque Saturnus et Mercurius coniunctis operibus hanc inquisitionem expedirent : cum & secundum Astrologos, illorum coniunctio aut benevola invicem radiatio . . . aut etiam intuitus beneuolentior, præ ceteris aliis significationibus ad ingenii et solertiæ contemplationisque profunditatem, laborisque invictam constantiam, plurimum conducere credantur." *Progym.*, p. 151.

[2] *Epist.*, p. 73. Rothmann was, therefore, not the inventor of this method of finding the meridian, as supposed by Wolf (*Geschichte*, pp. 374 and 598). Tycho had already used the Pole Star for azimuth in 1578, as appears from his MS. journals and *Observ. comet.*, p. 16.

Tycho, driven by necessity, had observed altitudes of the new star with a sextant, and as the planets never could attain an altitude above 60°, he found a sextant a convenient instrument for many purposes, and specially mentions that it was easily taken asunder and transported wherever required. Though Tycho believed himself to be the inventor of this instrument, he had been anticipated by the Arabs, as Al Chogandi in 992 at Bagdad erected a sextant (which is even said to have been of sixty feet radius) for measuring the inclination of the ecliptic.[1]

The sextant was with Tycho Brache a favourite instrument, which he had already constructed for Paul Hainzel in 1569 for measuring the angular distances of stars. At Hveen he constructed three large sextants for this purpose. One of these, which was placed in the great northern observatory, and was made entirely of brass, was on the same plan as the Augsburg instrument, the arc being attached to the end of one arm, the two arms being placed at the proper angle by a screw, the eye of the observer being at the hinge on which the arms turned.[2] The second was placed in one of the crypts of Stjerneborg, and was a solid sextant of wood, covered with painted canvas, and a brass arc 5½ feet radius, braced with stays and supported on a globe sheeted with copper, which enabled the two observers to place it in the plane through the two stars to be measured, while they steadied it in the position required by two long rods with pointed ends which rested on the floor. One of the observers sighted one star through a fixed sight at one end of the arc

---

[1] L. A. Sédillot, *Mémoire*, p. 204; *Matériaux pour servir à l'hist. des sciences chez les Grecs et les Orientaux*, i. p. 358. *Prolégomènes* (1847), p. xlii. Sarafedaula, who founded the Bagdad observatory, was not a Chaliph, as supposed by Bailly and Wolf, but Emir-ul-umara.

[2] *Sextans chalybeus*, used already in 1577; *Mechanica*, fol. E.; *De mundi aeth. rec. phen.*, p. 460. The sextant at Cassel (constructed from Wittich's description) also required one observer only, who placed his eye at the centre of the arc.

(C), while the other observer pointed to the other star through a sight at the end of a movable radius. Both observers em-

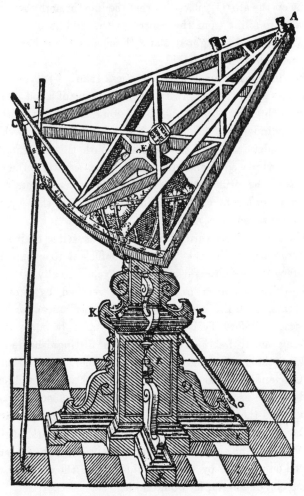

SEXTANS TRIGONICUS.

ployed the same object-sight, a small cylinder (A) at the centre of the arc and perpendicular to the plane of the sextant. As the observers when measuring very small distances would

get their heads too close together, there was for that purpose
a second cylinder on one of the end-radii (F), and a removable
sight on the arc (G), placed so that the line through them was
parallel to a line from the centre to the middle of the arc.
One observer then sighted along these, and the other along
the movable arm as usual.[1]

For measuring small distances (less than 30°) Tycho also
constructed an "arcus bipartitus," consisting of a rod $5\frac{1}{2}$ feet
long, with a cross-bar at one end, having at each extremity a
small cylinder, and two arcs of 30° at the other end, of which
the cylinders occupied the centres. With this instrument,
which was placed in the great northern observatory, the distances of the principal stars of Cassiopea *inter se* were measured in order to fix the position of the new star by the
measures taken in 1572–73.[2]

The size of these various instruments, as well as their solid
construction, would not have been sufficient to ensure the
accuracy in the observations which Tycho actually attained,
and which so much exceeded that reached by previous
observers, if he had not added special contrivances for that
purpose. Before Tycho's time there was only one way of
making small fractions of a degree distinguishable—by
making the instrument as large as possible. In addition to
Al Chogandi's 60-foot sextant, a quadrant of 21 feet radius is
said to have been constructed by Al Sagani (about the year
1000), and the value which the Arabs were obliged to attach
to large instruments was expressed in the remark of Ibn
Carfa, that if he were able to build a circle which was supported on one side by the Pyramids and on the other by the
Mocattam mountain, he would do it.[3]

---

[1] *Sextans trigonicus. Mech.*, fol. D. 5; *Progymn*, p. 248.

[2] *Arcus bipartitus. Mech.*, fol. D. 4.

[3] Sédillot, *Prolégomènes*, pp. lvii. and cxxix. The 180-foot quadrant of
Ulugh Bey was doubtless a kind of sundial, such as are also found in India.
Ibn Yunis quotes an observation of the autumnal equinox of 851 at Nisapur

The first to suggest a method of subdividing an arc of moderate dimensions was Pedro Nunez, whose work, *De Crepusculis,* was published in 1542. He proposed inside the graduated arc of a quadrant to draw 44 concentric arcs, and divide them respectively into 89, 88, 87 . . . 46 equal parts, so that the alidade in any position would (more or less accurately) touch a division mark on one of the 45 circles. The indication of this mark was multiplied by $\frac{90°}{n}$, where n is the number of divisions on the arc on which the mark touched by the alidade is. But however ingenious this proposal was, it was anything but a practical one, as it is not easy to divide an arc into 87 or 71 equal parts, and the observer would generally be in doubt which division was nearest the alidade.[1] Tycho Brahe tried this plan on three of the instruments first constructed at Hveen (the two smallest quadrants and a sextant), but abandoned it again as far inferior to the one he subsequently adopted.[2] By a strange misunderstanding, the name of Nonius is even at the present day often applied to the beautiful and practical invention of Vernier (1631), with which it has nothing whatever in common. A step towards the idea of Vernier was made by Christopher Clavius and the Vice-Chancellor Curtius, and the latter communicated this plan to Tycho in 1590, but it was not much more practical than that of Nunez, and was probably never carried out in practice.[3]

We have seen that Tycho Brahe in his youth followed the example of the Arabians by constructing a large quadrant at Augsburg, with a radius of 19 feet. But already

(Khorassan) with a great armilla which showed single minutes (Caussin, p. 148).

[1] The limited accuracy attainable is shown in tabular form by Delambre, *Moyen Age,* p. 404.

[2] *Mechanica,* fol. A. 2; *Epist.,* p. 62. The *quadrans mediocris* was, in addition to the arcs of Nonius, divided by transversals, and on the sextant Tycho removed the Nonian division altogether.

[3] *Mechanica,* fol. G. 5 ; compare *Chr. Clavii Opera* (1612), t. iii. p. 10.

before that time he had in 1564 obtained a cross-staff divided by transversals.[1] He says himself that Bartholomæus Scultetus had got the idea of this method of subdivision from his teacher Homilius, and now taught it to him; but in a letter to the Landgrave (of 1587) Tycho states that he was seventeen years of age when he at Leipzig learned the use of transversals for subdividing a straight line from Homilius.[2] The latter died, however, in July 1562, when Tycho was only $15\frac{1}{2}$ years old, and had only been a few months at Leipzig, and it is therefore more probable that Scultetus really was the means of imparting the idea to Tycho. At all events, Tycho did not attempt to claim the invention for himself, though it was afterwards often attributed to him. But whether Homilius really was the first inventor is more than doubtful, and Scultetus himself has even stated that the method was already known to Purbach and Regiomontanus.[3] We can, however, scarcely believe this to have been the case, as it would be difficult to explain why the method had never come to light, even though Walther notoriously guarded the belongings of Regiomontanus with a curious fear of their being known; and in the *Scripta* of Regiomontanus there is no trace of his having used so excellent a method. Curiously enough, there are two other names mentioned in connection with this invention. In his book *Alæ seu scalæ mathematicæ*[4] Digges states that transversals were first applied to the divisions on the cross-staff by the English instrument-maker Richard Chanzler, and Reymers Bär mentions that the

---

[1] *Mechanica*, fol. G., 2nd page; *Progymn.*, p. 671.

[2] *Epist.*, p. 62.

[3] "Von allerlei Solarien, das ist, himmlischen Circeln und Uhren . . . durch Bart. Scultetum, Görlitz, 1572," quoted by R. Wolf, *Astr. Mittheilungen*, xxxiii. p. 90.

[4] London, 1573, fol. I. 3, where there is a drawing of a rectilinear scale with transversals.

method was described in Puehler's Geometry, published in 1561.[1]

Under any circumstances, it was Tycho Brache who introduced the use of transversals on the graduated arcs of astronomical instruments. He did not use transversal lines, such as afterwards became universally used, but rows of dots, which were fully as convenient, and he showed that the error

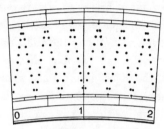

TRANSVERSAL DIVISIONS.

introduced by employing these rectilinear transversals for the division of arcs would not exceed 3″, which would be imperceptible.[2] When Wittich had brought the news of this contrivance to Cassel, Bürgi modified it a little by using lines instead of the rows of dots, and adding a scale on the alidade, the section of which with a transversal line showed the number of minutes to be added to the indication of the preceding division line, while on Tycho's instruments each of the dots corresponded to 1′.

But it was not sufficient to find means to read off the measured angle accurately; it was also of great importance to point the instrument to a star with greater precision than hitherto, and here Tycho had nobody to show the way. Up to his time an alidade had been furnished with two pinnules, one at each end, consisting of a brass plate with a small hole in the middle, and if this hole was made too small, a faint

[1] Kästner, *Gesch. der Math.*, iii. p. 479; Delambre, *Astr. mod.*, i. p. 299.

[2] *De mundi aeth. rec. phen.*, p. 461; *Mechanica*, fol. I. 2. According to R. Wolf (*l. c.*), Rothmann has in an unpublished MS. made the same investigation.

star could not be seen, while a larger hole made the observation too uncertain. To meet this difficulty Tycho introduced
a special pinnule at the eye-end of the alidade, consisting of
a square plate with a narrow slit close to the side next the
alidade, while there were three other slits between the three
other sides and small movable pieces of metal parallel to them.
By moving these pieces the slits could be made wide or narrow according as faint or bright stars were observed. At the
object-end was a small square plate exactly of the same size as
the plate at the eye-end. 'When the alidade was pointed to a
star, and the latter through the four slits was seen to touch
the three sides of the object-pinnule and shine through a slit
along the side next the alidade, the observer knew that the
alidade accurately and without any parallax represented the
straight line between his eye and the star. For observations of the sun there was in the centre of the objective
pinnule a round hole through which the light fell on a small
circle on the eye-pinnule, and the sunlight was generally
conducted "through a canal" to keep off extraneous light.
In many cases, Tycho (as we have already seen) modified
the arrangement by substituting a small cylinder (perpendicular on the alidade) for the objective pinnule. On
the armillæ this cylinder was placed in the centre of the
axis, while the eye-pinnules could slide along the graduated
circles.

Like the transversal divisions, the improved sights were
introduced at Cassel by Paul Wittich, and the value of
these improvements was found to be so great, that while
the observers could formerly scarcely observe within $2'$,
the attainable accuracy was now $\frac{1}{2}'$ or $\frac{1}{4}'$. It appears that
Wittich had not described the pinnules accurately, as he
had only given them two slits instead of four, which Rothmann (or probably Bürgi) soon found preferable.[1]

[1] *Epist.*, pp. 21, 28-29 ; *T. Brahei et doct. vir. Epist.*, p. 100.

Before Tycho settled at Hveen he had never regularly observed the sun,[1] but (as we have already mentioned) from his birthday, the 14th December 1576, he took regular observations of the meridian altitude of the sun, and later, when his stock of instruments increased, several quadrants were simultaneously employed for this purpose. Above all, he employed from March 1582 the great mural quadrant for observing the sun on the meridian, while the declination was also very frequently measured with the armillæ. These observations were made with the object of improving the theory of the sun's apparent motion. The equinoxes of the years 1584–88 were carefully determined, but owing to the difficulty of fixing the moment of solstice on account of the very slow change of declination at maximum and minimum, he did not make use of the solstices to find the position of the apogee and the amount of the excentricity of the orbit, but determined the time when the sun was 45° from the equinoxes, in the centre of the signs of Taurus and Leo. Copernicus had followed the same plan, but had made use of the signs of Scorpio and Aquarius, while Tycho objected to these that the sun was too low in the sky, and the influence of refraction and parallax too great. He found the longitude of the apogee = 95° 30′, with an annual motion of 45″ (should be 61″, Copernicus had only found 24″), and the excentricity of the solar orbit = 0.03584, the greatest equation of centre being 2° 3¼′. In the determination of the apogee he was more successful than Copernicus; but while the latter made the equation of the centre too small, Tycho made it too great. The length of the tropical year he found by combining some observations by Walther (reduced anew after determining the latitude of Nürnberg) with his own to be equal to 365^d 5^h 48^m 45^s, only about a second two small. With his new

[1] Only in 1574 he had at Heridsvad observed the meridian altitude of the sun on seven days in March and on two days in May.

numerical data he computed tables for the apparent motion of the sun, which he remarks are worthy of considerable confidence, as they depend on observations made with three or four large instruments made of metal, and capable of determining the position of the sun within 10″, or at most 20″; and by comparing the tables with observations by Regiomontanus, Walther, the Landgrave, and Hainzel, he shows that they represent the observed places within a small fraction of a minute, while the Alphonsine tables and those of Copernicus are often 15′ or 20′ in error.[1]

The solar observations at Uraniborg led to a result which Tycho does not seem to have anticipated. The colatitude, as found by the greatest and smallest altitude of the sun at the solstices, differed from that deduced from observations of the Pole Star by a considerable quantity, which sometimes amounted to 4′. Having ascertained that the discrepancy did not arise from instrumental errors, he was led to attribute it to the effect of refraction. As soon as the great armillæ at Stjerneborg were finished, he instituted systematic observations to prove this, and to determine the amount of refractions at various altitudes. Having first found, by following the sun throughout the day with the armillæ, that the declination apparently varied, as stated by Alhazen in his book on optics, he repeatedly in the years 1585 to 1589 devoted a whole day, generally in June, when the declination of the sun changed very slowly, to investigations on refraction. With an altazimuth quadrant he measured at frequent intervals the altitude and azimuth, and from the latitude of the observatory, the azimuth and the declination, he computed the altitude, which, deducted from the observed altitude, gave the amount of refraction. Another method was by observing simultaneously with the quadrant and with the armillæ. In the triangle between the pole and the true and

[1] *Progym.*, pp. 1–78.

apparent places of the sun, two sides (real and apparent declination) and one angle (180° minus the parallactic angle) were known, from which the third side could be computed, which was the effect of refraction in altitude. This is a most inconvenient and troublesome method, and must have given the computers plenty to do, if the observations were really extensively used in this way for the construction of his refraction tables. For these investigations he assumed the real declination (*i.e.*, corrected for refraction) as equal to the declination as observed on the meridian, as he thought the refraction at the meridian altitude at summer solstice ($57\frac{1}{2}°$) insensible. He assumed, in fact, that refraction disappeared already at $45°$, where it in reality amounts to $58''$. Unluckily Tycho spoiled the refraction table which he constructed from his solar observations by assuming, with all previous astronomers, from Ptolemy down to Copernicus, that the horizontal parallax of the sun was equal to $3'$. It is remarkably strange that Tycho should not have endeavoured to deduce this important constant from new observations which ought to have shown him that it was for his instruments insensible. This was the only astronomical quantity which he borrowed from his predecessors, and it was a wrong one.[1] The refractions, as given by him, must therefore be diminished by $3' \times$ cosine of altitude, and it is interesting to see that he was well aware of the fact that the refractions as found by the stars were different from those which he had mixed up with the imaginary solar parallax, as he gives a separate table of stellar refraction, in which the quantities are smaller than those in the solar refraction table by $4' 30''$; so that according to him refraction becomes insensible in the case of stars at $20°$ altitude (where it is in reality $2' 37''$). Possibly the refraction of stars was not as carefully looked into as that of the sun,

---

[1] Among Tycho's original observations there is at the beginning of the year 1581 a table of solar parallax for every degree (beginning with $2' 58''$ at $1°$) up to $60°$ altitude. Compare *Progymn.*, p. 80.

though the observations " pro refractionibus fixarum in-
dagandis " are numerous, particularly in the year 1589, and
are similar to those of the sun.[1]

Imperfect though Tycho's researches on refraction were,
they represent a great step forward, as he was the first to
determine from observations the actual amount of refraction,
and to correct his results for it. This was among the earlier
achievements at Uraniborg, and showed the great superiority
of the new instruments over the earlier ones. Though not
unknown to the ancients, and theoretically examined to some
extent by Alhazen and Vitello (whom Tycho quotes, though
he doubts whether they really carried out the experiments
mentioned by them, as their armillæ could not have been
large or accurate), the only astronomer who had practically
noticed the effect of refraction was Walther. He found, by
observing the sun when setting, by means of his zodiacal
armillæ, that the sun seemed to be outside the ecliptic, and
explained this as being caused by refraction ; but he thought
this could only be appreciable very near the horizon, and did
not attempt to investigate its laws, for which his instruments
would hardly have been accurate enough.

As to the cause of refraction, Tycho did not think that the
difference of density of the ether and the atmosphere was of
much importance, as he points out that in that case refraction
should not disappear except at the zenith, while he imagined
it to become insensible half-way towards the zenith. He
therefore ascribed refraction chiefly to atmospheric vapours,
though he believed that the atmosphere gradually decreased
in density and was essentially different from the ether, and
he naturally rejected as absurd the Aristotelean idea of a

[1] Table of solar refraction, *Progymn.*, p. 79. For comparison with modern
refractions (after deducting Tycho's parallax) see Delambre, *Astr. mod.*, i. p.
156. The table of stellar refraction in *Progymn.*, p. 280. On p. 124 Tycho gives
a table of lunar refraction, not differing much from the solar one. In Barretti
Hist. Cœl., p. 221, there is a table of refraction in A. R. and Decl. for the star
Spica Virginis.

sphere of fire encircling the earth. He had in his correspondence with Rothmann several times discussed questions connected with refraction, not only because the observer at Cassel only made the quantity of refraction about half as great as Tycho did, but also because Rothmann thought that there was no difference between the celestial ether and the air except density.[1]

Tycho recognised as an effect of neglected refraction various discrepancies between the elements of the solar orbit determined by Copernicus and his own results. We have already mentioned[2] that he sent one of his pupils to Frauenburg, and found that the latitude had been assumed $2\frac{3}{4}'$ too small, which, together with the neglect of refraction, accounted for the errors in Copernicus' determination of the obliquity and the other elements of the solar orbit, as the longitude concluded from the erroneous declinations would be as much as $13'$ in error at $45°$ from the equinox.

Among Tycho's " puerile and juvenile " observations there are very few indeed of the moon ; only now and then the approach of the moon to some bright star is mentioned, and the distance measured with the "radius" or sextant. At Hveen he gradually came to devote more attention to the moon, and from 1582 his lunar observations are very regular, and become year by year more numerous. They include distances from fixed stars, altitudes, declinations, and differences of right ascension from fixed stars, and as often as practicable the moon was observed in the nonagesimal or that point of her daily course in which the effect of parallax took place only in latitude. Eclipses were carefully attended to whenever they occurred;[3] but, unlike the ancient astronomers, Tycho did not confine himself to observing the moon

---

[1] *Progymn.*, p. 91 ; *Epist.*, pp. 83, 91, 106. Compare above, p. 206.

[2] See above, p. 123.

[3] The materials at Tycho's disposal included observations of twenty-one lunar and nine solar eclipses.

at the syzygies and quadratures, but followed her throughout
her monthly course year after year, determining her position
both on and off the meridian, and not forgetting to observe
her at apogee, or, as he called it, "in maxima remotione
utriusque epicycli." He thus succeeded in detecting the third
inequality in the motion in longitude, the, *variation*, which
reaches its maximum of 39'.5 (Tycho found 40'.5) in the
octants, when the difference of longitude of sun and moon is
45°, 135°, &c. But apart from this, he could not be satisfied
with the way in which Ptolemy had represented the motion
in longitude (by a deferent and one epicycle, the centre of
the former moving in a circle round the earth in a retrograde
direction), because it represented the apparent diameters of
the moon very badly. In fact, the moon ought, according to
the theory of Ptolemy, to appear nearly twice as great at
perigee as at apogee. This had not escaped Copernicus, who
avoided it by making the deferent concentric with the earth,
and adding a second epicycle with a motion twice as rapid
as the first one.[1] Tycho chose another way of representing
the motion in longitude. The deferent (radius $= 1$) according to him had its centre in a circle with radius 0.02174, in
the circumference of which the earth was placed, so that the
centre of the deferent was in the earth in the syzygies, and
farthest from it at the quadratures. There were two epicycles with radii 0.058 and 0.029, the period in the former
being the anomalistic month, and the moon moving in the
latter twice as rapidly and in the opposite direction, in such
a manner that at apogee the moon was 0.029 outside the
deferent, at perigee $0.058 + 0.029 = 0.087$ inside it. The
effect of the two epicycles gave the maximum of the first
inequality 4° 59' 30", while the circle through the earth gave

---

[1] For further details of Ptolemy's lunar theory, see, in particular, P. Kempf,
Untersuchung über die Ptolemäische Theorie der Mondbewegung, Inaugural
Dissertation, Berlin, 1878. Godfray's *Lunar Theory* (chap. viii.) gives short
sketches of Ptolemy's and Copernicus' theories.

an equation of 1° 14′ 45″ (evection), not differing much from Ptolemy's values, though somewhat more accurate.[1]

So far Tycho had not made much advance, but the discovery of the third and fourth inequalities was a very great step in advance. He probably thought that there were epicycles enough in his theory, and therefore he did not attempt to account for the variation by adding another. He merely let the centre of the first epicycle oscillate (*librate*) backwards and forwards on the deferent to the extent of 40′.5 on each side of its mean position, the latter moving along the deferent with the moon's mean motion in anomaly, and the centre of the epicycle being in its mean position at the syzygies and quadratures, and farthest from it at the octants, the period of a complete libration being half a synodical revolution.[2] At the same time Tycho's observations showed

---

[1] We have mentioned (p. 272) that Tycho had got part of the appendix on the lunar theory printed at Hamburg, but did not make use of the sheets thus printed, giving as reason that the printer had done his work badly. Tengnagel had given a copy to Magini, who in 1600 pointed out some discrepancies, the two first inequalities being stated to amount at most to 7° 41′ 15″, while the dimensions of the circles, so far as Magini could make out, gave 11′ or 14′ less. Tycho replied that the whole had been recast, partly at Wittenberg, partly in Bohemia, and that new tables had been calculated (Carteggio, pp. 232 and 238). In his *Astronomia Danica*, 2nd edit., Amstel., 1640, p. 242, Longomontanus talks of the lunar hypothesis described above as one "quam anno Salvatoris nostri 1600 apud Nobilissimum et omnium præstantissimum astronomum Dn. Tychonem Brahe invenimus."

[2] I shall not here enter into a discussion of the well-nigh thrashed out question whether Abul Wefa's *mohadzat* is the lunar variation or not, but only point out the utter absurdity of the suggestion of L. A. Sédillot *Matériaux*, i. p. 216) that Tycho might possibly have seen a translation of the Almegist from the Arabian, in which some abstract from Abul Wefa's book might have been given. If so, why has nobody else known this book until the present century? Tycho's discovery was not, as Sédillot believed, found among his papers and published by Kepler in 1610; it is distinctly announced in his *Mechanica* (fol. G. 2 *verso*), published in 1598, as a new inequality: "Nam & aliam quandam habet ea inæqualitatis insinuationem secundum Longitudinem, quam ab iis animadversum est." Kepler in many places mentions Tycho as the discoverer of the variation, and the insinuation that Tycho himself did not claim the discovery, but merely called his lunar theory "hypothesis redintegrata," is groundless, as Tycho used the same expression of his planetary system, which he most assuredly did claim as his own (*e.g.*, in a letter to Mästlin, *Kepleri Opera*, i. p. 45).

the existence of another inequality, the fourth one in longitude, of which the solar year was the period, so that the
observed place was behind the computed one, while the sun
moved from perigee to apogee, and before it during the other
six months. We have already mentioned that the solar and
lunar eclipses continued to be carefully observed by Tycho,
and at the latest during his stay at Wittenberg, he had
clearly grasped the peculiarity in the lunar motion just
described, since Herwart von Hohenburg wrote to Kepler
(in July 1600) that he had probably heard from Brahe himself how the latter in the paper he had printed at Wittenberg [1] had introduced a "circellum annuæ variationis, cujus
initium statuitur sole versante in principio Cancri, ita ut
in priori semicirculo hujus circelli verus locus Lunæ pro-
moveatur in consequentia, et in posteriori retrotrahatur in
præcedentia." Kepler also bears witness to the introduction of this *circellus* during Tycho's stay at Wittenberg.[2]
But the representation of the lunar motion had become so
complicated that Tycho shrank from introducing more circles
(for which reason he had adopted a mere libratory motion to
account for the variation), and the idea of a really unequal
motion was too much opposed to the time-honoured conception of uniform circular motion. He (or rather Longomontanus) therefore ultimately allowed for the annual equation
by using a separate equation of time for the moon, differing
by 8m. 13s. multiplied by sine of the solar anomaly from the
ordinary one, even though this left 5′ or 6′ of the irregularity
unaccounted for.[3] The correct amount of the equation (11′,

---

[1] "In dem deliquio Lunæ, so sie zu Wittenberg drucken lassen" (*Kep-
eri Opera*, iii. p. 28). We have seen (p. 272) that Tycho gave up the idea of
printing the lunar theory at Wittenberg.

[2] *Kepleri Opera*, viii. p. 627.

[3] Ibid. In a letter to Archduke Ferdinand, written early in July 1600,
Kepler gives an account of Tycho's researches on the moon, and alludes to
the annual equation in the following words (ii. p. 9) : "Solent ceteri astronomi
non experientia sed ratione moniti optima tempus aequare propter duas

and not 4'.5) was found by Horrox, but he applied it in the same manner as Tycho had done.

It is very interesting to see that Kepler had independently discovered the annual equation about the same time as Tycho did. In the calendar for 1598, which he had to prepare as provincial mathematician for Styria, Kepler had in detail described the solar eclipse of the 7th March (25th February) 1598, making use of Magini's tables.[1] But the phenomenon turned out very different from what he expected, as the eclipse not only was very far from being total (or nearly so), but occurred an hour and a half later than expected. As the only reservation taken by Kepler had been that the eclipse might possibly occur half an hour earlier, he had to say something about the cause of this error in the calendar for 1599. In this he therefore stated that the solar eclipse, as well as the lunar eclipse in February and the Paschal full moon, had been more than an hour late; but the lunar eclipse in August had been too early, and it appeared to him that one would have to assume "that a natural month or period of the moon with regard to the sun in winter, *ceteris paribus*, is a little longer and slower than in summer, and the fault is the moon's and not the sun's, as nothing can be reformed as to the latter without great confusion; but whether the inequality is to be applied to the moon itself or to the length of the day, and what cause it may have in nature and the Copernican philosophy, cannot be explained in a few words."[2] A letter to Mästlin of December 1598 shows that Kepler had not thought further about the matter, and

causas, primo propter inaequales partium signiferi ascensiones rectas, deinde propter motus Solis diurnos inaequales. Hanc posteriorem Tycho negligit, causam afferens experientiam, qua deprehendatur in collatione eclipsium aequalitatis rationem iniri non posse, nisi aut haec negligatur aequatio, aut annuus circellus tot epicyclis Lunae insuper adjiciatur." In 1603 Kepler had also to explain to Fabricius that experience had shown Tycho the necessity of omitting part of the equation of time in the lunar motion (ii. p. 96).

[1] *Opera*, i. p. 396.   [2] Ibid., i. p. 409.

merely threw out this solution because he thought it easier
to defend than one founded on corrections to the solar theory,
and he adds that his calendar was not written for learned
men, and would never be seen outside Styria.[1] It happened,
however, that the calendar was read by Herwart von Hohenburg, who in January 1599 requested Kepler to give him
further information about the solar eclipse. Being thus
obliged to consider the matter more fully, Kepler did so in
his reply, in which his wonderful genius displays itself by
the way in which he suggests that the moon might be retarded
in its motion by a force emanating from the sun, which
would be greatest in winter, when the moon and earth are
nearer to the sun than they are in the summer.[2] At the same
time he suggests that the phenomenon might also be caused
by an irregularity in the rotatory velocity of the earth, and
in after years he accepted this idea, and did not consider the
phenomenon as caused by an equation in the lunar motion.[3]

Tycho Brahe's discoveries as regards the lunar motion in
latitude were as important as those he made of inequalities
in longitude. The inclination of the lunar orbit to the
ecliptic had by Hipparchus been found to be 5°, which value
had been retained even by Copernicus. Several of the

---

[1] Ibid., pp. 409–411.

[2] Ibid., p. 412 et seq. Compare another letter from Kepler to Herwart
of April 1599, published in *Ungedruckte wissenschaftliche Correspondenz
zwischen J. Kepler und H. von Hohenburg*, 1599. Edirt von C. Anschütz.
Prag (Altenburg, S. A.), 1886.

[3] This idea is particularly developed in *Epitome Astr. Copern.*, Liber IV.
(*Opera*, vi. pp. 359 et seq.). See also an interesting paper by Anschütz in *Zeit-
schrift für Mathematik und Physik*, Jahrgang xxxi. and xxxii., 1886–87. In
this the author maintains that Tycho Brahe cannot be considered as the discoverer of the annual equation, because he did not distinctly announce it as a
separate inequality like the variation, but allowed for the effect of it by leaving
out part of the equation of time. I confess myself unable to follow this
reasoning. Tycho clearly perceived the effect of the annual equation, and
only adopted the peculiar dodge about the equation of time for fear of making
his theory too complicated. We might as well deny that Columbus discovered
America because he lived and died in the belief that he had merely come to
the eastern extremity of Asia.

Arabian astronomers had, however, noticed that this was not correct. Thus Abul Hassan Ali ben Amadjour early in the tenth century stated that he had often measured the greatest latitude of the moon, and found results greater than that of Hipparchus, but varying considerably and irregularly. Ibn Yunis, who quotes this, adds that he had himself found 5° 3' or 5° 8'. Other Arabians are, however, said to have found from 4° 45' to 4° 58', which does not speak well for the accuracy of their observations.[1] Tycho first began to suspect that the value of Hipparchus was wrong when examining an observation of the comet of 1577. On November 13 he had measured the distance of the comet from the moon, and found 18° 30', while the observed distances of the comet from stars by computation gave its distance from the moon equal to 18° 9', allowing for the lunar parallax. At first he attributed the difference to refraction, but in 1587, when the moon attained its greatest latitude about Cancer, so that neither errors in the parallax nor refraction could influence the result much, he found the lunar inclination to be 5° 15', and thought it might have increased since the days of Ptolemy, just as the obliquity of the ecliptic had diminished.[2] The examination of all his observations showed him, however, later, that the inclination varied between 4° 58' 30" and 5° 17' 30", while the retrograde motion of the nodes was found not to be uniform, so that the true places of the nodes were sometimes as much as 1° 46' before or behind the mean ones. This inequality of the nodes had not been detected by the ancients, because it disappears in the syzygies and quadratures, where they alone observed the moon. Tycho explained this and the change of inclination by assuming that the true pole

---

[1] L. A. Sédillot, *Matériaux pour servir, &c.*, t. i. p. 283 *et seq.* The sons of Musa ben Schaker (about 850) seem to have been the first to find a value differing from that of the ancients. Some Chinese observers found 5° 2'. *Copernicus, an Internat. Journal of Astronomy*, vol. ii. (1882), p. 128.

[2] *De mundi aeth. rec. phen.*, p. 40.

of the lunar orbit described a circle round the mean pole with
a radius of 9′ 30″, so that the inclination reaches its minimum
at syzygy and its maximum at quadrature.[1] He applied corrections separately to the latitude for equation of node and
for change of inclination, a form which was retained even by
Newton and Euler, until Tobias Mayer showed that the two
equations can be combined into one, varying with the double
distance of the moon from the sun, less the argument of
latitude of the moon.[2]

It would lead us too far if we were in this place to enter
into a description of Tycho's lunar tables, or of his precepts
for finding the longitude from his theory.[3] We shall only
mention that he was the first to tabulate the *reduction*, or
the difference between the moon's motion along its orbit,
and the same referred to the ecliptic. The table of parallax
makes this quantity vary between 66′ 6″ and 56′ 21″, the
apparent diameter varying from 32′ to 36′ at full moon,·
while he believed to have found from his observations of
eclipses that the diameter appears less at new moon (25′ 36″
to 28′ 48″), owing to the limb being "extenuated" by the
solar rays. He therefore denied the possibility of a total
solar eclipse, to some extent also misled by the accounts

---

[1] Copernicus had employed a similar construction to explain the trepidatio
or (imaginary) oscillation of the equinoxes.

[2] In Godfray's *Lunar Theory*, chap. viii., Tycho's hypothesis is described as if
he supposed the lunar pole to move in the small circle with double the synodical
velocity of the *node*. Though this, of course, is the correct representation of
the perturbations in latitude, it is not Tycho's idea, as he took no notice whatever of the position of the node with regard to the sun, but let the pole move
with double the synodical velocity of the *moon*. In the well-known term
$9' \sin (\mathbb{C} - 2 \odot + \mathcal{U})$, if we, instead of the quantity within the bracket, write
$2 (\mathbb{C} - \odot) - (\mathbb{C} - \mathcal{U})$, we get Tycho's period, as the inclination will vary by
$- 9' \cos 2 (\mathbb{C} - \odot)$. But if we put $(\mathbb{C} - \mathcal{U}) - 2 (\odot - \mathcal{U})$, the inclination
will vary by $+ 9' \cos 2 (\odot - \mathcal{U})$, and the period is 173 days. That Kepler
had remarked the importance of the position of the sun with regard to the
node may be seen from *Tab. Rudolph.*, pp. 89–90 ; *Opera*, vi. pp. 588 and 648.
Of modern authors, Montucla seems to be the only one who has remarked that
Tycho paid no attention to the node (*Histoire des Math.*, i. p. 666).

[3] For an account of these, see Delambre, *Hist. de l'Astr. mod.*, i. p. 164.

of the luminosity seen round the sun at the eclipses of 1560, 1567, and 1598.[1]

The planets had been favourite objects with Tycho from his youth. His very first attempts at observing had been sufficient to show him how imperfectly the existing theories of the planetary motions agreed with the actually observed positions of the planets, and throughout his life he never neglected to take regular observations of the five planets.[2] His early observations of planets were of course similar to those made by his predecessors. The ancients had generally fixed the position of a planet by mere alignment, or, if the distance from a star was small, by expressing it in lunar diameters, while conjunctions of planets *inter se* or near approaches to fixed stars were greatly valued as tests of theory. As long as Tycho only possessed few and small instruments, he naturally often had recourse to these old methods, but he commenced also very early to adopt the method, first used by Walther, of measuring the distance of a planet from two well-known fixed stars.[3] At Hveen he never quite gave up this method, but he chiefly depended on meridian altitudes and observations with the armillæ, and even the difficult planet Mercury was carefully watched for and observed on every opportunity.[4]

Though Tycho did not live long enough to try his hand seriously at the theory of the planetary motions, we have

[1] *Progym.*, p. 134; Kepler, *Ad Vitell. Paral.*, chap. viii. (*Opera*, ii. p. 309); Riccioli, *Almag. novum*, ii. p. 372. See also Tycho's letter to Mästlin in 1598 (*Opera*, i. p. 46). About Tycho's observations of the solar and lunar diameters, see above, chapter viii. p. 191.

[2] In a letter to Rothmann (*Epist.*, p. 114) Tycho expressed his regret that so little attention was paid to the planets at Cassel, since the positions of fixed stars were principally of interest by enabling an observer to follow the course of the planets.

[3] At first the youthful observer generally only measured the distance from one star; but from December 1564 two stars are often, and from 1569 always employed.

[4] The earliest observation of Mercury seems to be of April 17, 1574, at Heridsvad.

seen that he was at Benatky occupied with the theory of
Mars, and succeeded in representing the longitudes well
(Kepler says within 2′), while the latitudes gave more
trouble.[1] But already at Uraniborg he had not contented
himself with a mere accumulation of material, but had drawn
some conclusions from the comparison of his results with
the tabular places of the planets. We have seen that Tycho,
like Ptolemy and Copernicus, assumed the solar orbit to
be simply an excentric circle with uniform motion. But
already in 1591, he might have perceived from the motion
of Mars that this could not be sufficient, as he wrote to the
Landgrave that " it is evident that there is another inequality, arising from the solar excentricity, which insinuates
itself into the apparent motion of the planets, and is more
perceptible in the case of Mars, because his orbit is much
smaller than those of Jupiter and Saturn."[2] He concluded
(strangely enough) that his own planetary system alone
could account for this, and he can therefore not have had a
clear idea of the cause of the phenomenon. Again, in his
letter to Kepler of April 1, 1598, he mentioned that the
annual orbit of Mars (according to Copernicus) or the
epicycle of Ptolemy was not always of the same size with
regard to the excentric, but varied to the extent of 1° 40′.[3]
This eventually led Kepler to the discovery of the elliptic
orbits, but it showed him already in Tycho's lifetime that
the solar excentricity was only half as great as hitherto supposed, and that the remainder of the equation of centre
would have to be accounted for by a uniform motion round a
punctum æquans (that is, as long as only circular orbits were

[1] Above, p. 303.

[2] *Epist. Astr.*, p. 206. Magini had also noticed this apparent inequality in
Mars ; see above, chap. ix. p. 213. Tycho also alludes to it in *Mechanica*, fol.
G. 3 : "Tum quoque circuitum illum annuum, quem Copernicus per motum
Terræ in orbe magno, veteres secundum Epicyclos excusarunt, variationi
cuidam obnoxium esse perspeximus."

[3] *Opera*, i. p. 44, iii. p. 267 (*De Stella Martis*, xxii.).

admissible).[1] There was another important matter in which Kepler's suggestion was acted upon. Soon after his arrival at Benatky, he found that Tycho, like his predecessors, referred all the planetary motions to the mean place of the sun, while he had himself in his *Mysterium Cosmographicum* referred them to the actual place of the sun. He gave the impulse to this being done in the lunar theory by Longomontanus, and he mentions in the Appendix to the *Progymnasmata* that the necessity of this step had also become evident in the case of Mars.

With regard to Tycho's observations and researches on comets, we need only refer to Chapter VII., where they have been examined in sufficient detail. It is not among the least of Tycho's scientific merits that he finally proved comets to be celestial bodies.

That a new catalogue of accurate positions of fixed stars was urgently needed had early been felt by Tycho Brahe. The Ptolemean catalogue of stars was fourteen hundred years old, and was probably little more than a reproduction of the still older catalogue of Hipparchus. None of the Arabian astronomers had observed fixed stars, but had contented themselves with adding the precession to the longitudes of Ptolemy; and the only independent catalogue, that of Ulugh Beg, was not yet known in Europe. The co-ordinates of stars given in Ptolemy's catalogue were known at Tycho's time through the two Latin editions of the Almegist of 1515 and 1528 and the Greek edition of 1538; but to the original errors of observation had been added a goodly number of errors of copying, so that the discrepancies of the various editions *inter se* were numerous and large. The

---

[1] *Progymn.*, p. 821. In the *Tabulæ Rudolph.*, p. 57, Kepler says of Tycho: "De solis quidem Eccentrico simplici, cum videret, illum non tolerari ab observationibus planetarum caeterorum, desciscere ultimis temporibus cepit, eumque parem caeteris planetis concessit; quacunque ea res explicanda esset Hypothesi."

observations of the new star and of the successive comets
made Tycho feel the necessity of getting accurate places
for his stars of comparison, and when his observatory was
complete, he took up the work of forming a new star catalogue with great energy.

By Hipparchus the longitudes of stars had been deduced
from the longitude of the sun by using the moon as inter-
mediate link, which method is described by Ptolemy, who
gives a full account of the manipulation of the zodiacal
armillæ. Unfortunately Ptolemy does not say a word
about the manner in which the standard stars (Regulus
and Spica) were connected with the other stars, nor does
he give any details about the actual observations on which
the adopted places of the stars were founded. It is, therefore, not known whether every single star was connected
with a standard star, or whether he perhaps also made use
of conjunctions of stars with the moon (which had been of
great value for the deduction of stellar positions for earlier
epochs for the determination of the constant of precession),[1]
and nothing but a slight sketch of the method was handed
down to posterity. The Arabs, as already remarked, did
not observe fixed stars, and here, as in several other branches
of practical astronomy, Walther was the first to recommence
work. At the Nürnberg observatory he introduced a very
important improvement on the method of Hipparchus by
substituting Venus for the moon, as the small diameter, slow
motion, and very small parallax made the planet far more
suitable for the purpose than the moon. Among the observa-

---

[1] The statement by Copernicus (*De Revolut.*, lib. ii.), that Menelaus used
lunar conjunctions to determine a number of star-places, arises perhaps from
a mixing up of two circumstances, viz., the observations by Menelaus of two
conjunctions in A.D. 98 (recorded by Ptolemy, vii. cap. 3), and the tradition
mentioned by several authors, according to which Menelaus in the first year
of Trajan had compiled a star catalogue which Ptolemy had adopted, after
adding 25′ to the longitudes (Schjellerup's *Al Sûfi*, p. 42 ; Albohazzin, quoted
by Riccius, Delambre, *Moyen Age*, p. 380).

tions published in the *Scripta* of Regiomontanus, the first observation of this kind is from the 6th March 1489, and there are several from the following years; and as the book was published in 1544, Tycho Brahe has known Walther's plan, while the further development of it is due to himself.[1] The method recommended itself to Tycho because it did not involve the accurate knowledge of time by clocks or clepsydræ, while he made this objection to the method followed by the Landgrave of observing the altitude of stars, together with their transits over the meridian or a certain azimuth. The meridian method had been used by Tycho to determine the places of twelve stars observed with the comet of 1577, and for this purpose he made use of *a* Aquilæ as fundamental star, determining its right ascension by observing the meridian transits of it and the moon when not too far apart. He knew, therefore, by experience how undesirable it was to trust to the clocks.[2]

In the spring of 1582 Venus was most favourably situated, and from the 26th February it was for about six weeks clearly visible in full daylight even before it passed the meridian, so that it could be observed at a sufficient height above the horizon to make errors in the adopted refractions harmless.[3] With the *sextans trigonicus* two observers measured the distance between Venus and the sun, the shadow of the little cylinder at the centre of the arc falling on the movable pinnule; at the same time the

[1] Tycho does not allude to Walther, but mentions that Cardan had in 1537 determined the place of *a* Libræ by means of Venus (though apparently without reference to the sun), which he found absurd. Copernicus and Werner had determined the place of a few fixed stars (particularly of Spica) by measuring the declination, borrowing the latitude from the catalogue of Ptolemy, and from these calculating the longitude and right ascension. *Progymn.*, i. p. 146.

[2] *De mundi æth. rec. phen.*, p. 32.

[3] Therefore Tycho gladly turned from the morose Saturn and the deceitful Mercury (*i.e.*, from the use of timekeepers regulated by lead or mercury) to the charming Venus (*Progymn.*, p. 153).

altitudes of the two celestial bodies were measured, and
occasionally their azimuths, while their declinations were
observed by the armillæ, and their meridian altitudes as often
as opportunity offered.   After sunset the same sextant was
employed to measure the distance of Venus from certain
conspicuous stars near the zodiac (Aldebaran, Pollux, and
some others in the same constellations), while, as before,
altitudes and declinations were also observed.   In deducing
the positions of the stars observed, the motion of Venus and
the sun in the interval between the day and night observations was taken into account.   By simple trigonometrical
operations the difference of right ascension between the sun
and a zodiacal star was computed, and as the right ascension of the sun was known from the solar tables, the absolute
right ascension of the star was thus found from the
observations, while the declination was directly measured.
All the stars thus determined were connected by distance
measures with the star a Arietis, which he preferred to γ
Arietis, which by Copernicus had been adopted as principal
standard star, as being nearest to the vernal equinox, but
which Tycho found too faint to be conveniently observed by
moonlight.   Each observation thus gave a value for the right
ascension of a Arietis.   During the following six years
Tycho repeated these observations as often as an opportunity offered, and, in order to eliminate the effect of parallax
and refraction, he combined the results in groups of two, so
that one was founded on an observation of Venus while east
of the sun, the other on an observation of Venus west of the
sun ; while the observations were selected so that Venus and
the sun as far as possible had the same altitude, declination,
and distance from the earth in the two cases.   From the
observations of 1582 Tycho selects three single determinations, and from the years 1582–88 twelve results, each

being the mean of two results found in the manner just described. The fifteen values of the right ascension of *a* Arietis agree wonderfully well *inter se*, the probable error of the mean being only ±6″, but the twenty-four single results in the twelve groups show rather considerable discordances, the greatest and smallest differing by 16′ 30″. But anyhow the final mean adopted by Tycho is an exceedingly good one, agreeing well with the best modern determinations. He adopts for the end of the year 1585 26° 0′ 30″, the modern value for the same date being 26° 0′ 45″.[1]

From the absolute right ascension of *a* Arietis thus determined, and the directly observed declination, Tycho determined the co-ordinates of other stars by measuring the distance from *a* Arietis and the declination, after which the spherical triangle between the pole and the two stars (in which the three sides were known) gave the angle at the pole or the difference of right ascension. Proceeding thus from one star to another round the heavens, Tycho determined the right ascensions first for four, then for six, and finally for eight principal standard stars; and as the sums of the differences of right ascension in the three cases only differ a few seconds from 360°, he imagined that he had proved his results to be extremely accurate. It is needless to say that the accuracy cannot be so great as Tycho fondly hoped, as the errors of observation would be increased by neglect of refraction and by his ignorance of the existence of aberration and nutation. But it must be conceded that Tycho's results were an immense improvement on the positions of fixed stars as previously known, as the comparison with the best modern star-places for the nine stars reduced to the end of 1585 gives the probable error of Tycho's standard right ascensions equal to ±24″.1, and that of his

[1] For details see Note E.

standard declinations (after correcting them for refraction)
$= \pm 25''.9.$[1]

It is interesting to see that observations of absolute right
ascension were made at Cassel about the same time, and
by the same method, except that Jupiter was at first used
instead of Venus. As Jupiter could not be observed with
the sun above the horizon, this involved trusting to the rate
of the clocks for many hours, which perhaps was more
feasible at Cassel, where Bürgi introduced the use of the
pendulum for controlling the clocks. In 1587 Venus was,
however, made use of, the altitude and azimuth of the
sun, Venus, and Aldebaran being observed in succession.
The results thus found for the right ascension of the latter
star agreed well *inter se*, fixing it at 63° 10' for the beginning of 1586, or more than 6' greater than that found by
Tycho. This systematic error, with which all the right
ascensions determined by means of Aldebaran became
affected, and which also, with nearly the same amount,
entered into the longitudes, was discussed in several letters
between the Landgrave, Rothmann, and Tycho. The Landgrave thought 5' or 6' a very trifling quantity, not worth
mentioning, as nobody hitherto had been able to determine
longitudes with that accuracy.[2] Tycho at first suggested
that the discrepancy might be caused by an error in the solar
declination, caused by a faulty suspension of the plumb-
line which marked the zero point on the quadrant at Cassel,
and to which Rothmann had referred in a former letter.[3]
Afterwards he concluded that the error was caused by all
the observations being made in the evening, when refraction

---

[1] See Note E. Adopting the star-places given in Woldstedt's paper on
the comet of 1577, the probable errors in longitude and latitude of the stars
on p. 32 of *De mundi æth. rec. phen.* turn out to be $\pm 1'.18$ and $\pm 1'.25$.

[2] *Epist. Astron.*, p. 78.

[3] Ibid., p. 45 ; compare p. 33. Rothmann suggested that perhaps the appreciable size of Venus might have something to do with it (p. 88).

would tend to make the longitude of Venus appear greater.[1] It seems, however, that the real cause was the unlucky solar parallax of 3′ which Rothmann (like Tycho) had borrowed from the ancients, and which would act particularly injuriously on his results, as his observations were all made in winter, and at low altitudes of both the sun and Venus, and not combined, like those at Hveen, to eliminate errors as much as possible.[2]

On the basis of the nine standard stars and twelve additional stars near the zodiac, Tycho Brahe built up his star catalogue. Of a star to be determined, the declination was measured directly by the armillæ or a meridian quadrant, and the distance from a known star was measured with a sextant. This furnished, as before, a spherical triangle, with the three sides known, from which the angle at the pole or the difference of right ascension could be computed. Generally the star was connected with two known stars, one preceding and one following it, which gave two results for the right ascension as a control. Tycho communicates twelve examples of this double determination, the results always agreeing within a minute.[3] For stars in higher declinations the additional precaution was taken of connecting them with three stars, as in the case of the constellation of Cassiopea, in which Tycho was specially interested on account of the new star, and which he observed in 1578 and 1583. The other constellations were all observed in the years 1586 to 1591. It is needless to say that the

---

[1] *Progymn.*, p. 274.

[2] R. Wolf, *Astron. Mitth.*, xlv. (1878), p. 131. The Hessian star catalogue was to contain 1032 stars, but was never finished. In its incomplete state it is published in *Barretti Historia Cælestis*, under the year 1593, which the editor has erroneously assumed to be the epoch (instead of 1586), probably because the longitudes are about 6′ too great (as $7 \times 50'' = 5' 50''$). Compare Flamsteed's *Hist. Cæl. Brit.*, vol. iii., Proleg., p. 90, and p. 21 *et seq.*, where Tycho's and the Landgrave's star-places are given side by side.

[3] *Progymn.*, p. 224 *et seq.*

twenty-one standard stars were not sufficient, but that it became necessary to build further on the stars determined by them. Magnitudes were frequently noted, and in the final star catalogue they were entered, occasionally with two dots added (:) or one (.), to show that the star was slightly brighter or fainter than indicated by the figure. But these estimates of magnitude were probably not made with particular care, so that it would be risky to draw conclusions from a comparison of them with the more systematically made observations of relative brightness of Ptolemy, Al Sûfi, and astronomers of the nineteenth century.[1]

In reducing his observations, Tycho adopted $51''$ as the value of the constant of precession, which he deduced from a comparison of his own places for Regulus and Spica with those found by Hipparchus, Al Battani, and Copernicus.[2] Although the places of Spica recorded by Timocharis and Ptolemy gave respectively $49\frac{1}{4}''$ and $53\frac{1}{4}''$, he had sense enough to attribute this to the crudeness of earlier observations, and pointed out that these often erred very greatly as to the *relative* positions of stars which were supposed to have been well observed, so that there was no need of assuming any irregularity in the precession of the equinoxes in order to reconcile discrepancies in the absolute longitudes. The origin of this old idea, that the equinoxes did not recede with uniform velocity on the ecliptic, but were also subject to an oscillating motion, is shrouded in mystery. The name of Tabit ben Korra (who lived in the second half of the ninth century) is usually associated with this *trepidatio*, but the idea seems to be very old, and is first mentioned by Theon, the commentator of Ptolemy, according to whom "some ancient astrologers" had found that the stars had an oscillating

---

[1] Tycho's star catalogue was reprinted by Kepler in the *Tabulæ Rudolphinæ* (1627), and by Baily in *Memoirs R. Astron. Soc.*, vol. xiii.

[2] *Progymn.*, pp. 253-255.

motion 8° backwards and forwards in 672 years ; and according to Al Batraki (Alpetragius in the twelfth century), the
erroneous value of 36″ which Ptolemy had found for the
constant of precession, gave rise to the whole mischief, as his
successors could not believe that he had found an erroneous
value.   Al Battani was the only Arabian astronomer of note
who was not an implicit believer in trepidation, but from the
time of Al Zerkali of Cordova (about 1060) the theory of
this wholly imaginary phenomenon was developed minutely.
In the Alphonsine tables the period of the inequality of
precession was assumed to be 7000 years, though King
Alphonso personally seems to have believed precession to be
uniform.   From these tables and the Arabian authors the
theory was spread to Europe, and was further investigated
by Purbach and Regiomontanus, who assumed with Tabit
that the apparent equinox moved in a small circle with a
radius of 4° 18′ round the mean equinox, whereby the annual
precession was sometimes accelerated and sometimes retarded.
In the sixteenth century trepidation was made the subject
of two treatises by Johannes Werner of Nürnberg, and
in the third book of his great work Copernicus has also
examined it in detail, and showed how annual precession
had always varied from the time of Timocharis (300 B.C.)
till his own time.   It was a natural consequence of the
belief in the motion of the equinox on a small circle that
the obliquity of the ecliptic should also vary irregularly ; and
though it had been steadily diminishing since the days of
Eratosthenes, even Copernicus considered such irregularities
proved by the observations of the ancients and the Arabians.
The first to see that the obliquity of the ecliptic had always
diminished at a regular rate since the commencement of
history seems to have been Fracastoro (1538), after whom
the same was asserted by Egnazio Danti in 1578.[1]

[1] *Primo volume dell' Uso e Fabrica dell' Astrolabio e del Planisferio.* **Firenze,**
1578.

The authority of Tycho Brahe was so great, that the mere fact of his having ignored the phenomenon of trepidation was sufficient to lay this spectre, which had haunted the precincts of Urania for a thousand years, and possibly much longer. Though he had expressed himself somewhat guardedly (promising to discuss the matter further in the great work which he did not live to write), he had done enough by making his contemporaries aware of the vast difference between the accuracy of ancient observations and that of his own, and trepidation was never again heard of.[1]

It would not convey a correct idea of the accuracy which Tycho attained in his observations if we were to compare the positions of stars given in his catalogue with those resulting from modern observations. It would certainly be possible to reconstruct his catalogue from his original observations, but as this considerable labour would not benefit modern astronomy, for which a recurrence to Tycho Brahe's observations would hardly ever be of value except in very special cases, it is not likely to be undertaken. We are, however, able to form a conception of the accuracy of his results in other ways. First, the star of 1572 was, as we have seen, connected by distance measures with nine stars in Cassiopea. Computing the positions of these from modern data, Argelander found the probable error of one distance of the new star (with the sextant of 1572) to be $\pm$ 18″.2, while the distances between the stars of Cassiopea measured with the *arcus bipartitus* gave $\pm$ 41″.0.[2] The first result seems rather too small, but as we do not possess the original individual observations of Nova, we have no way of knowing how many such are embodied in the mean

---

[1] About the successive development of the ideas on trepidation, see Delambre, *Moyen Age, passim*, particularly pp. 53, 73, 186, 250, 264; Kästner, *Gesch. d. Math.*, ii. p. 60; *Mittheilungen des Coppernicus Vereins zu Thorn*, ii. (1880), p. 3 *et seq.*

[2] *Astron. Nachrichten*, lxii. p. 273 (1864).

results. From the distance measures of the comet of 1577 Woldstedt found the probable error of one observed distance = ± 4′.2,[1] but as he mixed the sextant measures with those obtained with the cross-staff, which Tycho always mentions as an untrustworthy instrument, this large probable error is not surprising. The most valuable investigation which we possess concerning Tycho's instruments is the discussion of the observations of the comet of 1585 by C. A. F. Peters.[2] When this comet appeared, Tycho's collection of instruments was complete, and we may assume that the observations are typical. Tycho states that his indications of time have been corrected by the observed hour-angles of stars, and by recomputing these the mean correction of + 22ˢ.5 was found, with a probable error of ± 37ˢ. This only shows, as Tycho merely gave the time in whole minutes, that the great armillæ of Stjerneborg were well adjusted. But a very much better proof of this is furnished by the observations. By the armillæ the comet was compared in right ascension with certain standard stars, while its declination was observed with the same instrument. From the total of these observations Peters found that the polar axis of the armillæ was inclined to the horizon by an angle which exceeded the latitude by only 65″ ± 33″, and formed with the meridian an angle of only 36″ ± 13″. The probable error of one observation of declination was ± 49″, that of one right ascension = 81″, and consequently that of one observed hour-angle = ± 57″. The error of collimation (or parallax, as Tycho called it) was − 30″.1, by which amount the observed declinations were too large. The comet was also observed with the *sextans trigonicus*, and the probable error of one observed distance was found

---

[1] F. Woldstedt, *De gradu praecisionis positionum cometæ* 1577. Helsingfors, 1844.

[2] *Astron. Nachr.*, xxix. p. 209 et seq. (1849).

equal to ± 45″, the collimation error being − 114″.6.[1]
These results are sufficient to show that Tycho's instruments
were really made with the great care which he declares he
had always bestowed on them,[2] and in connection with the
above results as to Tycho's standard stars, they exhibit the
vast stride forward which observing astronomy made at
Uraniborg, and which but for the invention of the telescope
could hardly have been much exceeded by his successors.[3]

It will not be out of place to say a few words here about
a time-honoured absurdity which has attributed great carelessness to Tycho Brahe in the adjustment of his instruments
in azimuth. In 1671, Picard, when determining the latitude of Uraniborg, measured the azimuths of the principal
church spires in Seeland and Scania visible from the site
of Uraniborg. At Copenhagen he found among Tycho's
manuscripts similar observations which showed considerable
differences from his own.[4] Picard did not lay any stress
on this discrepancy when mentioning it in the account of

---

[1] By computing the orbit from the sextant observations alone, Peters found
the probable error of one distance = 110″.5, which result, however, is less
certain than the one given above.

[2] "Plura enim hic quam ipsa magnitudo necessaria sunt. Nam et materiæ
soliditas, aëris mutationi nihil cedens, & preparationis concinnitas, diuisionum
subtilitas, pinnacidiorum atque perpendiculi iusta applicatio, firma fulcra,
debita dispositio, conueniens & obsecundans tractatio, accurata collimatio &
numeratio : & pleraque eiusmodi, adesse oportet. Quorum tamen vix omnia
instrumento ligneo, quantæcunque magnitudinis, competere, aut sane non diu
in eo sarta tecta perdurare possunt. Longe igitur præferendum censeo e
solida metallica materia confectum instrumentum." . . . *Progymn.*, p. 635.

[3] By using verniers, improved pinnules, &c., Hevelius (without using telescopes) reduced the probable error of a distance measure to 18″, to the
amazement both of contemporaries and of posterity (Lindelöf, *Ueber die
Genauigkeit der von Hevelius gemess. Sternabstände*, St. Petersburg Bulletin,
1853).

[4] They occur in a rough volume of observations, 1578–81, and are copied
into the volume for 1563–81, so often quoted above in Chapter ii. They are
entered at the end of the year 1578, but it is not stated when they were made.
There are also azimuths measured from a hill and from the church at Hveen,
probably with a cross-staff, and they are headed, " Observationes geographicæ
in insula Huena factæ."

his journey, probably because he saw from the MS. that
Tycho had merely measured these approximate azimuths
for the sole purpose of constructing a map of the island.
By others the matter was, however, misunderstood; and by
some the discrepancy was even supposed to prove a shifting
of the meridian line between the times of Tycho and Picard;
while others have pointed to Tycho as a blunderer in comparison with the builder of the Great Pyramid, who was
able to orient the sides of that remarkable structure with
considerable accuracy.[1]   It was, however, shown by a Danish
writer, Augustin, that Tycho and Picard had in two cases
pointed to different spires.   At Elsinore Tycho had pointed
to St. Mary's Church, while Picard had pointed to the taller
spire of the church of St. Olaus, built in 1614; and the
cathedral of Lund has two towers, of which Tycho had taken
the southern one, while Picard pointed midway between the
two.   This accounted for the most serious differences, and
the remaining measures would agree well by assuming an
error of 14′, by which amount Tycho's meridian line should
have deviated from the true south point towards the east.
Augustin even imagined that he had found in the printed
observations the proof that Tycho detected this error on
the 2nd November 1586.[2]   It is, however, evident from
the words used by Tycho that he must on this occasion
have referred to a *recent* readjustment (*in novo meridiano*)
of the instruments *at Stjerneborg only*, and not to some meridian line adopted since 1579, at which time (at the latest)
the azimuths of the church spires were measured.[3]   The

---

[1] In his *éloge* of Chazelles, Fontenelle had already in 1710 remarked the
absurdity of attributing such an error to Tycho, and Montucla had expressed
himself to the same effect. *Hist. des Math.*, i. p. 669.

[2] *Skrifter som udi det Kong. Videnskabernes Selskab ere fremlagte*, xii.,
1779, p. 191 *et seq.*; *résumé* in the *Connaissance des Temps pour l'an* 1820,
p. (385).   Compare *Corresp. astron. du Baron de Zach*, vol. i. p. 402.

[3] Tycho's words are (*Hist. Cœl.*, p. 170): "In novo meridiano monstrabant
armillæ 15 M. ante verum meridianum.   Quare omnia tempora hactenus

observations of the comet of 1585, as we have just seen, prove conclusively that in that year the great armillæ were in excellent adjustment, so that Tycho cannot have made use of any badly placed meridian mark. I have also computed a number of observed altitudes and azimuths of stars from 1582, and from these it is evident that the zero line of the azimuth circle was within 1′ of the meridian.[1] As Tycho never once alludes to the use of meridian marks or terrestrial azimuth marks (which he could not possibly have seen from the subterranean observatory, where stars near the horizon could only be observed with portable instruments in the open air), while he frequently states that he verified his instruments by observations, it is impossible that he can, even before 1586, have made a mistake of 14′ in azimuth in the adjustment of his numerous instruments.

The astronomical work in Tycho Brahe's observatory must have involved a considerable amount of computing, even though the great globe, no doubt, was very often used for the solution of spherical triangles. Trigonometry had made considerable advances in the fifteenth and sixteenth centuries, and Tycho could build on the labours of Purbach, Regiomontanus, Copernicus, and others, both as regards the solution of triangles and tables of sines and tangents. But

---

observata uno minuto tardiora sunt debito, non tamen ubique unius minuti est differentia, quia non semper eodem modo se habuit ; ubique dimidii." The instrument here referred to is the great equatorial at Stjerneborg ; the hour circle had probably been found to be set 15′ wrong. On p. 210 (same date) Tycho adds to some observations with the *quadrans volubilis* (also at Stjerneborg) the remark : "Azimutha sunt ex nova restitutione meridiani ante biduum facta." In the *Connaissance des Temps* for 1816, p. (230), Delambre quotes the note to the observation with the mural quadrant of 3rd December 1582 (*Hist. Cœl.*, p. 4), and assumes from this that Tycho in 1582 had found an error in his azimuths. The note in question has, however, nothing to do with this matter, as it only explains that the recently mounted quadrant had not yet been properly fixed to the wall.

[1] See Note F. at the end of this volume.

logarithms had not yet been invented, and great inconvenience was therefore felt whenever it became necessary to
multiply or divide trigonometrical quantities. To obviate
this difficulty a method was invented, the so-called Prostaphæresis,[1] by which addition and subtraction were substituted for multiplication and division, and in the history
of this invention, which was made independently by several
mathematicians, the name of Tycho is also mentioned. The
Arabs had had an idea of this method; at least, Ibn Yunis
makes use of the formula[2]

$$\cos A \cos B = \tfrac{1}{2} \left[\cos (A - B) + \cos (A + B) \right]$$

but, like many other discoveries of the Arabs, this formula
had to be deduced anew in Europe. It was found by Viète,
as well as the corresponding formula:

$$\sin A \sin B = \tfrac{1}{2} \left[\cos (A - B) - \cos (A + B) \right]$$

but as Viète's *Canon Mathematicus*, which was published in
1579, seems only to have been printed in a few copies at
his own expense, it is very possible that Tycho Brahe
never saw it, or at least that he had not seen it in 1580,
when, according to Longomontanus, he and Wittich invented
Prostaphæresis.[3] This was among the inventions which
Wittich a few years later brought to Cassel, where Bürgi
soon developed the method further. It appears that Wittich
merely had shown him the above formula for sin A sin B;
but Bürgi applied the principle to the formulæ of spherical
trigonometry, and ultimately was led to discover logarithms

---

[1] Astronomers need hardly be reminded that this word (formed from πρόσθεσις,
addition, and ἀφαίρεσις, subtraction) had originally signified the equation of
the centre, in which sense it was still used by Tycho.

[2] Delambre, *Astr. du Moyen Age*, pp. 112 and 164.

[3] Si autem de hujus compendii inventore quis quærat, nec Arabes aut
Joannem Regiomontanum fuisse, scripta eorum analemmatica declarent;
neminem certe habeo Tychone nostro & Vitichio Vratislaviensi antiquiorem :
quorum scilicet mutua opera primum anno 1582 [should be 1580] in Huæna,
sphærica quædam triangula tali pragmatiæ pro studiosis Vranicis sunt
subjecta."—*Longomontani Astr. Danica*, p. 8.

years before Napier did; but, as is always the case with that remarkable man, without securing the priority by a timely publication.[1] At Uraniborg the method did not make any progress after the departure of Wittich, and it is therefore more likely that it was he, and not Tycho, who was the inventor, as he is known to us (through the repeated testimony of Tycho) as an able mathematician. In 1591 a short treatise on plane and spherical trigonometry was drawn up at Uraniborg, but it does not indicate that Tycho had developed trigonometry in any way, as the rules are similar to those given in other treatises of that day, and are frequently expressed in even clumsier language than usual at that time.[2] The demand for the facilities offered by the Prostaphæresis was, however, so great, that Reymers Bär, Clavius, Joestelius, Magini, and others, with more or less success, continued to work in this direction, until the method was driven from the field by the discovery of Napier.

We have followed Tycho Brahe through his chequered career, and we have reviewed his scientific labours.   No doubt his contemporaries were not uninfluenced in their estimation of him by his princely residence, with its tasteful decoration and wonderful observatories, and also by its singular situation on the little island, which contributed to exhibit the noble astronomer in a romantic light.   But while these circumstances threw a halo over Tycho even before his works had become known beyond a limited circle, posterity has hardly been influenced by considerations like these when

[1] R. Wolf, *Astr. Mittheilungen*, No. 32; *Gesch. d. Astronomie*, p. 348 *et seq.*

[2] In the University Library at Prague, published in facsimile by Studnicka at Prague in 1886 : "[Tychonis Brahe] Triangulorum planorum et sphaericorum Praxis arithmetica." The original is written on twenty leaves, inserted at the end of a copy of *Rhetici Canon doctrinæ triangulorum.* Tycho has written his name under the title of the MS., but the handwriting of the remainder does not seem to be his.

affirming the judgment of his time.   He not only conceived the necessity of supplying materials for the discovery of the true motions of the heavenly bodies, and by his improvement of instruments and accumulation of observations made it possible for Kepler to reach this goal, but in almost all the branches of practical and spherical astronomy he opened new paths, and made the first serious advance since the days of the Alexandrian school.   Hereby he showed his superiority to the Landgrave ; for though the latter had perceived the necessity of systematic observations at least as early as Tycho did, he confined his attention almost entirely to the fixed stars, and had to borrow the improvements in instruments from Tycho, and let them be worked out by the great mechanical talents of his assistant, Bürgi, before his observations could rival those of Tycho in accuracy.   It was, therefore, not at Cassel, but at Uraniborg that the reform of practical astronomy was carried out, and posterity has not thought it an exaggeration when one of the greatest astronomers of the nineteenth century spoke of Tycho Brahe as a king among astronomers.[1]

[1] Bessel, *Populäre Vorlesungen*, p. 422.

APPENDIX.

———•———

IT was perhaps well for Tycho Brahe that his career in
Bohemia was cut short, for he would sooner or later have
been bitterly disappointed in the faith he had placed in
the Emperor's promises. His greatest wish had always
been that the observatory work should not cease at his
death, but that some competent person might be appointed
to carry it on; but though Kepler, two days after Tycho's
death, was informed by Barwitz that he was to be the new
Imperial mathematician, the observations with Tycho's instruments were not continued very long. The Emperor soon
agreed with Tycho's family to purchase the instruments for
the sum of 20,000 thaler; but when Tengnagel came back
to Prague in the summer of 1602, he assumed the position
of Tycho's scientific heir, promised the Emperor to have
the Rudolphine tables finished within four years, and though
Kepler had commenced to observe Mars, he was deprived
of the instruments, which were stored away in a vault under
Curtius' house. Kepler never got access to them again,
of which he complains repeatedly in his writings.[1] They
seem to have been preserved in this manner until the year
1619, when the Bohemians rose against the House of Habs-
burg and elected Frederic V., Elector Palatine, their king,
and during the disturbances which followed, some of the
rebels are said to have destroyed the instruments as Imperial

---

[1] *Opera*, ii. p. 760 (in the dedication to Hoffmann of the book on the star
in Cygnus), and p. 755. In the book on the star in Serpentarius (ibid., ii. 656)
Kepler quotes a few observations by Tengnagel, "in viridario Cæsaris, ubi
deposita habebantur instrumenta Braheana." Perhaps they had then (October
1604) been brought back to Ferdinand I.'s villa. In December 1601 and May
1603 Kepler used one of Tycho's clocks in observing two lunar eclipses (*Opera*,
ii. p. 3co).

property.[1] The great globe alone was saved, and was in
1632 found at Neisse, in Silesia, at the College of the
Jesuits, by Prince Ulrik, a son of King Christian IV. of
Denmark, who was in the service of the Elector of Saxony,
and had taken Neisse by storm. How or when the globe
had been sent there is not known, but Prince Ulrik now
sent it to Denmark, where it was first kept at the Castle
of Rosenborg, then at the University,[2] and afterwards in
a room of the Round Tower which had been erected in
Copenhagen to serve as a University observatory, and was
finished in 1656. An inscription, composed by Longomontanus, was attached to the globe or to the wall of the
room, and the beautiful monument of the great astronomer
remained at the Round Tower till October 1728, when it
was unfortunately destroyed in the great conflagration, in
which, among many other things, Ole Römer's unpublished
observations perished. At the present day there is neither
at Prague nor at Copenhagen the smallest vestige of Tycho's
celebrated instruments.[3]

Tycho's wife and children all remained in Bohemia, probably because they were honoured and respected there, while
the difficulty which they found in obtaining payment for
the instruments must also have tied them to Bohemia, as
they must have known well that they would have no chance
of getting their money unless they remained on the spot.
Tycho's widow died in 1604, and was buried beside her
husband, as we have already mentioned. A year or two
before her death she had purchased a country property
towards the Saxon frontier, and her eldest son, Tycho, had
in March 1604 married the widow of a country gentleman
in the same neighbourhood. He became the father of five
children, and died in 1627. His younger brother Jörgen
(George), died in 1640. Magdalene Brahe, Tycho's eldest
daughter, apparently never married; of the second daughter,
Sophia, nothing is known except that she became a Roman

---

[1] Gassendi, p. 216.'

[2] Where Huet saw it in 1652 (*Commentarius*, p. 81).

[3] The inscription is given by Gassendi, p. 217; Weistritz, i. p. 217. At
the Prague observatory (founded in 1751 in the Clementinum, far from Tycho
Brahe's observatory) there are two sextants and a clock showing the Tychonic
system, which are supposed to have belonged to Tycho Brahe; but they show
no sign of Tycho's refined workmanship, and the two sextants (the larger of
which is said to have been made for him in 1600 by Erasmus Habermel) have
not his peculiar pinnules. There is no proof of their ever having belonged
to Tycho Brahe (*Astron. meteorol. und magn. Beobachtungen an der Sternwarte
zu Prag im Jahre* 1880, p. iv.).

Catholic. The youngest daughter, Cecily, married a Swede, Baron Gustaff Sparre, colonel of a German regiment, and died at Krakau. In 1630 some of Tycho Brahe's nearest relations in Denmark, among whom was his sister Sophia, issued a declaration, stating that Christine, by the ancient laws of the kingdom, "on account of the open, unchanged, and honourable life of both of them, must be acknowledged as his wedded wife."[1]

Tengnagel very soon gave up the idea of working at the Rudolphine tables. He had probably only been a short time at Uraniborg (he is mentioned as an unpractised observer in September 1595), and there are no signs of his having occupied himself seriously with astronomy during Tycho's lifetime, so that probably it was only jealousy of Kepler which induced him to prevent the latter from taking up Tycho's work at once. He was in 1605, by the Emperor, made a Councillor of Appeal, and received a grant from the Benatky estate "for his astronomical observations," and he was also employed on various foreign embassies—among others, to England, whither he was accompanied by Eriksen, who also gave up astronomy.[2] Tengnagel was appointed Councillor to Rudolph's cousin, Leopold, Bishop of Passau, and afterwards became a privy Councillor to Ferdinand II. He was the only one of the family who was allowed to remain in Bohemia after the battle of Prague (November 1620), when the Protestants were driven from the Austrian possessions. His wife, Elizabeth Brahe, had died in 1613, leaving several children. He died in 1622.

Notwithstanding his connection with the two Emperors, Tengnagel had been unable to get the purchase-money for the instruments paid in full. From a letter which Magdalene Brahe wrote to Eske Bille in July 1602 we learn that, although the Emperor soon after Tycho's death had agreed to purchase the instruments for 20,000 thaler, he was, as

[1] The document was written in German, so that the children of Tycho could make use of it in Bohemia and Saxony, where they all lived. *Danske Magazin*, ii. p. 367; Weistritz, ii. p 375. Sophia Brahe died in 1643 at the age of eighty-seven. She was the only one of Tycho's brothers and sisters on whom some of his glory was reflected, and when she, nine years before her death, at Elsinore, met a French embassy, the secretary, Charles Oger, in the description which he afterwards wrote of his journey, mentioned the meeting with her among the most remarkable events.

[2] Eriksen observed the solar eclipse of October 1605 in London, and brought letters backwards and forwards between Kepler and Harriot. He had early in 1602 (with Tengnagel) visited Fabricius in Ostfriesland, and afterwards for some time assisted Kepler (*Opera*, iii. 533, ii. 432).

usual, without money, and had in vain tried to get the Bohemian Estates to pay the sum, which they declined to do, on the plea that this was a private matter of the Emperor's; and he attempted to persuade the heirs to accept some more or less doubtful securities instead of ready money.[1] This, however, they would not do, and in September 1603 they got 4000 thaler paid from the royal revenues. Owing to the disturbed state of Bohemia and the subsequent great war, the family apparently never received any part of the remaining 15,000 thaler, though they persevered for many years in their applications to the Government. The last time anything bearing on this matter is mentioned is in 1652, when it is stated that a married daughter of the younger Tycho had been six years in Bohemia on a safe-guard from the Elector of Saxony, endeavouring to get her share of the money due from the Treasury.[2]

The first piece of work which Kepler undertook after Tycho's death was to get the *Progymnasmata* published. The section about the lunar theory was not yet printed, but the woodcuts were ready and the text completed in manuscript. A postscript seemed desirable, explaining how the book had been written and printed by degrees, and Kepler at once wrote this appendix, which fills six pages.[3] He first explains how Tycho's anxiety that the book should contain the latest results of his investigations had made him push on with the printing before the whole manuscript was ready (it had been prepared in the years 1582–92). A few slight discrepancies are pointed out between these latest results and a few passages in the book, concerning the moon, but printed long before. It is also remarked that in the first chapter the planetary inequalities are referred to the sun's mean place, while it had recently been found in the case of the moon and Mars that it is the apparent

---

[1] *Danske Magazin*, ii. p. 361 *et seq.*; Weistritz, ii. p. 369 *et seq.* From this letter it appears that the family had previous to July 1602 left Curtius' house, and lived in the part of the city called Altstadt. They had commenced to remove the instruments to their new residence, as they had not yet received any payment; and even of Tycho's salary, which the Emperor had ordered to be paid up to the date of his death only, there were still a thousand florins owing to them. In April 1608 Magdalene wrote a letter to Longomontanus (ibid.) giving him information about the family.

[2] For full particulars about these transactions see the paper by Dvorsky, quoted above, p. 307.

[3] That Kepler is the author of this appendix is stated by himself in a letter to Magini (*Opera*, iii. p. 495; Carteggio, p. 331); it is reprinted in *Opera*, vi. p. 568.

place which enters into the equations, so that the same doubtless also was the case with the other planets. Lastly, the recently noticed fact that the solar excentricity is only half as great as formerly believed, is referred to. A dedication to the Emperor from Tycho's heirs, a short notice to the reader (stating that the author had intended to write a preface on the utility and dignity of astronomy), and the privileges of the Emperor and James VI. were also printed, and the book was published in the autumn of 1602. The title is: "Tychonis Brahe Dani Astronomiæ instauratæ Progymnasmata. Quorum hæc Prima Pars de restitutione motuum Solis & Lunæ, Stellarumque inerrantium tractat. Et præterea de admiranda noua Stella Anno 1572 exorta luculenter agit. Typis inchoata Vraniburgi Daniæ, absoluta Pragæ Bohemiæ MDCII " (some copies have MDCIII). The book seems to have been printed in 1500 copies,[1] and most of these appear to have been afterwards sold to Gottfried Tampach, a well-known bookseller in Frankfurt, who, in 1610, issued the book with a new title-page and the beginning as far as p. 16 reprinted.[2]

The second volume of the *Progymnasmata* had, as we have seen, been quite ready since 1588, though Tycho had only presented a few copies to correspondents, and had intended to add an appendix on Craig's allegations about the parallax of comets. This he had never done, and the book was now published in 1603 with a new title-page, and a dedication to Barwitz from Tengnagel, as well as a preface to the reader.[3] Like the first volume, it was re-issued by Tampach in 1610 with a new title-page, and the first seven leaves (including Tycho's own preface) and the two last pages reprinted.[4] The issue of 1610 is generally found bound together with the first (and only) volume of the *Epistolæ*, printed in 1596, furnished with a new title-page, but retaining the original colophon. Tampach had probably acquired the stock of copies of the *Epistolæ* on the death of Levin

---

[1] Tycho inquired in January 1600 if he could get the sheets yet wanting printed at Görlitz in 1500 copies. *Aus T. Brahe's Briefwechsel*, p. 16.

[2] The misprints so far are corrected by the list of errata at the end of the book.

[3] The title is the same as given above on p. 163, except that instead of the last sentence it has: "Typis inchoatus Vraniburgi Daniæ, absolutus Pragæ Bohemiæ." The colophon is the vignette "Despiciendo svspicio," and underneath : "Pragæ Bohemorum. Absolvebatur Typis Schumanianis. Anno Domini MDCIII."

[4] The two volumes were reprinted in Frankfurt in 1648 with the title *T. Brahei Opera Omnia*. This is a very poor edition with very small print.

Hulsius of Nürnberg, a well-known writer and publisher, to whom the heirs would seem to have sold them, as some copies have on the title-page: "Noribergæ, apud Levinum Hulsium MDCI."

The stock of copies of the *Astronomiæ instauratæ Mechanica* appears to have been exhausted, but most of the wood-cuts and copper-plates were in the possession of the heirs, who sold them to Levin Hulsius. He printed a new edition at Nürnberg in 1602, exactly like the original, but with narrower margins, and without the neat border which in the original runs round the pages. Paper and print are also somewhat inferior, and Tycho's portrait is on the title-page substituted for the vignette of the original.

On the state of Tycho's other manuscripts Kepler drew up a short report,[1] from which it appears that the printing of the second volume of letters had been commenced, and that Tycho had thought of adding some astronomical tables to the volume to make it more saleable. Kepler suggested that matter of astronomical interest occurring in the un-printed letters might be extracted and printed, so that the sheets already in print would not be wasted. This was, however, not done, and only a few of the letters have yet been published. For the third volume of *Progymnasmata* (on the comets of 1582, 1585, &c.), the materials were ready, but nothing was put into shape. As to the *Tabulæ Rodolpheæ*, Kepler stated that the materials were abundant, "nec deerunt ingenia, si Maecenates sint, et exiguum aliquid in certis pensionibus annuis in hunc usum erogetur." The *Theatrum astronomicum* (of which Tycho had sketched the plan in his letter to Peucer in 1588[2]) should contain the theory on which the tables were based, but nothing of it had been written.

It seems that Kepler received Tycho's observations, originals and copies, after signing a contract with Teng-nagel in 1604. He found them so indispensable to his studies that he never returned them, but it was not for-gotten that they were in his possession, and in November 1621 (when he had been obliged to stay more than a year in Würtemberg to watch the trial of his mother for witch-craft) Ferdinand II. wrote to the Duke of Würtemberg requesting him to command Kepler to return the manu-scripts.[3] Kepler was probably back again at Linz (where he

---

[1] *Kepleri Opera*, i. p. 191.   [2] See above p. 182.
[3] *Breve og Aktstykker*, p. 150.

had lived since 1612) when the letter arrived in Würtemberg, and it had no effect. After publishing the *Tabulæ
Rudolphinæ* in 1627, he thought the following year, while
living at Sagan, in Silesia, under Wallenstein's patronage, of
getting the observations printed. He wrote on the 17th
August 1628 to Jörgen Brahe that he hoped soon to commence the printing, but as he had found a selection from the
observations of the years 1600 and 1601 inserted in Snellius'
edition of the Landgrave's observations,[1] he inquired if Brahe
still had the originals for those two years, or whether Snellius
could have got hold of them, so that his widow might have
them still.[2] Nothing came, however, of the intended edition,
and the original observations remained in Kepler's possession,
and after his death in that of his son, the physician, Ludwig
Kepler.[3]

But while Kepler retained the originals as pledges for the
considerable arrears of salary due to him, a set of unfinished
copies in quarto volumes had remained in Austria.[4] These
volumes are alluded to by Tycho Brahe in his *Mechanica* (fol.
G. 2), where he mentions that the observations had been
entered in large volumes, and afterwards, for each year,
copied into separate volumes and sorted according to subject
—the sun, moon, planets (beginning with Saturn and ending
with Mercury), and the fixed stars. Albert Curtz, a Jesuit,
and Rector of the College of Dillingen, on the Danube, who
had corresponded with Kepler both on scientific and religious
subjects, conceived the idea of publishing Tycho's observa-

---

[1] "Coeli et siderum in eo errantium observationes Hassiacæ . . . et speci-
legium biennale ex observationibus Bohemicis V. N. Tychonis Brahe." Lugd.
Batav., 1618, 4to. Snellius had as a youth of twenty paid a short visit to
Prague in 1599 or 1600. See R. Wolf's *Astr. Mitth.*, lxxii.

[2] *Breve og Aktstykker*, p. 152. About Kepler's intention of publishing the
observations, see his *Opera*, vi. pp. 616, 621, vii. p. 215 (in his book *T.
Brahei Hyperaspistes*, Frankfurt, 1625, against Scipione Chiaramonte, who in his
book *Antitycho* had tried to prove from Tycho's own works that comets are
sublunary), also viii. p. 910. Gassendi, p. 207.

[3] On the outside of the cover of the volume for 1596-97 (bound in old
music-paper) there is pasted a slip of paper on which Kepler has written :
"Extract aus mein Johan Kepplers den Brahischen Erben zugestelter Erkle-
rungsschrift. Entlich und zum fünften sol auf einem jedem Tomum herab-
gemeldeten Observationen, sobalt ich nach Linz komme ein offene Zettel
auffgeleimet, und meinen Erben, darinnen von mir anbefohlen werden, dass
solche Bücher, da ich etwa Todtes verbliche, absobalden zu meinem Schatz
oder Kleinodien eingesperret und vor der Eröffnung Jhr. Kay. May. so wie
auch denen Brahischen Erben, umb weitere Vorsorg und Verwahrung dero-
selben angemeldet werden, damit also die Erben auch auff diesem Fall, de
abgesetzten ersten Puncts halben desto mehr versichert sein."

[4] Or perhaps they were purchased from Kepler's daughter, if the MS. account
printed by Kästner is authentic (*Geschichte d. Math.*, ii. p. 651 *et seq.*).

tions from these volumes. It is very strange that he should not have made any serious effort to obtain the originals, as he was engaged on the work already before 1647, in which year Gassendi heard of the undertaking; while Hevelius in the following year inquired how the rumour could be true that a Jesuit had got Tycho's observations from the Emperor and was about to publish them, since Hevelius with his own eyes had seen the original observations from 1564 to 1601 in Ludwig Kepler's house at Königsberg. Curtz himself seems, however, to have believed that the nineteen annual volumes for the years 1582–92 and 1594–1601, which he had before him, were originals and not copies, and though he suggests that they were the set of twenty-one volumes referred to by Tycho in the Appendix to the *Mechanica* (fol. H. 4), it does not seem to have occurred to him that in that case not only the volume for 1593, but five earlier volumes must have been lost, since Tycho wrote the *Mechanica* in 1597. The volumes which he used, and which he describes as being ornamented on the cover with Tycho's portrait and arms,[1] were therefore copied, and the observations of 1582 printed in 1656 as a specimen,[2] after which the complete *Historia Cœlestis* was published at Augsburg in 1666 in a handsome thick folio volume, on which the editor, instead of his own name, Albertus Curtius, has called himself anagrammatically Lucius Barrettus.[3]

The various astronomical observations, chiefly of eclipses anterior to Tycho's time, as well as the observations of the

---

[1] Like the presentation copies of the *Mechanica* (above, p. 261). The nineteen volumes are still in the Hofbibliothek at Vienna, where there is also a miscellaneous collection of loose leaves or small stitched books with computations, notes, or letters. Some of the letters have been published of late years (*Tychonis Brahei et ad eum doct. vir. Epist.*). A list of the MSS. was made by Friis in 1868, but owing to his complete ignorance of the subjects referred to in them, it is of little use (*Danske Samlinger*, iv., 1869, p. 250).

[2] " Sylloge Ferdinandea sive collectanea historiæ coelestis ex commentariis MS. obss. Tychonis Brahei ab anno 1582 ad annum 1601. Accessit epimetron ex obs. Hassiacis, Wirtenbergicis et aliis . . . vulgavit Lucius Barrettus, anno CIƆ IƆC LVI Viennæ Austriæ." Contains the preface and Liber prolegomenus and the observations of 1582, headed on every page : "Sylloge Ferdinandea." Different print from the *Hist. Cœl.* At the end of the volume is the colophon : "Operis Davidis Havtii, Bibl. Viennensis, anno 1657."

[3] "Historia Coelestis complectens Observationes astronomicas varias ad Historiam Coelestem spectantes Ill. viri Tychonis Brahe, Babylonicas, Græcas, Alexandrinas, Moestlini observationes Tubingenses, Hassiacas, &c. Aug. Vind. 1666" (also with title-page on which "Ratisbonæ, 1672"). Large plate with four emperors, more or less imaginary views of Uraniborg, Wandesburg, Benatky, Horti Cæsaris and Domus Curtii, most of Tycho Brahe's instruments, &c. cxxiv. + 977 pp. fol.

Landgrave, Mästlin, and others, which Curtz inserted in this volume, are not without value and interest, but Tycho Brahe's observations are presented in so mutilated and distorted a shape as to be well-nigh useless. Not only is there no explanation why the volume begins with the year 1582 (which has led many writers to believe that Tycho had not observed regularly until then), but it is evident that the copy at the disposal of Curtz had never been finished nor collated with the originals. Frequently several consecutive pages have been passed over, so that the volume is very far from containing a complete record even for the years it pretends to cover. For instance, at the end of 1584, half the observations made by Elias at Frauenburg and all those he made at Königsberg are omitted. Again, in 1589 and 1591 there are several large gaps in the observations of fixed stars, similarly in 1595 and 1597, while the omissions of one or two nights' work are very numerous indeed. But this is far from being the worst fault. There is scarcely a column which is not full of errors, figures misplaced or left out, words like *dexter* and *sinister, borealis* and *meridionalis,* are interchanged; sometimes the signs of the zodiac have even been mistaken for figures, so that the sign of Cancer becomes 69, &c. In short, the work is not far from being an Augean stable. Unfortunately there is no other edition of Tycho Brahe's observations except of the observations of planets made in 1593 [1] and of the observations of comets. The *Historia Cœlestis* gives the reader a fair idea of the general scope of Tycho's work, but it cannot be used for any scientific purpose.

In the meantime the original observations, of the existence of which Curtz was ignorant, were (at the latest in the beginning of 1662) by Ludwig Kepler sold to King Frederick III. of Denmark, who deposited them in the newly founded Royal Library at Copenhagen, where they are still preserved. King Frederick soon after decided to have them published under the direction of the mathematician Professor Erasmus Bartholin, under whom six students were employed in copying and collating, while the necessary pecuniary means were liberally supplied. A complete copy had been made and carefully read with the originals, when Bartholin heard of the publication of the *Historia Cœlestis,* and obtained a copy of it. As he found it extremely defective and erroneous, he pub-

---

[1] To account for the absence of these, Curtz invented a fable about the volume for 1593 having been sent to Cassel and thereby lost, which has been repeated by many writers.

lished in 1668 a critique of it, showing the errors in the observations of 1582, and announcing the forthcoming correct edition.[1] Bartholin furthermore compared a copy of the *Historia Cœlestis* (bound in two volumes) with the originals, and entered in it all the corrections and smaller omissions, while he in a third volume had the observations previous to 1582, the longer omissions, the year 1593, and the observations of comets, carefully copied and compared with the originals.[2] In 1669 he caused inquiries to be made from Blaev, in Amsterdam, about the printing of the new edition, which Blaev seemed disposed to undertake.[3]

Unfortunately, King Frederick III. died in 1670, and as his son and successor took no interest in literature or science, there was an end to the prospect of a correct edition of Tycho Brahe's observations. In the following year Picard came to Copenhagen to determine the geographical position of Uraniborg, and on learning how matters stood, he begged and obtained leave to take Bartholin's copy back with him to Paris to have it printed at the expense of Louis XIV.[4] For the sake of control during the printing, the originals were handed to Bartholin's assistant, Ole Römer, whom Picard had persuaded to go with him to France. The printing was commenced at Paris, but Louis XIV.'s wars required money, and the undertaking was eventually stopped.[5] Inquiries were made for the original manuscripts by the Danish Government in 1696, and they were found in charge

[1] "Specimen recognitionis nuper editarum observationum astronomicarum n. v. Tychonis Brahe, in quo recensentur insignes maxime errores in editione Augustana Historiæ Cœlestis a. 1582 ex collatione cum autographo . . . animadversi ab Erasmo Bartholino." Hafniæ, 1668, small 4to, 48 pp., of which 6 pp. are dedication to the King, 11 pp. introduction, and the remainder errata. Reviewed by Kästner, ii. p. 656.

[2] Strange enough, he did not copy the year 1581, but instead of it the year 1583, though this *is* in the *Hist. Cœl.* About this supplement, see Bugge, *Observationes astronomicæ*, 1781–83, Hafniæ, 1784, p. xviii., where a catalogue of Tycho's original MSS. is given. See also below, Note H.

[3] Werlauff, *Historiske Efterretninger om det store kongelige Bibliothek*, Kjöbenhavn, 1844, p. 411.

[4] Picard gave the following receipt for them (copied by Bartholin into the supplementary volume) : "Je confesse avoir recue de Mons. Erasme Bartholin les observations de Tycho Brahe escrites au net en cinq Volumes in folio depuis l'année 1563 jusques à 1601 avec les Observations des Cometes, à condition qu'ils seront imprimées à Paris au Louvre aux depens du Roy de France, & quant a la Dedication & Preface elles seront faites par le dit Mr. Bartholin. Je promets aussy, qu'incontinent après l'ouvrage achevé d'imprimer, il en sera fourny cinquante exemplaires, qui seront mis entre les mains de qui l'on voudra. Fait à Copenhague, le 2 Avril 1672. PICARD."

[5] Sixty-eight pages were printed (as far as 1582). See Lalande's *Astronomie* (2nd edit.), i. p. 198.

of La Hire at the Paris Observatory.[1] They were handed over to the Danish envoy in 1697, but were not sent back to Copenhagen till his return to Denmark in 1707. Being deposited in the Royal Library, they fortunately escaped the great fire of 1728, in which the University Library and the Observatory (with Römer's observations) were destroyed. In 1707 it was suggested by Dr. Arbuthnot, physician to Prince George of Denmark, that Tycho's observations ought to be printed in England, together with those of Flamsteed, and Newton drew up a letter to Römer on the subject, but nothing further came of it.[2] Bartholin's copy remained in Paris at the Académie des Sciences, and is now at the Paris Observatory.[3] La Hire had copied the observations of 1593 from the originals, and they were published in the *Mémoires de l'Académie* for 1757 and 1763. De l'Isle had made a copy of the whole series, translated into French, but with frequent omissions, which is now also deposited at the Paris Observatory. Pingré made extensive use of it for his *Cométographie.*[4]

While Tycho's observations were thus turned to lasting account, there is scarcely a trace left of the magnificent buildings he raised at Hveen.

> " Est in conspectu Tenedos, notissima fama,
> Insula dives opum, Priami dum regna manebant,
> Nunc tantum sinus et statio male fida carinis."

It would almost seem that Tycho did not build in a very substantial manner, for already in 1599 Eske Bille wrote to him that the farm buildings would soon tumble down, and that the forge was also in a very bad state, for which reason the clergyman wanted to know whether he might use the materials to repair the rectory, which already some years before had fallen into disrepair. Tycho answered that the clergyman had no claims on him, and had behaved very badly, and the peasants had been stealing building materials from the rectory. " As to the farm and the castle itself being in bad repair, I can only say, as I have done

---

[1] *Dänische Bibliothek,* viii. p. 684; Werlauff, *l. c.,* p. 57; *Observationes septem Cometarum* (1867), p. iii.
[2] Brewster's *Memoirs of Sir I. Newton,* ii. p. 168.
[3] M. Bossert of the Paris Observatory informs me that this copy is in six volumes (which agrees with Picard's receipt), in 4to, carefully written, the observations of comets being by themselves. The title is : *Tychonis Brahe Thesaurus observationum astronomicarum.*
[4] Pingré, i. p. 517 ; Lalande, i. p. 199.

before, that I do not intend to go to any further expense about it ; there was far too much spent on it before, and if I had the money back, it should hardly be so badly spent." [1]

Soon after Tycho's death, in May 1602, Cort Barleben received Hveen in fief, and shortly afterwards he was granted permission to pull down the forge. His successor was a mistress of the king's, Karen Andersdatter, who got the island in 1616, and was followed by her son, Hans Ulrik Gyldenlöve, who died in 1645, and seems to have been succeeded by some nobleman's widow. The destruction of Uraniborg had in the meantime gradually proceeded, as there was nobody to look after it. A new dwelling-house was erected on the site of Tycho's farm, called Kongsgaarden, which stood for about two hundred years, but has now disappeared, so that only some farm buildings remain. This Kongsgaard was built of the bricks and stones of Uraniborg, as a mason in 1623 was paid for 60,000 bricks which he had " pulled down and renovated from the old castle Oranienborg." [2] In 1645 Jörgen Brahe, a nephew of Tycho's, was granted permission to remove " any stones with inscriptions or other carved figures or characters " which might be found at Hveen.[3] Perhaps Tycho's nephew was anxious to secure some slight relic of his uncle before it was too late, for a couple of years later, when Gassendi inquired about the island, he was informed that there was only a field where Uraniborg had been.

In 1652 the island was for the first time after 1597 visited by a man of distinction. Pierre Daniel Huet, afterwards so well known as the editor of the classics "in usum Delphini," and sometime Bishop of Auranches in Normandy, was a young man of twenty-two when, in 1652, he accompanied the learned Bochart, who had been invited by Queen Christina to join the galaxy of learned foreigners at Stockholm. Passing through Copenhagen, Huet paid a visit to Hveen, and found scarcely a trace of the buildings. In his autobiography, which he did not draw up till more than sixty years later, when he says himself that both his senses and his memory were impaired after a serious illness, Huet adds the very absurd statement that neither the clergyman

---

[1] *Breve og Aktstykker*, pp. 53 and 104.

[2] Vandalism of that kind is not confined to any age or nation, but the destruction of many a fine old monastery or chapel in the heat of the Reformation had made people at that time particularly callous to the pulling down of historical relics.

[3] Friis, *Tyge Brahe*, p. 308.

nor the other inhabitants of Hveen had ever heard the name of Tycho Brahe, except one old man, who did not give a flattering account of him. Successive writers down to the present day have quoted this story without noticing the absurdity of the idea that a small community of a few hundred people should in the course of fifty years have quite forgotten the man who raised such fine and singular buildings and was visited by kings and princes.[1]

A few years after, Hveen ceased to belong to Denmark. In February 1658 the Danish king was forced to conclude the humiliating treaty of Roskilde, by which the provinces east of the Sound, which from before the dawn of history had been Danish, were handed over to Sweden. King Carl Gustav, who was not content with what he had got, but soon after made an ineffectual attempt to take the whole of Denmark, claimed Hveen as belonging to Scania, because the inhabitants were now under the jurisdiction of the court of Lund.[2] The spot where Tycho had lived and worked was thus torn from the country which had so little valued him, and, like Scania, the island soon became perfectly Swedish—a very natural consequence of the close affinity between the two nations, rivals for so many centuries, but now animated only by brotherly feelings.

In 1671 the Académie des Sciences sent Picard to Hveen to determine the geographical position of Uraniborg. The foundations were still easily recognised, and the earthen walls round Uraniborg untouched, except that a stone wall had been built across the enclosure, cutting off the north-eastern wall and a little of the two adjoining ones, and the parts thus cut off had been nearly obliterated by ploughing. Of Stjerneborg he saw nothing except a slight hollow in the ground, and he did not trouble himself about making exca-

[1] I have looked through the *Gazette de France* for 1652, in the hope that Huet might have sent a letter from Stockholm in which he might have described his trip to Hveen. But there is nothing from him or about him, so that the only account is the one given in his book *Petri Danieli Huetii, Episcopi Abrincensis, Commentarius de rebus ad eum pertinentibus,* Amsterdam, 1718, 12mo, p. 86 *et seq.* As perhaps hardly one of the writers who have copied Huet's story from Weidler's *Historia Astronomiæ* have seen it, I have quoted it in Note G. The proverbial "oldest man in the parish," from whom Huet got his information, attributed the destruction of the buildings to weather and the carelessness of Tycho's successors.

[2] Peder Winstrup, Bishop of Lund (son of the Bishop of Seeland of the same name), is said to have produced many arguments to support the claims of the Swedish king, who had the best possible argument—the sword. Hofmann, *Portraits historiques,* vi. partie, p. 8. About the change of jurisdiction, see above, p. 88.

vations.[1] During the eighteenth century the ruins were occasionally mentioned by travellers, but nobody seems to have explored them.[2] About 1740 a stone with a Latin inscription was removed from the site of Tycho's paper-mill to his old home at Knudstrup, from whence it was later brought to the museum at Lund. In 1747 a cellar was accidentally found on the site of the servants' dwelling, at the north angle of the wall enclosing Uraniborg. It is still to be seen, and if the statement on Braun's map be correct, that it was used as a gaol, it can certainly not have been a pleasant abode, though doubtless not worse than other dungeons of those days.

Within the present century the ruins at Hveen have been more thoroughly examined. They suffered a further desecration about eighty years ago, when the south-western enclosure wall round Uraniborg was broken through, in order to build a schoolhouse there. The Swedish antiquary Sjöborg visited the island in 1814, but was chiefly interested in the various slight antiquities from long before Tycho's time.[3] But in 1823 and 1824 the clergyman of Hveen, Ekdahl, examined the interesting spots carefully. At Uraniborg he found the deep well, which was easily cleaned out, and still gives excellent water; also some water-taps and pipes from the hydraulic works which had sent the water to various parts of the house. Parts of the foundation walls and some slight remains of the laboratory were also unearthed. At Stjerneborg Ekdahl was more successful, and found distinct traces of all the crypts, and one of them (F. on the plan, p. 106) in perfect preservation, with all the circular steps, and the low column in the middle, on which the large quadrant had formerly been fixed. The only ornament or inscription found was the stone with the words also put on Tycho's tomb: "Nec

---

[1] *Voyage d'Uranibourg par M. Picard*, Paris, 1680; also in *Receuil d'Observations*, 1693, and in Picard's *Ouvrages de Mathematiques*, Amsterdam, 1736. This little book is remarkable for containing the first distinct description of the phenomena of aberration and nutation.

[2] *Philos. Trans.*, xxii. p. 692, xxiii. p. 1407. Hell's *Reise nach Wardoe und seine Beobachtung des Venus-Durchganges*, Wien, 1835, p. 161. Hell and Sainovics were at Hveen in May 1770 on their return journey from Wardöhus. They give a rude diagram showing the ramparts, with a hole filled with water in the middle; also the site of Stjerneborg, the cellar found in 1747 (erroneously placed), and a hut where some Swedes had observed the transit of Venus.

[3] Sjöborg, *Samlingar för Nordens Fornälskare*, iii., Stockholm, 1830, 4to, p. 71 *et seq.*

fasces nec opes, sola artis sceptra perennant ; " so strikingly illustrated by the state of Tycho's works on earth and his labours in science.[1] Low stone walls form oval enclosures round the sites of Uraniborg and Stjerneborg, but otherwise the scanty remains of the buildings are quite unprotected, and will soon entirely disappear, being exposed to wind and weather. It is therefore well that they have been carefully described, first by the Danish poet, J. L. Heiberg, in 1845,[2] and by the distinguished astronomer D'Arrest in 1868,[3] both enthusiastic lovers of the memory of Tycho Brahe.

[1] " Fornlemningar af Tycho Brahes Stjerneborg och Uranienborg på Ön Hvén, uptäckte åren 1823 och 1824," Stockholm, 1824, 8vo. The inscription from the paper-mill given herein agrees with that of *Danske Magazin.* See above, p. 186.

[2] Urania, *Aarbog* for 1846. Af J. L. Heiberg, Copenhagen, 1846 (also in his *Prosaiske Skrifter*, vol. ix.), with fourteen plates, giving views of the island. On the 21st June 1846 a great festival in honour of the tercentenary of Tycho Brahe's birth was held at Hveen, attended by many thousand Scandinavians.

[3] *Astronomische Nachrichten*, vol. lxxii., No. 1718. The writer of the present work visited the island in 1874, at which time the ruins were still exactly in the state described by D'Arrest.

NOTES.

----◆----

A.—*SPECIMEN OF TYCHO'S EARLY OBSERVATIONS WITH THE CROSS-STAFF.*

1564 Oct. 20 mane, distantia inter Saturnum et Jovem.

1. Posito transversario in loco stato qui est partium 3500, reperi punctarum in transversario, a se invicem elongationem 1162, quibus juxta operationem proveniunt $398\frac{14}{35}$ cui numero in tabula gnomonica competunt 18 gr. 21 m. distantia syderum quæsita.

2. Collocato eodem in puncto 3700 pinnularum distantia reperiebatur 1233 quibus juxta operationem debentur $399\frac{33}{37}$ hisqve ex tabula respondent gradus ut antea, sed minuta numero 25 fere. Discrepat itaqve observatio in 3 [*sic*] tantum circiter minutis, quod admodum parum est. Erat autem sine dubio vera distantia 18^g 22′ infallibiliter. Stadius distantiam ponit 18^g 8, differentia 0^g 14′ parva (tantum erroris facit, amborum semidiameter additus 0^g 8 ita 6). Qvam Carellus constituit 15^g 20′ diff. 3^g 2′ Magna.

The above observations are calculated by the rule of Gemma Frisius (*De Radio astronomico et geometrico liber*, Antwerp, 1545, cap. 15): Multiplico igitur maximum tabulæ numerum, nempe 1200 per pinnularum interstitium, producuntur autem . . . , atqve hunc numerum divido per transversarii locum. In the first observation $\dfrac{1162 \times 1200}{3500} =$ $398\frac{14}{35}$; Gemma's *Tabula gnomonica G. Peurbachii* then gives the required angle thus :—

$$
\begin{array}{lll}
398 & 18° & 20′\ 57″ \\
399 & 18° & 23′\ 32″ \\
400 & 18° & 26′\ 7″
\end{array}
$$

B.—*LIST OF TYCHO BRAHE'S PUPILS AND ASSISTANTS.*

In the *Danske Magazin*, 4th Series, vol. ii. p. 32, the Rev. Dr. Rördam has communicated a list of Tycho Brahe's pupils, found on a loose sheet of paper (without heading or other description) in the Royal Library of Copenhagen. The handwriting seems to be that of Johannes

Aurifaber or Hans Crol, who had charge of Tycho's workshop, and whose writing frequently occurs in the volumes of original observations (*Danske Magazin*, 4th Series, vol. ii. p. 327). The list must have been written in 1588 or 1589, as Odd Einarsen (No. 13 on the list) became Bishop of Skalholt in 1588, and Longomontanus, who is not mentioned, arrived in 1589.

1. Petrus Jacobi.[1]
2. N.   Fionius.[2]
3. N.
4. Gellius Sascerides Hafniensis.
5. Andreas Wiburgensis, agit nunc pastorum siue ecclesiasten Wiburgi in Gutlandia.[3]
6. Jacobus Hegelius siue Hegelun, agit hypodidasculum in Selandia Sorae ; bene scribit et est bonus musicus, ingenij mercurialis.
7. Seuerinus N., turbator, phantasta.
8. Elias Olaj Cymber.
9. M. Nicolaus Norwegianus, præco diuini verbi Hafnie, in arce regia.[4]
10. Rudolphus   Groningensis, in primis non erat studiosus, creatus procuratione Tychonis in studiosum Hafnie, vbi deposuit Bacchanten, fuit apud Tychonem per 3 annos.[5]
11. Joannes Buck Cymber, natus non procul a Collingo, in pago Nabul, fuit optimi ingenij, occisus iactu lapidis ab altero quodam studioso Hafnie.
12. Andreas Jacobi Lemwicensis Gutlandus, siue Cymber, fortasse alicubi pastor in Gutlandia.
13. Otto Wislandus Islandus, episcopus in Islandia, est mediocris grammaticus aliasque non ignarus.[6]
14. Joannes Wardensis Cymber, sacellanus in quodam pago in Gutia.[7]
15.
16.
17.
18.
19.
20.

Hi 9 uno eodemque tempore accesserunt, menstruoque spacio vel vltra, petentes et obtinentes dimissionem, discesserunt.

[1] Peter Jacobsen Flemlöse.
[2] From the island of Fyen ; name evidently forgotten.
[3] His name occurs in the observations of February 1582 (*Hist. Cœl.*, p. 6) ; probably it was he who examined the pockets of Reymers (see above, p. 275).
[4] Niels Lauridsen Arctander died 1616 as Bishop of Viborg (Jutland).— *Rördam*.
[5] Observed, among other things, comet 1585 (*Obs. Comet.*, p. 63), occurs in the diary in 1586 and 1588.
[6] Odd Einarsen, Bishop of Skalholt in 1588.
[7] Perhaps this is Joannes Bernssön, "unus ex meis familiaribus," by whom Tycho sent a letter to H. Rantzov in 1585.

21. David Joannis Sascerides, Gellij frater, fuit hic per semissem anni.

22. Jacobus N. Malmogiensis, egit in hac insula diaconum, et fuit studiosus, is primus literas obligationis dedit ad triennium, nunc est sacellanus Malmogiæ, vel eius loci in vicinia, degit tamen Malmogiæ, non fuit autem hic diutius quam sesqui-annum.

23. Christiernus Joannis Ripensis,[1] accessit eodem tempore quo antedictus Jacobus N. Malmogiensis discessit, nempe anno 1586 circa finem Aprilis, videlicet 27 Aprilis.[2] Discedebat anno 90, 23 Aprilis.[3]

24. Petrus Richterus Haderslebiensis, fuit hic fere per semi annum.

25. Iuarus Hemmetensis Cymber, de pago Hemme in Cymbria, vbi pater eius pastorem agit, fuit hic per semestre, fuit poetici ingenij.[4]

26. Sebastianus, regiæ mensæ alumnus, fuit Borussus, vertit librum danicum in idioma germanicum, non dedit operam mathesi, fuit hic per mensem vnum vel alterum.

27. Joannes Hamon Dekent, Anglus nobilis, fuit hic studiosus, fuit hic per quadrantem anni, et ingenij mercurialis, musicus, et alias mediocriter eruditus.[5]

28. Joannes Joannis Wensaliensis Cymber, fuit hic per sesquianum, erat astrictus autem ad 3 annos.

29. Ego.[6]

30. M. Nicolaus Collingensis.

31. Martinus Ingelli Coronensis.[7]

32. Christiernus N. de Ebenthod oppido Cymbrie, huc missus a Falcone Göye,[8] fuit proximus qui post Christiernum Joannis Ripensem accessit, fuit . . .[9]

To supplement the above list I add the names of the other pupils or assistants, as far as their names are known, with references to places in this volume where further information about them is given.

Paul Wittich.

Longomontanus (1589–97).

---

[1] Afterwards Professor in the University; born 1567, died 1642 as Bishop of Aalborg. He was at Hveen 1586-90, observed the comet of 1593 at Zerbst in Anhalt, and the solar eclipse of 1598 in Jutland (*Hist. Cœl.*, p. 819).

[2] Agrees with diary.

[3] Remark about departure added afterwards (Rördam).

[4] Died 1629 as Bishop of Ribe in Jutland (Rördam).

[5] Probably *Dekent* should be *de Kent*. In the diary we read under 2nd November 1587: "Hamon abiit." Could he have been John Hammond, afterwards physician to Henry, Prince of Wales?

[6] Probably Hans Crol (see above and p. 211), died 30th November 1591.

[7] *i.e.*, from Landskrona.

[8] Falk Gjöe, a friend and kinsman of Tycho, to whom the latter addressed a Latin poem printed at Uraniborg.

[9] Paper worn at the corner; a few words lost (Rördam).

Cort Axelsen of Bergen (Conradus Aslacus), at Hveen 1590–93,
    afterwards Professor of Divinity at Copenhagen, edited Tycho's
    Oration on Astrology in 1610.

Joh. Isaacsen Pontanus, for three years at Hveen,[1] born at Elsinore
    of Dutch parents in 1571, some time Danish historiographer,
    died in Holland 1639.

Willem Janszoon Blaeu, p. 127.

Franz Gansneb Tengnagel von Camp, pp. 242, 301, &c.

Georg Ludwig Froben, p. 253.

Claus Mule, pp. 240 and 283.

At Wandsbeck and in Bohemia Tycho was assisted by Johannes
    from Hamburg (p. 280) ; Johannes Müller ; Johannes Eriksen ;
    Melchior Joestelius ; Ambrosius Rhodius ; Matthias Seyffart ;
    Paul Jensen Colding ; and Simon Marius.

C.—TYCHO'S OPINION ABOUT ASTROLOGICAL FORECASTS, FROM A LETTER TO HEINRICH BELOW, DATED THE 7TH DECEMBER 1587.

Meinen freuntlichen grues mitt wunschung alles guettes alzeitt beforr. Edler,
Ehrnvester, freundtlicher lieber Schwager vnd besonder vortraweter freundt.
Neben Dankfagung fur vielfeltige erzeigete wolthaten kan ich dir freundtlicher
wolmeinung nicht verhalten, das ich dein schreiben habe entpfangen vnd darinne
ein Copie des Durchleuchtigen Hochgebornen Fursten vnd Herren Hertzog Vlrichs
zu Mechelburg an Dir geschriebene brieffs, worauß ich erfahre, das ihr furstliche
Gnade begehrett von mir gnedichlich zu wissen, welcher meines erachtens von den
beiden Prognosticatoribus Tobia Moller vnd Andrea Rosa dem Zcill neher
zutrifft, ihndem daß der eine ihn diesem zukunfftigen 88 Jhar den Regenten des
Jhars Jovem vnd Venerem, der ander Saturnum vnd Martem.fetzet, darahn
fie nicht alleine keinstheils einig fein, fondern wie ihr Furftliche Gnade schreibet,
gahr widerwertiger meinung haben ; dan der eine macht beide beneficos Planetas,
der ander beide maleficos (wie fie die Astrologi nennen) zum Regenten im
felbigen Jhar, welchs gar contrarie bedeuttung bringett. Hierauff kan ich dir
freundlicher meinung nicht bergen, das wiewoll ich in die Aftrologische Sachen,
welche bedeuttung auß dem geftirn herholen vnd weiffagung tractiren, mich nicht
gerne einlaeße, dieweill darauff nicht vhill zu bawen ift, Sondern allein die Astronomiam, welche den wunderlichen lauff des geftirns erforschett, in einen gewitzen
vnd rechtmeffigen ordnung zu bringen mich etzliche Jhar her bemuhet, ban darahn
kan durch rechtgeschaffene Inftrumenten nach Geometrisch vnd Arithmetisch grundt-
vnd gewißheit die eigentliche warheitt durch langwirigen fleiß vnd arbeitt gefunden
werden, So habe ich doch nach ihrer Furftlicher gnaden begerung beide Prognostica, bie bu mihr zufchickeft, (bie ich boch nicht, wie bu gemeinett haft, zuvor

-----

[1] According to Pontoppidan, quoted in Bang's *Samlinger*, ii. p. 279 (Weistritz,
i. p. 73).

gehatt habe, dan ich niemals pflege solche practicen wider zu kauffen, noch zu lesen, ne bonas horas male collocem), durchgesehen, den mangel, worahn es hafftett, das sie so widerwertige Judicia stellen, daraus zu suchen, vnd befinde, das sie in ihre Rechnung gar vnterscheidliche fundament gebraucht haben ; dan der eine, nemlich Mollerus, bawett sein Calculation auf des hocherfarnen Copernici rechnung, der ander, Rosa, auff die alte durch des Königs Alphonsi in Hispanien liberalitet gemachte Tabeln, die man darumb die Alphonstinische nennett. Hirauß kömpt es, das der eine den anfang des Jhars in aequinoctio verno setzet ahm 10 tag Martii bey Neun vhr nach mittage, der ander ahm selbigen tag, aber vmb 2 Stunden nach der vorgehende Mitternacht, daß also zwischen beide ihre rechnung schier 19 gantze Stunden verlauffen, in welchen der himmel sich gar vil vorendern thutt, vnd kan gar ein ander Astrologisch Judicium darauß fallen, ebensowol als wen dar ein gantzes Jhar oder noch mehr zwischen wehre : Das darumb nicht zu vorwundern ist, das diese beide Astrologi in domino Anni nicht vberein stimmen, weil sie den auß der Figura Caeli introitus Solis in Arietem, wan daß vorbemelte aeqvinoctium vernum geschicht, pflegen herholen. Wiewol es auch lichtlich geschehn kan, das wan sie schon gleichmessige Tabeln vnd rechnungen volgeten, das sie gleichwoll in Dominis Anni vnd ihren gantzen weissagungen gar widerwertige meinungen können für- geben, das darauß leichtlich zu probiren ist, daß wan man hundertt der Progno- sticken lisset, so befindett sich doch gahr selten, das einer mitt dem andern concordirett, dan sie bawen nicht alle auff gleichformige grundt ihn ihren Judiciis vnd haben vntherscheidliche process vnd ahnleitungen. Es sein auch diese Astrologische weissagungen wie ein cothurnus, den man kan auff ein jeder Bein ziehen, gros vnd klein, wie man will, darumb ich auch niemals darvon ettwas Sonderlichs gehalten habe. Das ich aber Kong. Maj. meinen Gnedigsten Herren jhärlich ein solches Astrologisch Prognosticon vntherteniglich zustelle, mus ich in dem nach ihre May. befell vnd willen thun, wiewoll ich selbst nicht vill darauf halte vnd nicht gerne mitt solchen zweiffelhafftigen praedictionibus vmbgehe, darin man die eigenttliche warheitt nicht durchauß erforschen mag, wie sonst in Geometria vnd Arithmetica, darauff die Astronomia durch hulff der vleissigen observation ihm lauff des Himels gebawet wirtt. Dennoch dieweill ihr Furstliche Gnade gnediglich begert von mir zu wissen, welchen von den beiden ich beifellig sey, was den dominis Anni also widerwertiger weiß von ihnen gestellet thutt ahnlangen, So kan ich hierauff nicht anders sagen, dan das ihrer beiden rechnung vnd iudicium gehtt auß ein vormeinten vnrichtigen grundt ; dan weder die Alphon- sinischen noch die Copernianischen Tabeln, welchen sie folgen, geben den justen lauff der Sonnen, wie er ahn sich selbst am Himmel geschicht, vnd ist hierinne kein geringe vnterscheidt, wie auß meiner eigenen Restitution vnd vornewrung in Rechnung des lauffs der Sonnen zu sehen ist, welche ich auf etzliche vorgehende Jharen durch große vnd rechtmessige Instrumenten augenscheinlich vom Himel selbst her ab durch fleißige observation vnd warnemungen genomen habe, auß welchen sich befindet, das des Jhars Anfang in aequinoctio verno, darauß die Astrologi ihre vrtheil nehmen, geschicht ahm 10 tag Martii 8$\frac{2}{3}$ stunde nach der vorigen Mitternacht, welchs bey 7 stunden speder ist, als Andreas Rosa gesetzt hat, vnd 12 stunden fruer, als Tobias Mollerus meinett, darauß den vhil ein andere constitution des himels zur Zeitt des aequinoctii einfeltt, als ein jetzlicher von ihnen gefunden hat, worauff auch ein anber vrtheill folgett vnd auch wol andere domini Anni, wie sie es nennen, (darauß boch nicht vhil zu holen ist), mögen gesetzt werden. Was aber meine meinung sey ahnlangende Astrologische gitzung vber bis kunfftige 88 Jhar, habe ich Königl. May. meinen gnebigsten Herren ihn einen geschriebenen Prognostick vnterteniglich auffgezeignett, welchs ich auch ihre Furstliche Gnabe gerne wolte vnterteniglich mitteilen, aber ich hab keine vberige Exemplar darvon, wan ihr Furstliche Gnabe lasset bey ihr May. barumb ahnlangen, wirb ihr Furst. Gna. wol ein abschrifft bar von bekomen. Ich bin barinne gentzlich nicht ber meinung, baß solche gahr große vorenberunge in biesen negst= kunftigen Jhar vorhanben sein, als die Astrologi auß etzlichen alten reimen, bie sie den Regiomontano zumessen, furgeben, ban ich befinbe im Constitution bes Himels keine Sonberliche vrsachen barzu, sonbern achte, bas bis kunfftige Jhar wirb ben vorgehenben gleichmeßig sein vnb in zimlichen guten wesen in allen Sachen sich erzeigen, aleine wo zuuor krieg vnb vneinigkeit aufferweckt ist, bar möchte es wol ettwas weitter einreißen. Vnb kan ihr Furstliche gnabe meine meinung vom Astrologischen judicio vber bas gantze Jhar auß vorbemelten Prognostico, welchs ich Köng. May. meinen gnebigsten herren vnterteniglich habe zugestellt, weitter erfahren. Dis habe ich auff bein guttwilliges schreiben vnb beger nicht wöllen vntherlassen zu antworten. Bitte gar beinstlich, bu wollest vnbeschwert sein vnb mitt erster gelegenheitt ihr Furstliche Gnabe hierauff meine antwort vnb meinunge vnterteniglichen von meinettwegen zu vorstehn geben. Worin ich sonst ihre Furstliche gnaben zu willen vnb gefallen sein kan, bin ich alzeitt mitt aller beinstlichkeitt vnterteniglich erböttig.

[The remainder of the letter is about paper for Tycho's books; see above, p. 163, footnote.]

D.—KEPLER'S ACCOUNT OF TYCHO BRAHE'S LAST ILLNESS.

[Written in the volume of Observations for 1600-1601. Compare *Observationes Hassiacæ, publicante W. Snellio*, Lugd. Bat., 1618, p. 83.]

Die 13 Octobris Tycho Brahe Dominum a Mincowitz ad coenam Illustriss. Domini a Rosenberch comitatus, retenta præter morem urina consedit. Cum paulo largius biberetur sentiretque vesicæ tensionem valetudinem civilitati posthabuit. Domum reversus urinam reddere amplius haud potuit.

Erat hujus morbi initio ☽ in opposito ♄, & □to ♂ in ♉, & ♂ eodem loco quem sibi Tycho orientem gradum constituerat.

Transactis quinque diebus insomnibus, cum gravissimo cruciatu vix tandem urina processit, & nihilominus impedita. Insomnia sequebantur perpetua, febris interna & paulatim delirium, ratione victus, à qua prohiberi non poterat, malum exasperante. Ita die 24 Octobris cum delirium aliquot horis remisisset, victa natura inter consolationes, preces & suorum lachrymas placidissime expiravit.

Ab hoc itaque tempore series observationum Coelestium interrupta est, finisque impositus 38 annorum observationibus.

Nocte, quam ultimam habuit, per delirium, quo omnia suavissima fuere, creberrime hæc verba iteravit, quasi qui carmen texit.

Ne frustra vixisse videar.

Quem procul dubio ceu colophonem operibus suis addere, eaque his verbis Posteritatis memoriæ & usibus dedicare voluit.

E.—COMPARISON OF TYCHO BRAHE'S POSITIONS OF STANDARD STARS WITH MODERN RESULTS.

The following table gives the places for 1586 (anno 1585 completo) of Tycho's nine standard stars (*Progymn.*, p. 204), and their positions for the same epoch, computed from those of Bradley for 1755, with the proper motions of Auwers (*Neue Reduction der Bradley'schen Beobachtungen*, vol. ii.).

	Observed AR.	Observed Dec.	Computed AR.	Computed Dec.
α Arietis . .	26° 0′ 30″	+21° 28′ 30″	26° 0′ 44.″7	+21° 27′ 33″.0
α Tauri . .	63 3 45	+15 36 15	63 4 11.2	+15 35 35.6
μ Geminorum	89 29 10	+22 38 30	90 28 41.0	+22 37 48.5
β Geminorum	109 58 0	+28 57 45	109 57 47.9	+28 56 19.1
α Leonis . .	146 32 45	+13 57 45	146 33 19.7	+13 56 56.9
α Virginis .	195 52 10	− 8 56 20	195 52 36.4	− 8 58 22.0
δ Ophiuchi .	238 11 20	− 2 33 15	238 11 9.8	− 2 33 21.5
α Aquilæ . .	292 37 20	+ 7 51 20	292 38 36.2	+ 7 50 35.8
α Pegasi . .	341 2 30	+13 0 40	341 3 1.3	+12 59 47.1

The differences between Bradley and Tycho are shown in the next table. The effect of refraction (which Tycho thought insensible above Zen. Dist. 70° or Declination − 14° for Uraniborg), is seen at a glance, and I have therefore in the last column given the difference of declination corrected for mean refraction.

	BRADLEY MINUS TYCHO.			
	Δ α	Δ α cos δ	Δ δ	Δ δ′
α Arietis . .	+ 14″.7	+ 13″.7	− 57″.0	− 17″.4
α Tauri . .	+ 26.2	+ 25.2	− 39.4	+ 9.6
μ Geminorum .	− 29.0	− 26.7	− 41.5	− 3.6
β Geminorum .	− 12.1	− 10.6	− 85.9	− 56.5
α Leonis . .	+ 34.7	+ 33.7	− 48.1	+ 4.4
α Virginis . .	+ 26.4	+ 26.1	− 122.0	+ 0.6
δ Ophiuchi . .	− 10.2	− 10.2	− 6.5	+ 88.1
α Aquilæ . .	+ 76.2	+ 75.4	− 44.2	+ 19.5
α Pegasi . .	+ 31.3	+ 30.4	− 52.9	− 0.1

From the third and fifth columns we find the probable errors of Tycho's standard right ascensions = ± 24″.1, and of his standard declinations = ± 25″.9. Of course the accuracy of most of his star-places must be much less, as they were neither as often nor as carefully determined as those of the standard stars, and were vitiated not only by refraction, but also by aberration and nutation. See above, pp. 351 and 353.

F.—ON THE ALLEGED ERROR OF TYCHO'S MERIDIAN LINE.

In order to remove any doubt which the reader might have as to the correctness of the opinion set forth above on p. 360, that the azimuth error of 15′ detected in some of the instruments in November 1586 does not prove all Tycho's instruments to have been erroneously placed during all the years previous to 1586, I have tested the matter by calculation. Azimuths were never very extensively observed at Hveen, and the only series of observations sufficient for my purpose was that made in the beginning of 1582, when a number of altitudes and azimuths of bright stars were measured. Many of these were taken too near the meridian to be of any use in this case, but there are many observations made in the prime vertical, by which a great error in the assumed zero of the azimuth circle would easily be detected. I have made use of all the observations taken in or near the prime vertical during March 1582 (except of a few which were vitiated by some great error of copying, observing, or in the identification of the star), and of one observation of α Tauri in azimuth 69° on February 25.[1] The following mean declinations for 1582 were computed from Auwers' *Bradley* the reductions to apparent declinations for 1st March being appended :—

α Tauri	.	.	.	.	+ 15° 35′ 0″	+ 4″
β Herculis	.	.	.	.	+ 22 27 50	− 20
κ Ophiuchi	.	.	.	.	+ 10 5 45	− 17
α Herculis	.	.	.	.	+ 14 56 20	− 12
α Ophiuchi	.	.	.	.	+ 12 56 14	− 18
α Lyræ	.	.	.	.	+ 38 26 50	− 19
β Cygni	.	.	.	.	+ 27 8 32	− 10
ι Cygni	.	.	.	.	+ 50 52 36	− 18
γ Cygni	.	.	.	.	+ 38 57 48	− 12

Assuming the observed altitudes to be correct, the azimuth corresponding to each altitude was computed by the formula

$$\cos A = \frac{\sin \phi \sin h - \sin \delta}{\cos \phi \cos h}$$

---

[1] *Historia Cœlestis*, p. 34 *et seq.* My copy of this work formerly belonged to Professor Schjellerup of Copenhagen, who corrected a great many of the errors in the observations of fixed stars (by comparing with Bartholin's copy) including fortunately those of 1582.

in which δ is the apparent declination. h the observed altitude corrected for refraction, and ϕ the latitude of Uraniborg, $55° 54' 26''$. The computed and observed azimuths were found to be in excellent agreement, the differences only in nine cases exceeding $3'$. Thirty-six observations of nine stars gave the correction to the observed azimuths (counted from $0°$ to $360°$ from south through west) $= -0'.4 + 0'.30$. Of course this result is affected by the errors of observation in the altitudes, but at any rate it shows most conclusively that Tycho Brahe's azimuth circles were adjusted with the same care which we have seen he bestowed on his other instruments ; and the assertion that Tycho from 1579 to 1586 had all his instruments set $15'$ wrong in azimuth is hereby proved to have been utterly without foundation.

G.—HUET'S ACCOUNT OF THE STATE OF HVEEN IN 1652.

[From P. D. Huetii *Commentarius de rebus ad eum pertinentibus*, Amsterdam, 1718, 12mo. After describing the great celestial globe, the author relates how he started for Hveen, with only one companion, on the 26th May 1652, and after telling the story about the origin of the name of Insula Scarlatina and praising the hospitable reception by the Lutheran clergyman, he continues on p. 86.]

Humaniter ergo excepti, postquam paululum interquievimus, multa ab hospite nostro, aliisque adstantibus Huenæ incolis percontatus sum de Tychone, deque Vraniburgica arce, cujus visendæ gratia istuc venissem : quodque mirere, incognita sibi esse hæc nomina uno ore professi sunt omnes, nec quicquam se de iis accepisse. At cum virum quemdam valde senem superesse rescivissem in insula, misi qui eum ad me adducerent. Rogatus ille a me numquid de Tychone Braheo inaudisset aliquando, deque condita olim ab eo istic arce quam Vraniburgi nomine decorasset, & per annos unum & viginti incoluisset ; se vero respondit non Tychonem modo & Vraniburgum nosse, sed & in Tychonis famulatu fuisse per aliquod tempus, & operam quoque suam ad hanc exstruendam arcem contulisse. Tychonem autem referebat hominem fuisse violentum, impotentem, iracundum, famulos & villicos male multantem, ebriosum quoque et mulierosum ; cum uxorem duxisset ex infima natalis villæ suæ Knudstrupii plebe rustica, ex qua liberos multos suscepisset, atque hac affinitate, velut sibi probrosa, magnopere offensam fuisse illustrem Braheorum gentem. Tum subjecit bonus ille vir, si istuc venissem spectandi Vraniburgi causa, irritum me laborem suscepisse, quippe solo æquata omnia, vixque superesse parietinarum vestigia. Cujus rei causas cum ab eo, & jam ante Hafniæ a viris doctis exquisivissem, varias ac plane discrepantes commemorabant. Hi enim Tychonem ipsum, e Dania excedentem opus suum diruisse dicebant, cum satis tamen constet res suas Huenicas & Vraniburgicas colono & famulis aliquot procurandas reliquisse ; quippe fundi hujus fructus in totum vitæ tempus fuisse ei a Frederico Rege concessos. Erant qui dicerent hostiles Suecorum copias pervasisse in insulam, stragemque hanc edidisse ; quod non ignorasset profecto vetus ille Huenæ incola qui ad turbidas Freti Sundici tempestates & rapidissimos ventos rei causam referebat ; quibus facile concussæ fuerant ædes leviter materiatæ : cum præsertim de tuendis sartis tectis Astronomicæ domus parum omnino laborarent aulici viri, quibus clientelæ jure, post Tychonem, Regis beneficio obtigerat.

H.—CATALOGUE OF THE VOLUMES OF MANUSCRIPT OBSERVATIONS OF TYCHO BRAHE IN THE ROYAL LIBRARY, COPENHAGEN.

[Compiled by means of the list in Bugge's *Observationes astronomicæ* (1784), compared with the library catalogues and the volumes themselves, with the kind assistance of Dr. C. Bruun, chief librarian. The MSS. form part of the so-called Old Royal collections, *Gamle Kongelige Samlinger*, the numbers of which are added in brackets.]

1. *Original Observations.*

Observationes astronomicæ a solstitio Hiberno anni 1577 ad annum 1582 ; maxima ex parte autographum, 4to, 197 ff. [No. 1825, 4to.]

Observationes ab anno 1582 usque ad annum 1587 ; ex parte autogr. In folio, 500 ff. [No. 311, fol.]

Observationes astronomicæ ab anno 1587 ad 1590 usque ad Jan. 26. In folio, 344 ff. [No. 312, fol.]

Observationes astronomicæ ab anno 1590 usque ad a. 1595. In folio, 340 ff. [No. 313 fol.]

Observationes planetarum super annum 1595 institutæ. In folio, 173 ff. [No. 314 fol.]

Observationes cometæ anni 1590, nec non cometæ 1596 apparentis. [No. 315 fol.]

Observationes planetarum et fixarum anno 1596 et 1597 ad d. 15 Mart. ; narratio de occasione interruptarum observationum et discessus mei ; obs. continuatæ Wandesburgi a 20 Oct. 1597. Adjectus est in fine catalogus fixarum ad annum 1588 completum.[1] In folio, 151 ff [No. 316 fol.]

Liber observationum Tychonis Brahe pro 1598 Wandesburgi, nec non observationes astronomicæ factæ Wittebergæ a solstitio hyberno præcedente annum 1599, et variæ aliæ in Arce Benatica factæ eodem anno. In folio, 167 ff. [No. 317 fol.]

Observationes astronomicæ annorum 1600 et 1601 in Arce Caesarea regni Bohemici Benatica & Pragæ habitæ per instrumenta Tychonis Brahe. Accedunt observationes, quas anno 1600 factas Keplerus e Stiria Tychoni misit. In folio, 171 ff. [No. 318 fol.]

Observationes [autographæ] cometarum a. 1577 et a. 1580 et observa-

---

[1] The catalogue has columns for RA, Decl., Long., Lat., and Magn., but very little has been filled in.

tiones [apographæ] cometarum qui annis 1582, 1585, 1590, 1593 et 1596 apparuerunt.[1] 4to. [No. 1826, 4to.]

2. *Copies of Observations, Miscellaneous Computations, &c.*

Observationes Lipsiæ annis 1563–65, Rostochii 1568, Augustæ Vindelicorum 1569–70, Heridsvadi, Hafniæ et in insula Hvena 1573–81 factæ, præmissis excerptis ex observationibus G. Frisii et Regiomontani.[2] 4to, 280 ff. modern binding. [No. 1824. 4to.]

Observationes septem Cometarum 1577–96. 4to. [No. 1827, 4to.]

Fasciculus continens theoriam Solis per observationes Lunæ ad illum et fixas, Lunaria 1582, 1586, 1590, 1593. Observationes Veneris 1586, Jovis 1593, Saturni 1590. 4to, 145 fol., modern binding. [No. 1830, 4to.]

Computatio observationum Solis, Lunæ et planetarum 1599 et 1600 & comparatio cum tabulis. 4to. [No. 1829, 4to.]

Fasciculus observationum Brahei, variorum annorum ex quibus 1590 1598, 1599 et 1600 expressi sunt. Addita sunt excerpta ex quodam libro MSo ex collegio Pragensi a M. Bacchatio Tychon, communicata. 4to, 478 ff., modern binding. [No. 1828, 4to.] [3]

Tychonis Brahei stellarum octavi orbis inerrantium accurata restitutio. [No. 306, fol., presentation copy to King Christian IV.]

Idem liber [No. 307, fol., presented to Johannes Adolph, Bishop of Lübeck.]

Observationes quædam astronomicæ habitæ 1589 per Quadrantem orichalcicum Tychonis Brahe, divisionum satis capacem, in diversis locis Daniæ ad eruendas eorum Poli elevationes. 4to. [No. 1831, 4to, by Elias Olai, see above p. 123, footnote.]

Collectio observationum Tychonis Brahei in L. Baretti Historia Coelesti omissarum per Erasmum Bartholinum. In Folio.[4] [No. 310, fol., includes the observations of comets.]

---

[1] The original observations of the comets of 1582 and 1585 are in the folio volume for 1582–86.

[2] These excerpts fill two leaves. There are also some notes on the comets after the year 1500, and among the observations are astrological notes and comparisons with ephemerides. In several places the observations have afterwards been verified or reduced by means of the great globe, *e.g.*, under 1570, 5th March, where there is written in the margin : "Manu Christiani Severini Longomontani recentius scriptum;" and on the next page : "Manu Tychonis, sed recentius, scriptum." This volume has often been quoted above in Chapters ii. to v. It is the volume referred to in *Connaissance des Temps* for 1820, p. 386.

[3] I have not seen this volume myself. Dr. Bruun informs me that it has originally consisted of two bundles, the first begins "Observationes ☉ Die 27 Februarii," and consists of 260 leaves ; the second is headed "Astronomicæ Observationes anni 1598," and consists of 218 leaves. It is written in the same hand as No. 1829, 4to. Bugge's No. 6 (p. xvii. last two lines) could not be identified, and the last and third last items on Bartholin's list (Werlauff's *Historiske Efterretninger om det store kgl. Bibliothek*, 1844, p. 54) are not now at the Royal Library. One of these is an 8vo volume of observations, 1563–74 (originals?), which probably was lost in Paris, or on the way back in 1707.

[4] Bartholin's corrected copy of Barrettus is mentioned above p. 374.

About the copies of observations in the Hofbibliothek at Vienna and at the Paris Observatory, see above, p. 372 *et seq.* The three astrological MSS. [No. 1820-21, 1822, 1823, 4to] are described above, p. 145 *et seq.*

I.—*BIBLIOGRAPHICAL SUMMARY.*

The titles of Tycho Brahe's own publications (including the posthumous *Oratio de disciplinis mathematicis* and the observations published by Snellius and in the *Historia Cœlestis*) have already been given in full in the text. The following is a complete chronological catalogue of books and memoirs containing biographical details or investigations of Tycho's scientific work, omitting popular accounts in which nothing new occurs.

Tychonis Brahei, Equitis Dani, Astronomorum Coryphæi, Vita. Authore Petro Gassendo, Regio Matheseos Professore. Accessit Nicolai Copernici, G. Peurbachii & J. Regiomontani, Astronomorum celebrium Vita. Editio secunda, auctior & correctior. Hagæ Comitum MDCLV. 4to, lx. + 384 pp. [of which 287 pp. about Tycho Brahe; 1st edition published in 1654; pagination of the two editions agree. About Gassendi's sources, see his letters in his *Opera*, vol. vi. pp. 518-527].

Encomion Regni Daniae. Det er Danmarckes Riges Lof. Af J. L. Wolf. Kjöbenhavn, 1654. 4to [pp. 525-529 about the island of Hveen].

Le Grand Atlas ou cosmographie Blaviane, en laquelle est exactment descritte la terre, la mer et le ciel. A Amsterdam, 1663, fol. [Large map of Hveen, with letterpress in vol. i. p. 61 *et seq.*, and figures of Tycho's instruments.]

P. J. Resenii Inscriptiones Hafnienses, Stellæburgenses, Uraniburgenses, &c. Hafniæ, 1668. 4to [p. 310 *et seq.* are given a number of inscriptions at Uraniborg and Stjerneborg, and a letter from Tycho Brahe to Peucer, all of which are reprinted by Weistritz in his first volume].

Voyage d'Uranibourg ou Observations astronomiques faites en Danemarck, par M. Picard. Paris, 1681, fol. [Reprinted in the Recueil des observations faites en plusieurs voyages, Paris, 1693, fol.; and in Picard's Ouvrages de mathematiques, Amsterdam, 1736, 4to, pp. 61-99.] .

Epistolæ ad Joannem Kepplerum scriptæ. Ed. M. G. Hanschius. Lipsiae, 1718, fol. [On pp. 102 *et seq.* letters from Tycho, see above, p. 290 *et seq.*]

Pet. Dan. Huetii, Episcopi Abrincensis, Commentarius de rebus ad eum pertinentibus. Amstelodami, 1719. 12mo [pp. 81-89 about Tycho Brahe and Hveen].

Tychonis Brahe relatio de statu suo post discessum ex patria in Ger-

maniam et Bohemiam ad M. Andr. Velleium ex manuscripto
edita a G. B. Casseburg. Jenae, 1730. 4to, 23 pp. [The same
is more correctly printed in Dänische Bibliothec oder Sammlung
von alten und neuen gelehrten Sachen aus Dänemarck, III.
Stück. Copenhagen, 1740. 8vo, p. 177 *et seq.*]

Singularia historico-literaria Lusatica, oder historische und gelehrte
Merckwürdigkeiten derer beyden Marggrafthümer Ober- und
Nieder-Lausitz, 27ste Sammlung, 1743. 8vo [contains, p. 177 *et
seq.*, four letters from Tycho to Scultetus, see above, p. 133].

Cimbria Literata. A Joh. Mollero. Hafniæ, 1744. Tom. ii., fol.
[contains on pp. 103–118 a biography of Tycho Brahe, with some
interesting details].

Samling af adskillige nyttige og opbyggelige Materier, saa vel gamle
som nye. Kjöbenhavn, 1745. 8vo, v.–vii. Stykke [vol. ii.] ; Den
store vidtberömte Danske Astronomus . . . Tyge Brahe . . .
Hans Liv og Endeligt. [By Mag. Malthe and the editor, O.
Bang, fills about 183 pp., chiefly founded on Gassendi, but also
containing a few new details].

Portraits historiques des hommes illustres de Danemark. Par T.
de Hofman. Sixième Partie. Copenhague, 1746. 4to [pp. 2–30
life of Tycho Brahe, some new documents].

Danske Magazin, indeholdende allehaande Smaa-Stykker og Anmærk-
ninger til Historiens og Sprogets Oplysning. Andet Bind.
Kjöbenhavn, 1746. 4to. [Nos. 18 to 24, pp. 161–372, "Rare og
utrykte Efterretninger om Tyge Brahe," by Jacob Langebek.
In vol. iii., 1747, there is an account of Tycho's sister Sophia, pp.
12-32 and 43-53, in which occurs the only known Danish poem
by Tycho Brahe. The engravings in vol. ii. are borrowed from
Hofman's *Portraits*.]

Lebensbeschreibung des berühmten und gelehrten Dänischen Stern-
sehers, Tycho von Brahe, aus der Dänischen Sprache in die
Deutsche übersetzt von Philander von der Weistritz. Kopen-
hagen und Leipzig, 1756. 2 vols. 8vo. [Vol. i. is a translation of
the biography in Bang's *Samling*, vol. ii. of the papers in *Danske
Magazin*, vol. ii. ; the Danish original mentioned in Houzeau
and Lancaster's *Bibliographie*, vol. ii., No. 6193, never existed.
The portrait is borrowed from Hofman. The translator's name
was C. G. Mengel.]

Observations de Mars faites en 1593 par T. B., tirées des manuscrits par J. de Lalande. Mémoires de l'Acad. des sc. de Paris,
1757.

Observations de Saturne et de Jupiter en 1593 par Tycho. *Ibid*,
1763. [See above, p. 375.]

Om Forskiellen imellem Tycho Brahes og Picards Meridian af
Uraniborg. Ved [J. S.] Augustin. Skrifter, som udi det Kgl-
Videnskabernes Selskab ere fremlagde og nu til Trykken befor-
drede. XII. Bind. Kjöbenhavn, 1779. 4to, pp. 191–216.

Histoire de l'Astronomie moderne. Par M. Bailly. Tome i. Paris,

1779. 4to [pp. 377-389, 398-424 contain an account of Tycho's scientific work].

Sammenligning mellem de 1672 af Hr. Picard og de nyere i Skaane gjorte Observationer og Opmaalinger for at bestemme de tre Punkters, Uraniborgs, Rundetaarns og Lund's Observatoriers Situation. Af C. C. Lous. Nye Samling af det kgl. Danske Videnskabernes Selskabs Skrifter. I., 1781. 4to, pp. 142-155.

Observationes astronomicae annis 1781, 1782, & 1783 institutæ in observatorio Regio Hafniensi, auctore Thoma Bugge. Hauniæ, 1784. 4to [pp. xiii.-xix. about Tycho's MS. observations].

De meritis Tychonis Brahe in Astronomiam Mechanicam. O. Schilling et A. P. Weller. Upsaliae, 1792. 4to, 13 pp. [Merely short descriptions of Tycho's instruments, and a plate with twelve small figures of them.]

Geschichte der Mathematik. Von A. G. Kästner. Zweiter Band. Göttingen, 1797. 8vo [pp. 377-416 life of Tycho, pp. 613-660 reviews of his books ; compare vol. iii. p. 469].

Histoire des Mathematiques par J. F. Montucla. 2ᵐᵉ edition. 1798. Tome i. 4to [pp. 653-674 about Tycho and his scientific work].

Geographische Länge und Breite von Benatek wo Tycho Brahe vor 203 Jahren beobachtet hat . . . Von Aloys David. Abhandlungen der Königl. Böhm. Gesellschaft der Wissenschaften. Prag, 1802. [Résumé in Zach's Monatliche Correspondenz, vi., 1802, p. 468 et seq.]

Trigonometrische Vermessungen zur Verbindung der K. Prager Sternwarte mit dem Lorenzberg, und zur Bestimmung der geogr. Breite und Länge des Ortes auf dem Hradschin, wo Tycho Brahe beobachtet. Abhandlungen, &c. Prag, 1805. [Résumé in Monatl. Corr., xii. p. 248].

Méridienne d'Uranibourg. Par M. Delambre. Connaissance des Tems pour l'an 1816. Paris, 1813, pp. 229-239.

Extrait d'une lettre de M. Schumacher. Connaissance des Tems pour l'an 1820, pp. 385-387 [also about the meridian, compare Zach's Correspondence astronomique, Tome i., pp. 338 and 402].

Histoire de l'Astronomie moderne par M. Delambre. Tome i. Paris, 1821. 4to [pp. 148-260 reviews of Tycho's principal books].

Astronomische Nachrichten, No. 63 (Band iii.). [On an album belonging to Tycho's son, and on a portrait of Tycho in watercolours (above, p. 263), by Biela.]

Fornlemningar af Tycho Brahes Stjerneborg och Uraniborg på Öen Hvén, aftäckte åren 1823 och 1824. Stockholm, 1824. 8vo, 27 pp., 1 plate.

Samlingar för Nordens Fornälskare. Tredje Tomen. Af N. H. Sjöborg. Stockholm, 1830. 4to [pp. 71-83 about the state of Hveen in 1814, with 2 plates].

Tycho de Brahe als Homöopath. Von W. Olbers. Schumacher's Jahrbuch für 1836, pp. 98-100.

Letter from Tycho Brahe to Thomas Savelle. A collection of letters

illustrative of the progress of science in England . . . edited by
J. O. Halliwell. London, 1841. 8vo, pp. 32–33.

Historiske Efterretninger om det store kongelige Bibliothek i Kjöbenhavn. Ved E. C. Werlauff. 2^{den} Udgave. Kjöbenhavn, 1844.
8vo [52–60 and p. 411, about Tycho's MSS. at Copenhagen].

Observationes cometae anni 1585 Uraniburgi habitæ a Tychone Brahe.
Edidit H. C. Schumacher. Altonæ, 1845. 4to, 32 pp., 2 plates.

Urania, Aarbog for 1846, udgiven af J. L. Heiberg. Kjöbenhavn.
8vo [pp. 55–170, Hveen tilforn Danmarks Observatorium, on the
state of the ruins at Hveen in 1845, with 14 plates].

T. B. Mikowec : T. Brahe ; ziwotopisni nastin, ku 300 leté památce
jeho narozeni. Prag, 1847. 8vo [contains some information about
the fate of Tycho's relations after his death].

Bestimmung der Bahn des Cometen von 1585 nach den . . . Origi-
nalbeobachtungen Tycho's. Astronomische Nachrichten, Band
xxix. pp. 2c9–276. [By C. A. F. Peters, the most important
investigation of the accuracy of Tycho's observations. About
the orbits of the other comets, see above, p. 159 et seq.]

De hellige tre Konger Kapel i Roskilde Domkirke. Af E. C.
Werlauff. Kjöbenhavn, 1849 [p. 17 et seq. about Tycho's pre-
bend].

Historiske Efterretninger om Anders Sörensen Vedel. Af C. F.
Wegener, 291 pp. fol. [Appendix to Den Danske Krönike af
A. S. Vedel, trykt paany. Kjöbenhavn, 1851.]

Danske Magazin. Tredie Række. 3^{die} Bind [pp. 79–80, Letter
from Tycho Brahe to Valkendorf of 1598]. 4de Bind [pp. 263–
264, Royal Letter to Tycho Brahe about the tenant Rasmus
Pedersen, of 17 November 1592].

Ueber den neuen Stern vom Jahre 1572. Von Prof. Argelander.
Astronomische Nachrichten, vol. lxii. pp. 273–278.

Tychonis Brahe Dani Observationes septem Cometarum. Ex libris
manuscriptis qui Havniæ in magna bibliotheca Regia adservan-
tur nunc primum edidit F. R. Friis. Havniæ, 1867. 4to, viii. +
120 pp., 5 plates.

Die Ruinen von Uranienborg und Stjerneborg im Jahre 1868. Von
H. L. d'Arrest. Astronomische Nachrichten, lxxii. pp. 209–224.

Tycho Brahe und seine Verhältnisse zu Meklenburg. Von G. C. F.
Lisch. Jahrbücher des Vereins für meklenburgische Geschichte,
xxxiv. 20 pp., 8vo.

Minder om Tyge Brahes Ophold i Böhmen. Af F. R. Friis. Dansk
Tidsskrift, i., 1869, 8vo, pp. 225–236 and 257–269.

Tyge Brahe's Haandskrifter i Wien og Prag. Af F. R. Friis.
Danske Samlinger, iv., 1869, 8vo, pp. 250–268.

Danske Magazin. Fjerde Række. 2^{det} Bind. Bidrag til Tyge
Brahes Historie, af H. J. Rördam, pp. 16–34 ; Bidrag . . . af
F. R. Friis, pp. 324–328. [About Gellius ; list of some of
Tycho's pupils ; the sale of Knudstrup ; Parsbjerg's complaint
of Jessenius, above, p. 311, footnote.]

Tyge Brahe. 'En historisk Fremstilling efter trykte og utrykte Kilder af F. R. Friis. Kjöbenhavn, 1871. 8vo, 386 pp.

"Tyge Brahe, en historisk . . . af F. R. Friis," kritisk betragtet af J. Dreyer. Kjöbenhavn, 1871. 8vo, 19 pp.

Joannis Kepleri Astronomi Opera Omnia. Ed. C. Frisch, 8 vols. Frankofurti et Erlangæ, 1858–71. 8vo. [Contains many abstracts from letters between Tycho and Kepler, and the Vita Kepleri in vol. viii. is a valuable source for the last years of Tycho's life.]

Tycho Brahe und J. Kepler in Prag. Eine Studie von Dr. J. von Hasner. Prag, 1872. 8vo, 47 pp.

Breve og Aktstykker angaaende Tyge Brahe og hans Slægtninge. Samlede og udgivne af F. R. Friis. Kjöbenhavn, 1875. 8vo 169 pp.

Tyge Brahes meteorologiscke Dagbog holdt paa Uraniborg for Aarene 1582–1597. Udgiven af det kgl. danske Videnskabernes Selskab. Kjöbenhavn, 1876. 8vo, iv. + 263 + lxxv. pp.

On a portrait of Tycho Brahe. By Samuel Crompton. London, 1878 (Mem. Lit. Phil. Soc. of Manchester, vol. vi. pp. 77–81 [compare *Nature*, xv. p. 406 and xvi. p. 501].

Tychonis Brahei et ad eum doctorum virorum Epistolae ab anno 1568 ad annum 1587 nunc primum collectae et editae a F. R. Friis. Havniæ, 1876–86. 4to, 112 pp., 2 plates.

Astronomische, meteor. und magn. Beobachtungen an der K.K. Sternwarte zu Prag im Jahre 1880. 4to [on pp. iii.–iv. about some instruments at Prague alleged to have been Tycho's].

F. Dvorsky : Nové zpravy o Tychonu Brahovi a jeho rodine. Published in the Časopis Musea Královstvi Českeho, vol. lvii., 1883, pp. 60–77.

Carteggio inedito di Ticone Brahe, Giovanni Keplero e di altri celebri astronomi e matematici dei secoli xvi. e xvii. con Giovanni Antonio Magini. Tratto dall' Archivio Malvezzi de' Medici in Bologna, pubblicato ed illustrato da Antonio Favaro. Bologna, 1886. 8vo, 522 pp.

Tychonis Brahe Triangulorum planorum et sphaëricorum Praxis Arithmetica. Qua maximus eorum, praesertim in Astronomicis usus compendiose explicatur. Nunc primum edidit F. I. Studnička. Pragae, 1886, 4to, 2 + 20 ff. [Facsimile.]

Aus Tycho Brahe's Briefwechsel. Von F. Burckhardt. Wissenschaftliche Beilage zum Bericht über das Gymnasium 1886-87. Basel, 1887. 4to, 28 pp.

Elias Olsen Morsing og hans Observationer. Ved F. R. Friis. Kjöbenhavn, 1889. 8vo, 28 pp. and 4 plates.

INDEX.